BRE Digests   **Services and Environmental Engineering**

BRE Digests

Services and Environmental Engineering

Essential information from
the Building Research Establishment

THE CONSTRUCTION PRESS

LANCASTER   LONDON   NEW YORK

Published in 1977
by
The Construction Press Ltd
a company within the Longman Group.

The Construction Press Ltd
Lunesdale House
Hornby
Lancaster LA2 8NB
England.

Longman Group Ltd
5 Bentinck Street
London W1M 5RN.

Longman Inc.
New York.

Printed in Great Britain.

ISBN 0 904406 44 X

Foreword

This book and its companion volumes bring together in a bound and edited form the monthly publications of the Building Research Establishment known as the BRE Digests. Respected throughout the construction industry, the BRE Digests have earned a reputation for authoritative information presented in a readily understood manner. Each Digest takes a subject of particular building concern and provides an analysis of the most important and useful relevant information available about it.

The first edition of this series of volumes presented the BRE Digests in a format which has proved to be outstandingly useful and successful. They were originally published in 1973 with a somewhat uncertain expectation of likely demand, but three printings were required to fill the orders that started to arrive from almost every country in the world.

Since the publication of the first edition, however, about 40 new Digests have appeared and many of the earlier ones have been revised and updated. This new edition therefore will meet a very real need in the many offices and colleges where it is essential to be up to date. These volumes contain all current Digests up to and including that issued in February 1977, with the exception of No. 149.

The editorial format of the first edition has been retained, with the series contained in four volumes on Construction, Materials, Services, and Defects and Maintenance respectively. Each volume includes a comprehensive index to the whole series, thereby making easy cross-reference from one volume to another.

New BRE Digests are published at monthly intervals by HMSO and these may be purchased either singly or on a subscription basis through any supplier of HMSO publications or direct from their London bookshop:

HMSO
PO Box 569
London, SE1 9NH.

ACKNOWLEDGEMENT

We have pleasure in acknowledging the co-operation of both the Building Research Establishment and Her Majesty's Stationery Office in granting us permission to publish the BRE Digests in this volume.

Contents

1 Piped Services and Drainage

Soil and waste pipe systems for housing

This digest describes briefly the problem of maintaining under all conditions the seals of traps to sanitary appliances in dwellings and makes detailed recommendations for the design of single-stack soil and waste pipe systems to meet this requirement in low-rise and multi-storey dwellings.

Importance of sealing

For any plumbing system to operate satisfactorily, the traps to the various appliances must remain sealed in all conditions of use; otherwise there is a risk of odours contaminating rooms in different parts of the dwelling. A seal will be broken if the pressure changes in the branch pipe are of sufficient size and duration to overcome the head of water in the trap itself. Such pressure changes can be brought about by:

(a) Induced siphonage: water flow down main stack; e.g. WC discharge; also water flow from other appliances through combined branches.

(b) Back pressure: water flow down main stack in conjunction with sharp bend at foot of main stack.

(c) Self-siphonage: liquid flow in waste branch pipe to main stack.

Both the old two-pipe system and the fully vented one-pipe system virtually eliminated the risk of seal breakage, the first by interrupting the continuity of the system at hopper head and gulley, the second by ensuring with the help of vent pipes that the pressure in branch pipes never deviated appreciably from atmospheric.

The single-stack system relies on the soil stack alone for venting and experience has shown that given good design and workmanship the scope of such a system is considerable. Using this as a basis, the usefulness can be extended by providing partial additional venting to either the stack or its branches, or both. The early recommendations have been steadily extended to multi-storey installations of increasing load.

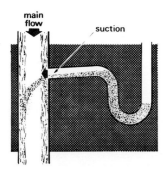

(a) Induced siphonage

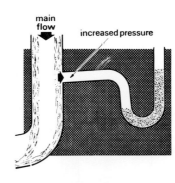

(b) Back pressure

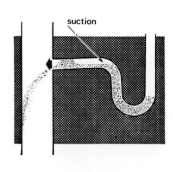

(c) Self-siphonage

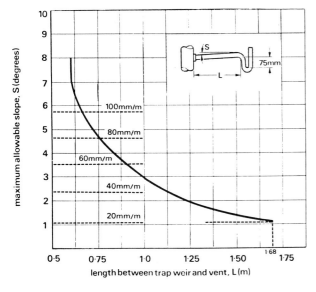

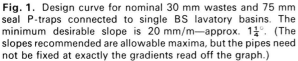

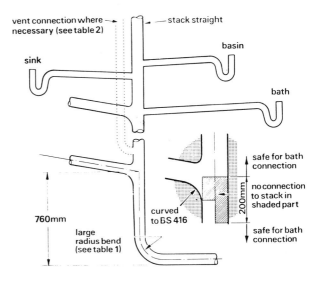

Fig. 1. Design curve for nominal 30 mm wastes and 75 mm seal P-traps connected to single BS lavatory basins. The minimum desirable slope is 20 mm/m—approx. $1\frac{1}{4}°$. (The slopes recommended are allowable maxima, but the pipes need not be fixed at exactly the gradients read off the graph.)

Fig. 2. Design of single-stack system—main features.

Branch pipes to lavatory basins

For lavatory basins the risk of self-siphonage is much more closely dependent on the design of the branch connection to the stack than it is for other appliances. Research has shown that the bore, length and fall of the waste pipe are all important factors. Basin branches of the usual nominal 30 mm bore normally run full and so cause suction on the trap at the end of the discharge. One remedy is to limit the length and, if necessary, the slope of the pipe. Fig. 1 gives maximum allowable slopes for varying lengths of

30 mm nominal bore waste pipe, used in conjunction with a 75 mm seal P-trap. When the basin is more than 1·68 m from the stack, a 30 mm nominal bore trap, with a short 30 mm tail pipe can be discharged into a 40 mm or 50 mm waste pipe. This should prevent the waste pipe from running full (which is sometimes noisy) and thus reduce the suction on the trap. It also reduces the risk of blockage by the deposit of sediment from the waste water. Any bends in the waste pipe should be of large radius.

Table 1. Design of single branches and fittings

Component	Action to be guarded against	Design recommendations
Lavatory basin waste	Self-siphonage	75 mm seal P-traps to be used. The maximum slope of a nominal 30 mm waste pipe to be determined from Fig. 1. Any bends to be not less than 75 mm radius to centre-line.
		Waste pipes longer than the recommended maximum length of 1·68 m should be vented, or a larger diameter waste pipe or suitable resealing trap should be used.
Bath and sink wastes nominal 40 mm trap and 40 mm waste pipe	Self-siphonage	75 mm seal traps to be used. Self-siphonage problems unlikely. Length and slope of waste branch not critical, but sediment may accumulate in long waste pipes and access for cleaning should be provided.
	Backing up of discharge from WC branch into bath branch	Position of entry of bath waste into stack to be as in Fig. 2.
Soil branch connection to stack	Induced siphonage lower in the stack when WC is discharged	WC connections should be swept in the direction of flow. Fittings should have a minimum sweep of at least 50 mm radius (but see also note on straight WC inlets in text). WC branches up to 6 m long have been used successfully.
Bend at foot of stack (Fig. 2)	Back pressure at lowest branch. Build-up of detergent foam	Bend to be of large radius. Two large radius bends (similar to the long one-eighth bends shown in Fig. 5, BS 65 and 540: Pt 1: 1971) should be used for 150 mm and also preferably for 100 mm. The minimum recommended bend radius for 100 mm pipes is 200 mm. Vertical distance between the lowest branch connection and invert of drain to be at least 760 mm (460 mm for up to three-storey houses with a 100 mm stack).
Offsets in stack	Back pressure above offset	There should be no offsets in stacks below the topmost appliances unless venting is provided to relieve any back pressure. Offsets above the topmost appliances are of no significance.

Another means of avoiding self-siphonage in long waste branches is the use of special re-sealing traps, but these may require periodic cleaning to perform well. Designs vary, and each must be judged on its merits.

S-traps produce particularly severe self-siphonage conditions, and venting or a special resealing trap is normally needed.

Combined lavatory basin and bath wastes

Some arrangements perform satisfactorily, but it is not possible to set down limits on design. Tests are needed to assess the behaviour of particular arrangements.

Branch pipes to WCs

Water closet branches do not run full and so there is no risk of self-siphonage whatever their lengths. However, the shape of the WC branch connection to the stack is important because it influences the amount of induced siphonage acting upon branches to other appliances lower in the stack. Thus the recommendations of Tables 1 and 2 assume that WC branch connections are swept in the direction of flow (Fig. 2).

If straight-inlet WC branches are used, more venting or a larger diameter stack may be necessary (see footnote to Table 2).

Design of main pipework

Back pressure and detergent foam

A sharp bend at the base of the stack can cause back pressure to affect the seals of the lowest branch connections. It can also cause a build-up of detergent foam. These difficulties can best be avoided by ensuring that this bend is of large radius (as described in Table 1). A further precaution is to connect ground-floor appliances directly to the drain or manhole or to use a bend at the base of the stack which is one size larger than the stack itself.

Venting

Induced siphonage can be reduced by providing cross vent connections between the soil and vent stacks, either at every floor or at alternate floors. To prevent cross-flow, the vent connection should slope upwards from the drainage stack at an angle not less than 135°. Table 2 shows the vents required for various loads on 90 mm to 150 mm stacks.

Offsets

An offset in the stack above the topmost connection to the stack has little effect on the performance of the system. Offsets below the topmost connection should be avoided, or extra vent pipes may be required to prevent large pressure fluctuations in the stack.

Table 2. Minimum stack sizes and vents required for various loading conditions

Type	Stack diameter mm	Requirements	
Houses (one-family)			
Up to 3 storeys	90	Single-stack	
Flats		Stack serving one group* on each floor	Stack serving two groups* on each floor
Up to 10 storeys	100	Single-stack	Single-stack.
11 to 15	100	50 mm vent stack with one connection on alternate floors	50 mm vent stack with one connection on each floor.
16 to 20	100	65 mm vent stack with one connection on alternate floors	65 mm vent stack with one connection on each floor.
Up to 12 storeys	125	Single-stack	Single-stack.
12 to 15	125	Single-stack	50 mm vent stack with one connection on alternate floors.
Up to 25 storeys	150	Single-stack	Single-stack.
Maisonettes		Stack serving one group on alternate floors	Stack serving two groups on alternate floors
Up to 10 storeys	100	Single-stack	Single-stack.
11 to 15	100	Single-stack	50 mm vent stack with one connection on alternate (bathroom) floors.
16 to 20	100	50 mm vent stack with one connection on alternate (bathroom) floors	65 mm vent stack with one connection on alternate (bathroom) floors.

* Each group consists of a WC, bath, basin and sink. Where dwellings contain more appliances, it may be necessary to provide more vents.

N.B. The above recommendations apply to systems with swept-inlet WC branches. With straight-inlet branches, a 100 mm stack with no vents has been found satisfactory up to 4 storeys; a 150 mm stack with no vents has been found satisfactory up to 15 storeys.

Surcharging of the drain

If the drain to which a stack is connected is surcharged, the normal flow of air down the stack during discharge can be interrupted and very high back pressures can result. If this condition is likely, a vent pipe should be connected to the base of the stack above the likely flood level. The vent pipe bore should be at least 75 mm for a 100 mm bore stack, or 100 mm for a 150 mm bore stack; it should be as short as possible.

Drains with intercepting traps

An intercepting trap fitted in a drain serving a discharge stack can restrict the flow of air down the stack and cause high back pressures unless the base of the stack is vented. The vent pipe should be of the size described above for drain surcharging conditions.

Removal of rainwater

The highest likely rainwater flow from the roof of a small house or flat (with an area of approx. 40 m²) is equivalent to the maximum discharge from a lavatory basin. For this size of roof the rainwater can normally be ignored in the calculation of hydraulic loading on a soil stack. Where easy access to the roof exists, the roof gulley linking the roof to the stack should be trapped.

Wind effects

Suction caused by wind blowing across the tops of stacks on tall buildings has been known to result in loss of water from seals. Normal trap venting is no remedy but a protective cowl at the top of the stack is of some help. Suction is greatest nearer the corners of roofs and the edges of parapets and if possible the tops of stacks should be sited away from these positions.

Further reading

British Standard Code of Practice CP304
Sanitary pipework above ground

Soil and waste pipe systems for office buildings

Substantial information now exists on the use of simplified drainage systems in low-rise and multi-storey domestic buildings, and much of this is summarised in Digest 80, Soil and waste pipe systems for housing. *The present digest deals with the design of soil and waste pipe systems for office buildings. Ranges of appliances, each individually vented to a separate vent stack, are typical features in a traditional layout. BRS work has shown that much of this vent piping is unnecessary or can be reduced in diameter. For an eight-storey installation, using copper pipework, simplified design has been shown to effect a 40 per cent reduction in cost.*

Introduction

A drainage system should carry away waste materials quietly, without blockage and without the escape of foul air into the building. To ensure that an adequate water seal is retained at each appliance, air-pressure fluctuations within the pipework must be limited.

The recommendations given in this digest are not intended to be comprehensive and reference should be made to BS CP304 : 1968. For situations outside those covered by the digest and the code—especially where venting requirements for long ranges of appliances are being considered—a test on a mock-up is recommended.

Pressure effects

The pressure effects that may occur at the traps are:
Self siphonage: suction due to full-bore flow in horizontal drainage pipework (Fig. 1)

Induced siphonage: the suction normally associated with water flow down the drainage stack (Fig. 2)

Back pressure: also normally associated with water flow down the drainage stack (Fig. 2). Conditions near the base of the stack—bends and offsets—influence the pressure.

To limit pressure fluctuations, vent piping is traditionally employed. The requirements for venting crucially affect the overall cost of the system and, in

the housing field, major economies have been made by the omission of vent piping and the introduction of single-stack drainage, which is now well established and embodied in CP 304. A traditional situation for office buildings is shown in Fig. 3, where the venting nearly doubles the length of piping needed. Design of the traditional system for offices is covered by the code, with drainage stack sizes based on water-flow considerations, not related to air-pressure fluctuations in the stack, and with venting requirements determined in an arbitrary way. For the situation in Fig. 3, with a 100 mm drainage stack, the vent stack and branch vents to the WC ranges would be 50 mm in diameter, whilst the branch vents to the basin ranges would be typically 30 mm in diameter.

Designs in this digest are based on the experience that some loss of seal is acceptable and that stacks may be designed to restrict pressure fluctuations to $\pm$ 375 N/m². A negative pressure of this magnitude corresponds roughly to 25 mm loss of seal from a washdown WC.

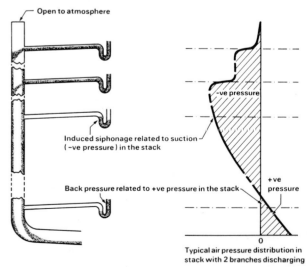

Open to atmosphere

Induced siphonage related to suction (-ve pressure) in the stack

-ve pressure

Back pressure related to +ve pressure in the stack

+ve pressure

Typical air pressure distribution in stack with 2 branches discharging

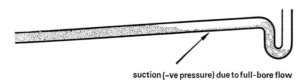

suction (-ve pressure) due to full-bore flow

Fig. 1 Self siphonage

Fig. 2 Air pressures in a stack

The frequency of use of sanitary appliances has also been considered since it is unreasonable to design drainage systems on the assumption that all or indeed most of the appliances discharge at the same time. The principles may be applied using normal pipe sizes and the normal components and materials but in these initial recommendations the drainage stack is assumed to be cast-iron, i.e. the pipe and fittings conform to BS 416. In addition, the recommendations assume that washdown WCs to BS 1213, having a 50 mm seal, are used with a 9-litre flushing cistern.

Pipework for appliances on individual floors

The recommendations for individual WC and basin branches in Digest 80 and CP 304 (1968) are equally applicable to offices and are not repeated here. This section of the digest gives recommendations for pipework to ranges of appliances, and is particularly concerned to limit the effects of suction within the combined branches serving such ranges. The recommendations are summarised in Fig. 4. Where venting is recommended, the vent pipes may be run separately or be connected to a vent stack. The connection of vent pipe to stack should be above the spill-over level of the highest fitting it serves and the pipe should be installed so that there is continuous fall back to the appliance.

(a) Ranges of WCs
Branch pipes serving ranges of WCs are normally 100 mm diameter and do not run full. Hence there is usually no need for branch venting. This has been checked in the laboratory for up to eight WCs in a range with a straight branch at an angle of $2\frac{1}{2}$ deg; it is likely that this number of WCs could be exceeded. Field studies have shown that the angle is not critical. Where there are bends in the pipe it may be necessary to fit a vent pipe to the appliance farthest from the stack. From a general performance standpoint, it is an advantage for the WC connections to the common branch to be swept in the direction of flow (as provided, for example, in the appropriate BS 416 fittings).

(b) Ranges of lavatory basins
With the discharge rates that commonly occur when lavatory basins are emptied, and with the sizes of branch waste normally used, full-bore flow and hence a suction effect in the branch is much more likely than with WC ranges. In addition to discharge rate and pipe size, the occurrence of full-bore flow depends upon factors such as the length and slope of the branch and the shape of tee connections. Full-bore flow is unlikely if washing is done under a running tap. As the latter is common in offices, this gives a safety margin, since the following recommendations are based on filled basins.

The recommendations in Fig. 4 relate to British Standard traps with 75 mm seal used on BS 1188 basins.

Spray-tap wastes do not normally run full even when the branch is only 30 mm diameter, and trials have shown that up to eight basins may be connected to this size of pipe without venting. Spray-tap waste pipes are, however, likely to become blocked by sedimentation and regular cleaning is usually necessary.

Another way to avoid siphonage in basin ranges is to use special resealing traps but these may be noisy and require periodic cleaning to perform well. Designs vary and each should be judged on its merits. The above venting considerations do not apply when running traps are used.

(c) Ranges of urinals
On flow considerations alone, venting is not normally necessary for a branch pipe to a range of urinal stalls. Information is lacking on the way in which venting may affect the build-up of deposits in

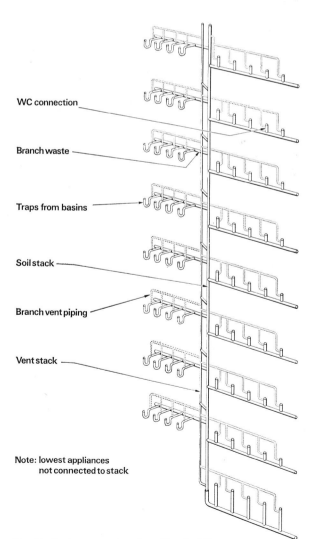

Fig. 3 Drainage services for office buildings

The example shows both drainage and vent piping. The latter (broken lines) may be largely omitted when design is in accordance with the digest.

urinal branches. As with wastes for spray-tap installations, regular cleaning of urinal branches may be necessary, especially in hard water areas where scaling may be severe.

Drainage and vent stacks

In vertical pipework serving appliances on different floors, means must be found to limit the effects of induced siphonage and back-pressure. These effects depend on such factors as the number and distribution of appliances, pattern of use, height of building and dimensions of pipes and fittings. The recommendations for stacks are related to ranges of not more than five appliances on each floor. Diameters of 100 mm and 150 mm are commonly used for drainage stacks, and these sizes have therefore been taken as the starting point for recommendations. Table 1 covers 100 mm pipes used up to 12 floors and 150 mm up to 24 floors. It gives the minimum vent stack sizes recommended for use with vertical drainage stacks serving equal ranges of WCs and basins. For example, section (a) of the table shows

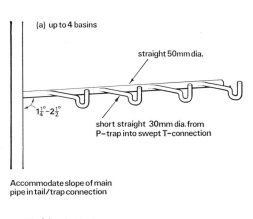

Accommodate slope of main pipe in tail/trap connection

Accommodate slope of main pipe in tail/trap connection
The 50mm dia. pipe may contain a bend in the 'horizontal' plane

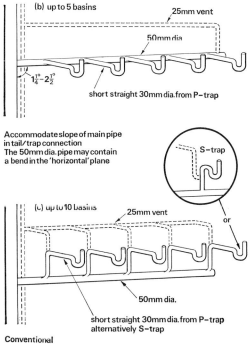

Conventional

Fig. 4 Design recommendations for ranges of appliances

that with a 100 mm drainage stack serving five WCs and five basins on each of eight floors, a 40 mm diameter vent stack is needed. The effect of the flow from wash-basins on induced siphonage is small and hence the results apply without much over-sizing when the number of basins is less than the number of WCs. Stacks serving basins only are usually less than 100 mm in diameter and Table 1 does not therefore apply.

The research has so far not covered the sizing of vertical pipes serving basins only, nor is information complete on the sizing of stacks accepting rainwater from the roof.

At this stage in the development of the sizing tables, it is convenient to use the provisions of two Statutory Instruments* made under the Offices, Shops and Railway Premises Act 1963 to equate wc/urinal/basin combinations to wc/basin combinations. Four examples, that may be taken as hydraulically equivalent, are shown in a separate panel of Table 1.

Attention is drawn to the main assumptions given in Table 1. The intervals of use of 10 minutes and 5 minutes correspond to the 'public' and 'peak' headings given in the Code of Practice. Whilst the 'public' use (10 min) values should be sufficient for most purposes, the 'peak' use (5 min) section is included to cover special situations where a concentrated peak use may be expected. The Table does not cover offsets in the 'wet' part of the stack, nor a series of changes of direction between the lowest branch connection to the stack and the sewer, which may increase pressure fluctuations above that likely in the simpler situation of Fig. 3. In such cases, a vent connection should be made to the stack at or close to appliance branches nearest to the offsets or changes in direction. This is especially so where Table 1 shows that no vent stack is needed. A 50 mm pipe for a 100 mm stack and 75 mm for a 150 mm stack are likely to be sufficient. If the drain connection at the foot of the stack is likely to be surcharged then larger vent stacks should be used: a minimum of 75 mm for a 100 mm drainage stack, or 100 mm for a 150 mm stack. Intercepting traps in the drain connections can also require similar vent sizes, but generally these traps should be avoided.

The use of a large-radius bend at the foot of the stack is noted in Table 1; the minimum radius to the root should be 150 mm but two 135-degree large-radius bends are preferred. With a 100 mm stack, a bend and drain of 150 mm diameter is recommended for the situation shown in Fig. 3. This figure also shows one way in which a vent stack may be connected to a drainage stack at each floor. An alternative method is to connect the vent stack to the WC branch on each floor. Cross-connecting pipes should slope upwards from the drainage stack into the vent system. The vent stack should join the drainage

* SI No. 965 The Washing Facilities Regulations 1964
  SI No. 966 The Sanitary Conveniences Regulations 1964

stack just above the bend at its base to help relieve back pressure. Ground floor appliances connected directly to the drain are shown in Fig. 3; this is recommended for large multi-storey systems. Vent and drainage stacks should terminate well away from the corners of roofs and from parapets, as suction due to wind effects in these areas may cause loss of seal which cannot be remedied by normal venting.

Application

Assuming a building of four floors with four WCs on each, the individual ranges may be served by 100 mm straight branches without venting. Table 1 shows that the four branches may discharge into a 100 mm drainage stack with no vent stack. Consider now the effect of four wash-basins on each floor also connected to the drainage stack. With the arrangement shown in Fig. 4(a) the individual branches do not need venting and Table 1 confirms that no vent stack is needed. If the basins are installed with S-traps, as Fig. 4(c), the individual ranges require venting and this would normally require a vent stack to link this vent piping together

and with the outside air. A vent stack of 30 mm diameter would probably be sufficient; its position is not critical. A vent stack remains unnecessary as far as Table 1 is concerned; limitation of induced siphonage and back pressure associated with flow in the drainage stack itself is virtually independent of the way in which the basins are installed.

As a further example, consider 12 floors with four WCs and four basins on each, all discharging into the same drainage stack. The individual WC ranges do not need venting, nor do the basins if served by P-traps as shown in Fig. 4(a) but Table 1 shows that a 40 mm vent stack is required. If the basins are fitted with S-traps, venting is required as in Fig. 4(c) and this could be linked by the vent stack derived from Table 1.

Such considerations give the basic pipe sizes and the design may be completed by reference to CP 304.

Further reading
Lillywhite, M S T and Wise, A F E *Towards a general method for the design of drainage systems in large buildings,* BRE Current Paper CP 27/69 (free on application).

Table 1 Vent stack sizes (dia in mm) for office buildings

Diameter of drainage stack:		100 mm			150 mm				
Number of floors:		4	8	12	8	12	16	20	24
	WC/Basin								
(a) 10-min interval	1+1	0	0	30	0	0	0	0	0
	2+2	0	0	30	0	0	0	0	0
'Public'	3+3	0	30	40	0	0	0	0	0
use	4+4	0	40	40	0				
	5+5	0	40	see note	0		see note		
(b) 5-min interval	1+1	0	0	30	0	0	0	0	0
	2+2	0	50	50	0	0	0	0	0
'Peak'	3+3	0	50		0	0	0	50	65
use	4+4	30		see note	0				
	5+5	30			0				

Note: for situations outside the range of the table, refer to CP 304

0 *means no vent stack needed*

Main assumptions:
Cast-iron drainage stack and fittings to BS 416
Washdown WC to BS 1213, with 9-litre flush
No offset in 'wet' part of drainage stack
Large-radius bend at foot of stack
Surcharging of the drain connection at the base of the stack is unlikely
No intercepting trap in the drain adjacent to the stack

WC/Urinal/Basin					WC/Basin		
2	+	1	+	2	equivalent to 2	+	2
2	+	2	+	3	3	+	3
3	+	3	+	4	4	+	4
4	+	4	+	5	5	+	5

Pipes and fittings for domestic water supply

This Digest deals with the plumbing installation for cold water supply for houses, but much of what is said here is valid also for larger buildings. Soil and waste plumbing is dealt with in Digest 80 and plumbing applications of plastics in Digest 69. The installation is considered in its logical order, from the mains supply to the various types of outlet. In the second part, experience with the 'Garston' ball valve is reviewed and remedies for some early defects are described.

Imperial/metric units

Some of the dimensions quoted in this Digest are statutory requirements for which rationalised metric equivalents would not be appropriate, e.g. minimum depth of external pipes—2 ft 6 in., overflow pipes (Scottish) $1\frac{1}{4}$ in. diameter; others are nominal pipe bore sizes, still retained in metric standards. Imperial units have therefore been used throughout the text, with precise metric equivalents.

For domestic supply, water is not sold by measure; it is provided as a service, the cost of which is recovered by means of the water rate. As there is no incentive for the consumer to limit his consumption, the suppliers need to exercise some control. This they do through their byelaws, which are generally based on the Model Water Byelaws (1966 Edition) of the Ministry of Housing and Local Government and make provision against waste, undue consumption, misuse and contamination of the water supplied. The byelaws lay down in some detail the kind of materials and fittings to be used, the protection to be provided, etc., the emphasis being on avoiding waste while maintaining the amenity. Nevertheless, they afford considerable latitude in interpretation; this sometimes results in a wide variation between different suppliers' specific requirements and consequently leads to difficulties for designers and manufacturers.

Besides byelaw requirements there are other considerations, mainly of interest to the designer and consumer, affecting the design and construction of a water system. Durability, ease of maintenance, good appearance and quiet operation are treated extensively in the BS Codes of Practice for Water Supply (CP.310:1965), for Frost Precautions (CP.99: 1965) and for Sanitary Appliances (CP.305: 1952). The notes that follow cover various other factors, not mentioned in the byelaws and codes and commonly disregarded in traditional practice, that can make all the difference between an installation that causes repeated annoyance and an efficient one.

Design

It is common practice to provide the plumbing contractor with copies of drawings prepared in compliance with the local authority's requirements for approval under general building byelaws, and on them merely to indicate the positions of sink, bath, basin, etc. These drawings (which need not be to a scale larger than 1 : 100 or in some cases 1 : 200) are not intended to determine the system in all its technical details, and the resulting installation, governed mainly by costs and by the need to comply with the water authority's requirements, may not be wholly satisfactory to the user. It cannot be emphasised too strongly that an installation, to be free of faults, and dependable, must be functionally designed in relation to the building it is to serve, and that this calls for considerable understanding of the design principles involved. The builder or plumber should be provided with working drawings showing clearly, on an appropriately large scale, the position and full description of appliances, pipe runs, valves and other fittings, methods of fixing, protection and all other information which may affect the functioning of the system.

Distribution system

Water is delivered to the building from the main by way of a service pipe, from which branches may be taken to the various supply points or to the storage cistern, if any. Byelaws require one such branch at least to supply drinking water direct to the tap over the sink, and in some districts all other cold water supply points are required to be fed from a storage cistern in the roof space (or at least as high as possible above the topmost tap). The boiler for a hot water system is fed either from the storage cistern or from its own feed cistern.

Service pipes

Byelaws generally require underground service pipes outside buildings to be laid at a minimum depth of 2 ft 6 in. (0.762 m) as a precaution against frost. There have been many instances of frost damage, however, where pipes, although at the required depth as far as the footings, have been bent up to take them through the wall above the footings. This should not be permitted; the service pipe should enter the house at a depth of at least 2 ft 6 in. (0.762 m) and it should not rise until it has reached a point at least 2 ft (0.61 m) away from the inner face of any external wall. Where the pipe, on emerging, passes through the ventilated air space under a suspended floor, it should be protected with a 2 in (50 mm) lagging secured by galvanised wire netting or copper wire.

Within the building the service pipe should be placed on an internal wall and it should run straight to the storage cistern, with the drinking-tap branch taken from it where needed. This keeps the pipe away from the eaves, where it would be more likely to freeze up than anywhere else in the loft. All water byelaws require the service pipe to have a stop-valve of the 'screw-down' type where it enters the house, and it is advisable to have immediately above this a draw-off tap (with a detachable key) for emptying the pipe if the house is to be left unheated in the winter. The stop-valve is a very important fitting; its use in an emergency may avert much damage, and it should, therefore, be placed where it is easily accessible and where it is not liable to be cluttered up with household articles—when water is dripping through the ceiling, time becomes a very important factor. At least once a year the valve should be closed and opened to ensure that it does not become too stiff to operate. To avoid possible waterhammer it should have a fixed washer plate.

Materials for pipes

The composition of water varies between different areas, and some waters tend to corrode certain metals, used alone or in combination with other metals. The characteristics of the water in a particular locality are known by the local water engineer and, in the absence of past experience, he should be consulted on the choice of materials before any new work is put in hand. In general it is advisable to keep to the same metal throughout the system, as far as possible (*see* Digest 98).

Recently, tubes made of polythene under various trade names (e.g. Alkathene, Telcothene, Vulcathene) have come to be used for conveying water, especially on farms and in other developments where this material, when used in long lengths, has certain advantages over metal pipe. When polythene is used for internal plumbing, certain points should be borne in mind. Experience with the material is required for the making of good joints of either the fusion or the compression type; the former, when well made, seem to be the more reliable. If compression type joints are used, no jointing compound should be used; it may lead to cracking of the polythene tube. Tubes must be supported at short intervals (every 8 in. (200 mm) for ½-in. (nominal) tube on horizontal runs). For this reason, and also because the material does not readily take paint, it is preferable to place the pipework out of sight. Because of the slight permeability of polythene to gases, including coal gas, particular care should be taken when siting polythene tubing.

Polythene has a very high coefficient of thermal expansion; in the past, this has caused tubes to pull out at joints on cooling. As a precaution, long lengths of tube in unheated spaces should be 'snaked'. Because polythene tube is less resistant to physical damage than metal pipes, it should not be left exposed in places where it is liable to be damaged through accident or mischief. It resists damage by frost better than metal tube, but when frozen it must not be thawed out by means of a blow-lamp as the flame would damage it.

The above remarks apply primarily to low-density polythene tube (BS.1972: 1967). A new, more rigid type of polythene pipe (type 710) is covered by BS.3284: 1967) which has an Appendix giving notes on its use. For use of polythene and PVC pipework, Digest 69 should be consulted.

Although pipework of any kind is usually concealed as far as possible, the need for ready access to the pipe runs for repair should not be overlooked.

Storage cisterns

From the water supplier's point of view, the purpose of the storage cistern is to provide a reserve of water and to reduce the peak hour demand on the mains. To the user, the cistern gives the advantage of supplying all cold water draw-off fittings (except the drinking-water tap mentioned earlier) with water at a pressure much lower than that in the mains. This is very desirable, because the mains pressure in pipes and fittings causes noise and damages the seatings of taps and valves, in so far as these are still made to the traditional designs, developed at the time when mains pressures were much lower.

The life of a galvanised steel cistern may be considerably extended by an inside coating of a non-toxic bituminous composition that does not impart any taste to the water. Before a new galvanised cistern is put into use it should be carefully cleared of any metallic filings and scraps of metal; if these are allowed to remain they may set up rapid local corrosion through electrolytic action and cause a leak within a matter of months. Cisterns made of non-metallic materials, such as asbestos cement or glass-fibre reinforced plastics, though more expensive, have their obvious advantages. In some areas, because of the properties of the water, their use is compulsory.

Glass-fibre reinforced plastics and polythene cisterns are discussed in more detail in Digest 69.

It is sometimes practicable to place the cistern below the top floor ceiling, to give it full protection from frost and make it easily accessible. If the cistern is placed in the roof space, it should be located centrally and cased in timber or rigid sheet material, leaving a 50 mm gap for a layer of insulating material. Instead of casing in the cistern, pre-formed insulating slabs, made to fit cisterns of different sizes, are available. Finally the cistern should be provided with a well-fitting cover, similarly insulated. Trap doors to lofts should be large enough to pass the cistern in case it has to be replaced. For further details of frost precautions, see CP.99 : 1965.

The ball valve

Ball valves, which control the supply to the storage and flushing cisterns, can cause more trouble and annoyance than any other part of the water system, through failure of their moving parts. A ball valve may either fail to close, causing an overflow or flooding, or, less commonly, it may 'stick' in the closed position, cutting off the supply.

A valve may fail to close for any of the following reasons: perforated float, eroded seating, defective washer, or the presence of grit or lime deposit. Copper floats, especially when soldered, are likely to become corroded; sometimes they simply drop off the boss attaching them to the valve arm, or they become perforated and filled with water. Floats made of plastics—solid cellular or hollow—will not fail in this way.

High-velocity discharge of water from a ball valve may erode the seating, producing radial channels which cause the valve to leak. The remedy here is either a new seating, preferably made of nylon (if the valve is of the British Standard pattern), or re-seating the valve by means of a special tool sold by builders' merchants.

Worn washers should be replaced. It is advisable to inspect the seating at the same time, because damage to the washer is often caused by an eroded seating.

In hard water districts the combined effect of corrosion and lime deposition on the piston and the valve body can cause the piston to stick in the open position, causing an overflow or flooding. In areas where the trouble is known to recur, the valve should be periodically dismantled and cleaned, and the piston greased. Grit in the water is a similar source of trouble, especially in newly-built houses if the water system has not been thoroughly flushed, as it should be, after installation; it also occurs in a few districts where solid matter is formed in galvanised pipes by the action of certain types of water.

Sticking of the valve in the closed position tends to occur when the house has been unoccupied for some time, and lime and dirt have dried out on the working parts; it can be remedied by working the float arm up and down a few times or, better still, by taking the valve to pieces and cleaning it thoroughly.

Where the supply pressure is high, say about 350,000 N/m², ball valves of traditional construction make a high-pitched noise for which there is no cure. The splashing noise can, however, be reduced by fitting a silencing tube to the valve.

The 'Garston' ball valve, developed at the Building Research Station, was designed to overcome some of the troubles mentioned above. The experience gained in its field use is reviewed in the second part of this digest.

Overflow pipes

As a precaution against overflow in the event of failure of the ball valve, the cistern is provided near the top edge with a pipe led outside the building. In byelaws, this pipe is termed 'the warning pipe', emphasising its primary purpose. In England, the bore of this pipe when a $\frac{1}{2}$-in. ball valve feeds the cistern must be not less than $\frac{3}{4}$ in. (19 mm), and this is the usual size. Flushing cisterns for wcs are normally made to take this size of overflow pipe. If the ball valve leaks only slightly this safeguard may be sufficient; if the valve becomes stuck in the fully open position, however, the capacity of the warning pipe will be quite inadequate and water will overflow into the house. Scottish byelaws treat this 'warning pipe' as an overflow pipe in the full sense of the term and, for storage cisterns, require it to have a bore of not less than $1\frac{1}{4}$ in. (32 mm) or twice the diameter of the inlet pipe. A pipe of this size, costing perhaps £1 more than a $\frac{3}{4}$-in. (19 mm) pipe, can under similar conditions carry away three times as much water and thus give much greater protection. Whatever the size of pipe used—and for greater safety the larger size is advisable on a storage cistern—it can take away water more quickly the greater its fall.

Stop valves

When a fitting needs repair it is necessary to be able to stop the flow of water to it. Each pipe carrying water from the storage cistern should therefore be separately controlled by means of a stop valve. In the part of the system fed from a storage cistern, valves of the 'full-way' type should be used because 'screw-down' stop valves offer too much resistance to flow. The traditional practice is to place these valves as near as possible to the cistern, to bring the greatest length of piping under control. However, if the cistern is in the roof space, such an arrangement makes it difficult to reach the valve quickly in an emergency, while the control over a short additional length of pipe is of little practical value. It is better to fit the valves at a height at which they can be reached from the floor, but out of reach of children. If they are fitted in an airing cupboard, they should be positioned so as to be easily accessible without removing the contents of the cupboard. A label on each stop valve indicating the outlet it controls may be of great help in an emergency.

Water closets

It is often not realised that noise and inefficient flushing of wc suites are avoidable. Most annoyance is caused by the high-pitched hissing of the ball valve as the flushing cistern fills. There will be less noise if the ball valve is fed from a storage cistern or, if the supply has to be under mains pressure, if a 'Garston' valve is fitted.

A flushing cistern should fill reasonably quickly, say, within 2 minutes, and, to ensure this, the size of valve orifice should be chosen to suit the supply pressure. As a rule a $\frac{1}{8}$-in. (3 mm) orifice is used on mains supply where the pressure is above, say, 40 lb/in.^{2} (276,000 N/m^2), and a $\frac{3}{16}$-in. (5 mm), or $\frac{1}{4}$-in. (6 mm) orifice, depending on the available head, where the supply is taken from a storage tank. With a very small head, a $\frac{3}{8}$-in. (10 mm) orifice may be found to be necessary to give the required flow. Persistent slow filling may also be due to partial blockage of the valve by foreign matter, which should, in that case, be removed and cleaned.

There are several reasons why an apparently good wc suite may not consistently clear the pan at one flush. First, the cistern may be delivering less than the statutory 2 gallons; accordingly, the water level in the cistern should be checked, when the ball valve shuts off, against the water line marked on the inside back wall. If the level is below this line, the ball valve arm should be carefully bent upwards until the cistern fills to the designed level. The 'Garston' valve has a special screw adjustment for this purpose.

Some cisterns, particularly cast-iron ones, are operated by means of a cast-iron bell which is lifted by the chain pull and then allowed to drop with a clanging noise. These in time become corroded and deliver a slow ineffectual flush. Only too often they are replaced by one of the same type for no better reason than to avoid altering the position of the supporting brackets to fit the dimensions of a better cistern. If a change has to be made, this small amount of extra work should not prevent the installation of a better type of cistern with a piston-actuated syphon. Cisterns of this type work with much less noise, require less force to operate and are more reliable. For greater quietness, a thick-walled cistern is to be preferred.

Inefficient flushing may also be due to a fault in the joint between the flush pipe and pan. The waterway may be obstructed by putty which is still sometimes used as a jointing material, or the pipe may enter the pan socket at an angle. In both cases the rate of flow is reduced, and in the latter case the designed flow distribution in the pan may be disturbed as well. The remedies here are obvious but they may be difficult to apply if, through lack of space between pan and wall, the flush pipe has been forced into the pan socket at an angle and needs to be shortened. Experiments have shown that, other things being equal, low-level cisterns with flush pipes larger than the common $1\frac{3}{8}$-in. (35 mm) bore give a more efficient flush. A bad joint between the flush pipe and

flushing cistern may reduce the flushing efficiency of a suite very considerably. Such a joint is difficult to detect because high velocity flow inside the pipe will cause air to be sucked in *without* water leaking out.

Finally, the trouble may lie with the pan itself. The action of the more common 'wash-down' (as against syphonic) type of pan depends on the weight and velocity of the flush water pushing the contents into the soil pipe, and its efficiency depends largely on the extent to which the shape of the pan usefully employs the available momentum of the two gallons of water flowing into the pan. It is not easy to tell its efficiency from the looks of the pan, but the way in which water flows into it during flushing may give some guidance. In a good 'wash-down' pan a strong jet of water at the back of the pan should discharge into the well of the pan near its back wall. The stream of water coming from the front and formed by the joining of the flow in both sides of the rim should be central and also directed into the well of the pan.

Pans with syphonic discharge are more positive in their action because the contents are pulled out by the suction created behind the trap. In some syphonic pans, particularly those with a single trap, the syphonic action is initiated by means of a slight constriction in the waterway. In normal use this causes no trouble, but an obstruction will cause a blockage more readily than would occur in a 'wash-down' pan, and it may be more difficult to clear. When a syphonic pan is installed it is even more important than with others that the flushing cistern made for the particular type of pan should be used. For maximum quietness and efficiency, a close-coupled double-trap syphonic suite fitted with a 'Garston' ball valve is recommended. Its cost may, however, be £10 above that of the cheapest suite.

Wash-basins

Wash-basins and wc suites for domestic use are made of vitreous china.

The present British Standard for ceramic basins (BS1188: 1965) provides for only two sizes, namely 25 in. × 18 in. (635 mm × 457 mm) and 22 in. × 16 in. (550 mm × 406 mm). Although basins of other sizes do not comply with these requirements, this certainly does not mean that they are necessarily of a lower quality. British Standard basins will be found, however, to be priced lower than others.

The Standard requires basins to be provided with an overflow to prevent flooding, so designed that it can be easily cleaned. In practice this is very difficult to achieve, and it is virtually impossible to clean the overflow on most basins; when the waste plug is opened, the dirty water rises in the overflow passage leaving behind a coating of scum, which produces an unpleasant smell, particularly in warm weather when various bacteria and fungi thrive in the damp and dark environment. Moreover, an overflow of the specified dimensions cannot take away the full discharge of a tap in a plugged basin, and its effectiveness on the whole is doubtful. These considerations have led manufacturers in U.S.A. and on the Continent to make basins and baths without overflows, a practice which is slowly gaining ground in this country also; one large housing authority specifies basins, as well as baths and sinks, without overflows for its multi-storey housing schemes.

Taps

Leaking taps waste a lot of water, besides being a nuisance, particularly on baths, where the drip often makes a stain which is difficult to remove. Taps leak when the washer is worn out, or when the usual metal seating on which the washer bears has become eroded. As with ball valves, erosion produces radial channels through which the water leaks; the channelled metal also tends to damage the washer, making matters worse. Whenever a washer is replaced, therefore, the seating should also be examined; if necessary, it can be re-faced by means of a special tap reseating tool sold by builders' merchants.

Of the many types of washer available, not all give satisfactory service. Some are suitable for cold water taps, but not for hot and vice versa. Black synthetic rubber washers, as used by some of the largest water suppliers, are suitable for either cold or hot taps, and last a long time.

The 'Garston' ball valve in use

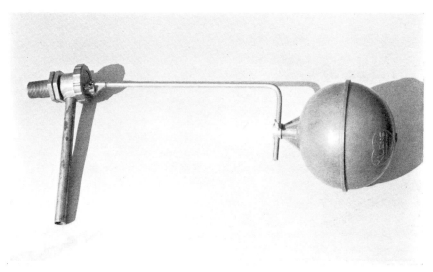

The 'Garston' ball valve

The 'Garston' ball valve developed at the Building Research Station has now been on the market for over ten years and about half a million valves of this type are in use in various parts of the country.

They are sold under the following trade names:

'Chem-Garston'—C. H. Edwards Ltd., Millfield, Wolverhampton, Staffs.
'Kingley—B.R.S.'—Kings Langley Engineering Co. Ltd., Kings Langley, Herts.

The 'Garston' valve was designed to overcome three disadvantages of other types of valve in common use: (a) early damage to the seating caused by cavitation, (b) 'sticking' of the piston, and (c) excessive noise. The main features of the valve are a nylon nozzle, shaped to reduce cavitation, and a rubber diaphragm which stops the flow of water when pressed against the nozzle by a plunger. The diaphragm keeps the moving parts of the valve dry and free from corrosion and incrustation. Movement of the plunger is con-trolled by the usual float on a hinged arm which, at its free end, is bent down at a right-angle. By means of a thumbscrew the float can be clamped on to this part of the arm at any height depending on the water level required in the cistern.

Reports on the performance of the valves in field use show that in general they are proving satisfactory. Experience in the installation and assembly of any new type of fitting often leads to minor changes in procedure and the 'Garston' valve is no exception. Three different causes of defects have been identified and the following review of these should help present and future users of the valve to obtain the full benefit of the performance it is designed to give.

Leakage past the nozzle thread

In the early days of manufacture, loosening of the nylon nozzle in the tapped hole of the valve body, due to creep of the material, allowed water to leak past the nozzle thread. Experiments showed that this

Exploded view

tended to occur a few days after the nozzle had been screwed home, and that the nozzle could then be screwed in by another quarter of a turn. After the second tightening there was no further movement and the valve remained watertight indefinitely.

After this defect was first reported and investigated, the manufacturers modified their method of assembly which now ensures a permanently watertight joint at the nozzle, and they have replaced the faulty valves. They report no complaints of leakage from this cause in valves made since that time.

Valves failing to close

Failure of valves to close has been found to occur mainly in newly built houses. There, when the valves were taken apart, brick and mortar grit and large grains of sand were always found embedded in the diaphragms. All ball valves are liable to leak when grit lodges on the rubber washer or diaphragm, but the 'Garston' valve is more easily affected than the 'Portsmouth' type valve because of the small clearance between nozzle and diaphragm, which is one of the design features ensuring quiet operation.

The flushing of water supply systems in newly completed houses, a normal practice on a well-managed site, is particularly important where 'Garston' valves are fitted. The valves should be thoroughly flushed with the diaphragms removed and all grit adhering to the diaphragm and inside the valve cleaned off. If much grit has been allowed to enter the system a second flushing may be necessary later on. Reports from a large building scheme where several thousands of 'Garston' valves are in use have confirmed that once the water system has been cleaned of grit the valves need no further attention.

In this country there are three or four areas where grit is inherent in the water supply. It may be carried in the mains or it sometimes forms in the pipework in houses owing to the action of a particular type of water on the material of the pipes. In these areas all types of ball valves are liable to leak, but for the reasons already given, the 'Garston' valve is more likely to do so unless protected by a filter at the valve inlet.

Faulty fitting and assembly

In a number of instances, leakage was found to be due to faulty fitting and assembly of the valve, including, in a few cases, valves fitted at an angle or even with diaphragms missing. A more common fault is failure to ensure engagement between the locating lug on the end plate and the slot in the rim of the valve body, when assembling the valve. When this happens a gap is formed between the end plate and diaphragm and the plunger cannot reach the diaphragm to stop the flow. For the benefit of plumbers not yet familiar with the construction of the 'Garston' valve it may be worth while pointing out that the easiest way to assemble the valve is, with the cap nut placed over the end plate to start the cap nut on the valve body thread without trying to locate correctly the end plate. The cap nut is screwed up until a resistance is felt and then the arm, hinged to the end plate, is slowly twisted towards its correct position until the end plate lug is felt to click into the slot. A gentle twist is then given to the arm to make sure of the correct engagement, and, if it does not move, the cap nut is screwed up hand tight. *Grips or pipe wrenches must not be used.*

Although noise in operation has not been the subject of complaint, it has been noticed that occasionally the valves are fitted without silencing tubes. One of the main advantages of the 'Garston' valve is its quiet operation, and while it is true that most of the noise is eliminated by the shape of the nozzle, the full effect of its performance cannot be obtained unless it is fitted with a proper silencing tube. It is essential that there should be a small hole either in the body of the valve or at the upper end of the silencing tube. The absence of this hole may be responsible for a humming noise.

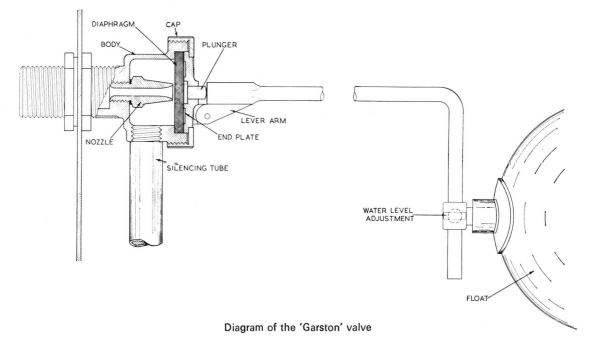

Diagram of the 'Garston' valve

Modernising plumbing systems

Part of the work involved in the modernisation and conversion of old dwellings is usually the improvement of water services including, in some cases, the addition of hot water space heating. When undertaking such modernisation, particularly when the extension of an existing system is involved, it is essential to be aware of the problems of corrosion and scaling that may arise.

The modernisation and conversion of old dwellings usually include the provision of improved hot and cold water services which may qualify for grants to meet part of the cost. As an adjunct to the improvement of a hot water system, it may be convenient to consider extending it to provide hot water central heating.

Before choosing a particular type of hot water system, criteria other than the physical design and performance of the system must be considered. Whilst the initial objective is to produce hot water, it must be known if it is intended in the future to extend the system to include space heating, and to know the effects of the local water on the durability of the metal components.

Choice of system—direct or indirect
In a direct system (Fig 1), hot water circulating between the boiler and storage tank or cylinder is drawn off as required for domestic use and replaced by fresh, cold water, fed directly into the same circuit.

In an indirect system (Fig 1), hot water also circulates between the boiler and storage vessel, but the storage vessel is so designed that the hot water in the primary circuit from the boiler is used only to raise the temperature of the stored water; it does not mix with it nor is it drawn off for domestic use. Hot water for domestic use is drawn from the secondary side of the system (which may be a complete circuit) and is replaced by cold water fed into the secondary circuit. Apart from replacement of water lost by expansion, the same water circulates continuously in the primary circuit.

Although both systems can be extended to serve hot water radiators for space heating there are disadvantages with the direct system: radiators tend to run cold when domestic water is drawn from the tank and the rate of circulation of heated water is slow. Hot water for space heating in an indirect system is taken from the primary circuit and is less likely to cool when water is drawn from the secondary side of the hot water cylinder.

If they are supplying only domestic hot water, both direct and indirect systems usually rely on gravity circulation to transfer heat from the boiler to the storage cylinder. The cylinder must then be located at a higher level than the boiler and the pipes must be of adequate size to maintain the flow. Where an indirect system serves a separate space-heating circuit, the water is normally pumped through the circuit to achieve a higher rate of initial heating and to ensure that the rate of water flow through the radiators is sufficient to maintain radiant surface temperatures.

Single-feed or self-priming indirect cylinders are available in which the calorifier is designed to allow filling of both primary and secondary circuits from one cold feed, via the secondary circuit. When full, the two circuits are separated by an air-bubble.

Chemical and physical effects of water

The composition of the water supplied in UK varies between individual local water authorities and is dependent on the area from which it is supplied and the treatment it receives before distribution (see Digest 98). As a general rule, however, it contains dissolved salts and oxygen, both of which affect the durability of metal. Both have a chemical effect (in stimulating corrosion) and the dissolved salts may also have a physical action (in the formation of scale).

Calcium bicarbonate, which causes temporary hardness, is decomposed most rapidly where the water is hottest, in the boiler water jacket and near the outlet, often producing heavy calcium carbonate scale at these points. Note that once the discharge of salts has occurred, however, it will not recur unless fresh water is introduced. Scale can also be built up over a long period by the precipitation of salts which cause permanent hardness in the water, but this process is slow and scaling which leads to eventual blockage from compounds producing permanent hardness is unlikely.

The problem of scaling and furring is therefore confined mainly to the hottest parts of a system in which fresh water with moderate to high temporary hardness is being regularly replaced, ie in a direct system. Where the same water is being recirculated, ie in an indirect system, there will be an initial production of scale but the formation will not be maintained.

Corrosion of pipes, tanks, cylinders, boilers and radiators usually requires the presence of oxygen. The severity of attack depends upon the susceptibility of each metal to aqueous attack and the protection afforded by scale, but it may be accelerated by bimetallic contact.

The corrosion risk produced by dissimilar or bimetallic contact varies with different metals; the most usual contacts in plumbing systems are as follows: copper/brass, copper/galvanised steel, brass/

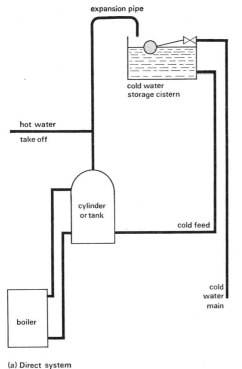

(a) Direct system

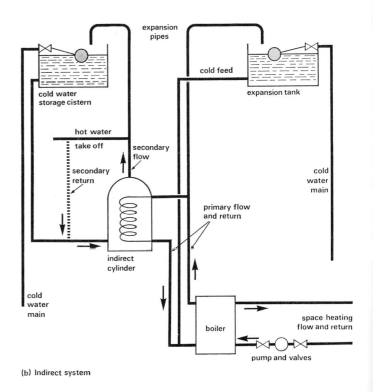

(b) Indirect system

Fig 1 Diagrammatic representation of systems

galvanised steel, lead/galvanised steel and lead/brass or copper. The most aggressive of these is the copper/galvanised steel combination in which the copper acts as a cathode to promote corrosion on the adjoining zinc layer, and the risk of corrosion is intensified when there is either a high concentration of oxygen in the water, or where the water contains high levels of chloride ions in solution. The problem associated with the copper/galvanised steel combination can be reduced by placing an intermediate, less aggressive metal, or plastics, coupling between the copper and the galvanised steel. If the plumbing system is used as an earth for the electrical circuit, it must be ensured that the continuity of the earth is not broken by the use of a plastics pipe. Suitable materials for a metal coupling are brass, lead and gun-metal.

Pitting corrosion on zinc and steel can also be stimulated by a plentiful supply of oxygen and will be further accelerated by the presence of copper dissolved in the water. This is a particular risk when new copper components are used in waters known to be cuprosolvent and especially when the new components are being incorporated into an old galvanised steel system. Small amounts of copper dissolved from the surface of the new tubing produce trace quantities of copper in solution. If this copper-bearing water comes into contact with galvanised steel, serious pitting corrosion can occur, especially if the water has not laid down a protective scale. To minimise the risk in extensions to old systems the flow of water should be from the galvanised part of the system to the new copper part.

The problem of bimetallic corrosion can be severe in direct plumbing systems where fresh quantities of both dissolved salts and oxygen are regularly introduced, but in indirect systems the low oxygen content of the primary water reduces the risk of bimetallic and pitting corrosion. In well-designed and installed, properly used systems few problems should be encountered even where steel radiators are used with a cast-iron boiler and a copper calorifier using copper pipe in the primary circuit. During the commissioning period of a steel/cast iron/copper system it is likely that a small amount of gas will be produced and will collect in the top of the radiators. If the evolution of gas continues for several months the system should be checked for leaks and points at which air can enter, and a check kept on the amount of fresh water used for topping up as this introduces further supplies of dissolved gases.

Any flow of water through the expansion pipe into the expansion tank can be a further cause of aeration of the water; oxygen can also be introduced by fresh water entering the system during cooling or by a pump operating at too high a pressure. If, on expansion, hot water overflows from the expansion tank it will be replaced on cooling by fresh water through the valve. The expansion tank should therefore be designed with sufficient spare capacity to accommodate all the water expansion that will occur during normal conditions. The ball should be positioned in the tank so that, when cold, the water level is above the cold feed pipe to the boiler and, when hot, does not overflow.

Similarly, with self-priming or single-feed cylinders care must be taken to accommodate the expansion and contraction of the water on heating and cooling. The number of radiators, and hence the volume of water used in this type of system, must be kept within the manufacturer's recommendations; if too much water is used, its expansion during heating may be sufficient to allow water to pass from the primary to the secondary circuit; on cooling the flow will be reversed, with the result that fresh water is introduced into the system.

20

In all cases the continuing introduction of fresh water into the primary circuits of indirect systems can lead to corrosion problems, often manifest as persistent air or gas locking in radiators.

However, when all checks have been made on a system consistently producing gas and there has been no obvious introduction of fresh water into the primary circuits it is likely that a small quantity of copper dissolved in the water is stimulating the breakdown of the iron corrosion products releasing hydrogen gas. This problem may be eased by the use of a corrosion inhibitor (preferably containing a biocide) in the primary water. Sodium benzoate-nitrite corrosion inhibitors will reduce chemical or electro-chemical attack on steel and benzo-triazole will reduce the activity of dissolved copper. The biocide is used to kill fungal growths which might feed on the sodium benzoate. Some inhibitor/biocide systems may be harmful to health and it is therefore essential that no transfer of these chemicals is possible from the primary to the secondary circuit or to the water supply by back siphonage. Their use should therefore be restricted to closed primary circuits, and the local water authority should be consulted before the decision to use them is made.

Before improving or extending a hot water system, account must be taken of the condition of the system and the long-term requirements. In general, if the system is to provide only domestic hot water and is in an area where the water has low temporary hardness, the best choice will probably be a direct system, provided the metals used are mutually compatible. In areas of moderate to high temporary hardness, an indirect system will be preferable, as it will in all water conditions if the system is to include space heating.

Further reading
BRS Digest 98. Durability of metals in
natural waters.

Plumbing with stainless steel

Recent improvements in manufacturing techniques have led to the supply, at competitive prices, of small-bore stainless steel tube for plumbing and for domestic heating purposes. Such products are claimed to be stronger than copper and as versatile and corrosion resistant. This digest deals with the properties of stainless steels and their suitability for plumbing installations.

For any plumbing application the resistance to corrosion is the most important single factor determining durability. There are three main kinds of stainless steel whose corrosion resistance varies according to their composition. Table 1 summarises this.

The carbon content of steels is an important factor in determining durability; for welded steels under corrosive conditions a low carbon content is required. Three types of corrosion may be distinguished: pitting, stress corrosion cracking and weld decay, although nowadays the problems associated with weld decay are negligible for plumbing tubes.

Table 1 Classification of stainless steels

Type	Typical constituents		Heat treatment
Austenitic	Chromium 18%		Softening by cooling in air or by quenching from 1000 to 1100°C
	Nickel 8%		Cannot be hardened by heat treatment
Ferritic	Chromium 16% to 22%		Softened by cooling in air from 750 to 820°C
Martensitic	Chromium 12% to 20%		Hardened by oil quenching at 950–1000°C, followed by tempering at temperatures up to 750°C
	Nickel	Nil to 3%	Carbon content of up to 1 per cent

Note: Titanium or niobium are added to stainless steels, principally of the austenitic variety, to improve their welding properties; molybdenum is added to improve corrosion resistance.

Pitting

Water containing high chloride concentrations can be highly corrosive. The chlorides interfere with the natural repair of the oxide film on the surface of stainless steel; increasing the chromium content of the alloy helps to offset the tendency to pitting and it is for this reason that the austenitic 18/8 steels are the most popular for use in water systems.

Stress corrosion cracking

This occurs with austenitic stainless steels through the combined action of a corrosive environment and stress. It does not occur below a certain minimum temperature, not yet specifically determined but thought to be of the order of 80°C. In water services the only examples of stress corrosion failure have been with stainless steel back boilers, and a warning has been inserted in BS 3377 that such boilers should not be used in hard water areas and that particular care should be taken to ensure that no air space remains when the boiler is filled with water. Chlorides concentrate behind hard water scale and by water evaporating in the air spaces. This leads to stress corrosion cracking and failure of the boiler.

Stainless steel tubes for domestic purposes

Steel used for domestic tubes is of the austenitic type (see Table 1) designated 302 S 17 in BS 1449 : Part 4. It contains 17·0–19·0 per cent chromium, 8·0–11·0 per cent nickel, 0·50–2·00 per cent manganese, 0·20–1·00 per cent silicon and with a maximum carbon content of 0·08 per cent.

British Standards

Table 2 compares the dimensions of stainless steel tubes to BS 4127 : Part 2 *Light gauge stainless steel tubes,* and copper tubes to BS 2871 : Part 1 *Copper tubes for water, gas and sanitation.* It shows that the outside diameters are the same for both materials; the tolerances on

Table 2 Dimensions of stainless steel tube to BS 4127 : Part 2 and copper tube to BS 2871 : Part 1

| | | Nominal thicknesses (2) | | | |
| | | Copper to BS 2871 : Part 1 | | | Stainless steel To BS 4127: Part 2 |
Nominal size mm	Maximum outside dia mm (1)	Table X Half hard light gauge mm	Table Y Half hard and annealed mm	Table Z(3) Half hard thin wall mm	mm
6	6·045	0·6	0·8	0·5	0·6
8	8·045	0·6	0·8	0·5	0·6
10	10·045	0·6	0·8	0·5	0·6
12	12·045	0·6	0·8	0·5	0·6
15	15·045	0·7	1·0	0·5	0·6
18	18·045	0·8	1·0	0·6	0·7
22	22·055	0·9	1·2	0·6	0·7
28	28·055	0·9	1·2	0·6	0·8
35	35·07	1·2	1·5	0·7	1·0
42	42·07	1·2	1·5	0·8	1·1

(1) Minimum outside dia is 0·08 mm less for copper and 0·105 mm less for stainless steel
(2) Thickness shall not vary by more than ±10 per cent.
(3) Bending this tube is not recommended.

thickness of stainless steel tube walls are within those specified for light gauge copper tubes. The minimum hydraulic pressures that stainless steel tubes are required to withstand are equal to or higher than those for copper. Stainless steel's tensile strength of 510 N/mm², compared with 250 N/mm² specified for copper, ensures greater protection from accidental damage and necessitates fewer fixing clips.

The ductility test specified in BS 4127 requires that stainless steel can be expanded 25 per cent of its original diameter compared with 30 per cent called for in BS 2871.

Working the tube

Capillary and Types A and B compression fittings to BS 864 are suitable for use with stainless steel tube to BS 4127 and have been approved by the British Waterworks Association. For the metric sizes shown in Table 2, Part 2 of each of these standards is applicable. Part 1 of each standard applies to imperial sizes. Tubes so joined must meet the hydraulic test requirements of 20 bar or 300 lb/in², respectively. Many of the fittings can be obtained in a nickel-plated or satin chrome finish which blends well with the finish of the tube. With a suitable flux, it is as easy to join stainless steel by capillary fittings as it is copper tube. However, as stainless steel has a lower thermal conductivity than copper it is advisable to use a smaller flame and to direct it at the fitting, not at the tube.

Experience has shown that acid chloride fluxes have caused severe pitting corrosion of stainless steel plumbing tube, particularly where liquid fluxes have been used in excessive amount and where a long period—weeks or even months—has elapsed between making the joint and admitting water. To overcome this risk, safer fluxes have been developed and some are available in liquid or paste form based on phosphoric acid. Safe fluxes based on phosphoric acid should always be used for soft soldering; those in paste form are preferable. By their very nature in cleaning material to take solder, all fluxes are corrosive to some degree and should not be left in contact with metal for long periods. Manufacturers of fluxes often give recommendations for their use in soldering and these should be observed.

The tube can readily be bent by either an internal spring or machine. More force is needed than for copper, and spring bending of tubes above 12 mm diameter is not recommended. Spring bending of smaller diameter tubes is satisfactory but the springs used should be

Fig 1 Applying the flux   **Fig 2** Making the joint

Table 3 Manufacturers or suppliers

Stainless steel tube	Bending machines	Phosphoric acid or acid phosphate based fluxes
Permatube Ltd Carriers Close Charles Avenue Canley, Coventry	C and J Hampton Ltd Parkway Works Sheffield S9 3BL	*T R Bonnyman Son and Co Inchcape Chemical Works 121 Millerfield Road Glasgow SE
TI Stainless Tubes Ltd Green Lane Walsall WS2 7BW	Hilmor Ltd Caxton Way Stevenage, Herts	A Levermore and Co Ltd Broadway House Broadway, London SW19
Wednesbury Tube Company Oxford Street Bilston, Staffs	W Kennedy Ltd West Drayton Middlesex Tubela Engineering Co Ltd Fowler Road Hainault, Essex	*Southern Cross Fluxes 103 Cannon Street London EC4 Welding Improvements Ltd Lady's Lane, Northampton
These lists are presented for information only; they do not claim to be complete or to comment in any way upon the relative merits of products.		*For use only with solder ring fittings

those suitable for thin-walled copper tube, to BS 2871. Satisfactory bends are produced by machine bending, irrespective of diameter and position of the weld.

Acceptability

The tube has been accepted by the British Waterworks Association and most local Water Authorities in the United Kingdom for use with compression (Types A and B) and capillary fittings.

Tests have shown that stainless steel can be safely incorporated into existing copper systems without risk of bimetallic corrosion. In areas where dezincification of brass occurs, gunmetal or copper fittings must be used.

Back boilers

Stainless steel back boilers give satisfactory service in soft-water areas. They should be installed in accordance with BS 3377, correctly aligned so that no air space remains when the boiler is filled with water.

Hospital sanitary services: some design and maintenance problems

Some of the lessons to be learned from a preliminary study by the Building Research Station of hospital sanitary services are instructive in showing how critically the design and use of these installations determine subsequent maintenance needs and the ease with which these needs can be met. The main observations contained in this digest are directed principally to those involved in the design and maintenance of hospital services, but they have a wider relevance for the design of complex installations in general.

The advantages of single-stack plumbing for housing, the subject of the previous digest, are best realised when there is a close grouping of sanitary appliances round a vertical service duct. When appliances are more widely spaced apart, as they often are in hospital wards, it may be impracticable or even impossible to use a single-stack method.

In wards it is common practice for a drain to be run at a gradual slope beneath the floor and to receive the discharge from many appliances on its way to a stack. While this more or less horizontal arrangement may be economic and may fit in with architectural requirements, it has disadvantages in that the risk of blockage is likely to be greater than in vertical pipes, particularly if the drain has bends and junctions in addition to a shallow slope. There is a lack of information about the performance of these systems and of their anti-siphonage requirements. For these reasons the Station, with the encouragement of the Ministry of Health, began to study sanitary services in hospitals. The first stage of this work, involving the examination of several existing and proposed schemes, is reported below.

Pipework in existing hospitals

In several important respects the design of drainage systems in modern hospitals shows little advance over that of a hundred years ago. Admittedly, most piping is now installed internally rather than externally, and there is greater provision for anti-siphonage devices, but there are still many different drainage layouts, including vertical and horizontal main drains within the building, and pipe systems which take tortuous paths through the structure (*see* Fig. 1). The main reason for this seems to be lack of co-ordinated planning of the different services, both with each other and with the main structure. The result is a network of piping which is spread throughout hospital buildings and which reveals how little thought has been given to the need for adequate access to the pipework.

Fig. 1.   ...tortuous path through the structure...

Fig. 2. Three short-radius bends close together

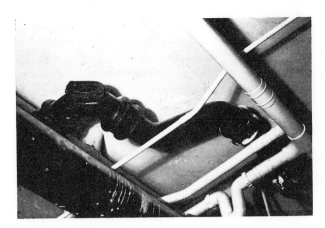

Fig. 3. An unnecessary offset

Layout of pipework in relation to stoppages

The complexity of pipe layout directly affects the number of stoppages which have to be cleared—expensive and often troublesome operations. Not only do they disrupt the work of the hospital, they may be a possible cause of cross-infection. Most stoppages seem to occur where the pipework is complicated by knuckle bends and sharp offsets and $92\frac{1}{2}°$ junctions. None was reported where straight lengths of 100 and 150 mm pipes sloped gradually. The kind of problem is well illustrated by Figs. 2 and 3. No excuse is offered for saying again that better design, planning and overall integration are called for.

Misuse of system during construction

Stoppages may result also from misuse of the building while under construction. Builders' rubbish was frequently found in the drainage system above and below ground. The open ends of soil branch connections are often to be seen on site stuffed with sacks or cement bags to keep out rubbish (*see* Fig. 4)—an unreliable method if these materials have to be removed as rubbish themselves from the completed system after handover.

While this kind of misuse occurs with all types of buildings, in hospitals it seems to have been particularly prevalent, and is of course particularly unfortunate in its effects. Plastic caps have been recommended for closing the ends of heating pipes during installation and there is much to be said for using this type of protective cover on soil and waste pipes also. In general, improved site supervision of the contract could help a lot in preventing the carelessness which causes trouble later.

Misuse of system during occupation

Special appliances such as sluices, slop hoppers and cleaners' sinks are available to ward staff for the disposal of materials unsuited to removal by conventional drainage systems. However, some misuse of these appliances was reported and dressings, sanitary towels, floor cloths, mop heads, syringes, spatulas and rubber gloves are known to have been among articles removed from the systems surveyed; cases were also reported of a blanket, a stainless steel bowl, false teeth and tea cloths contributing to stoppages. In addition, traps are readily blocked by match sticks and cigarette filter tips.

There is a special problem concerning goods such as plastic syringes and spatulas. Sometimes it is not appreciated by hospital staff that these 'disposables' should be incinerated rather than put into the normal drainage system. In other cases it appears that ward incinerators are much less efficient than they need to be and this leads to the misuse of drainage systems as the first alternative to hand.

Kitchen refuse also causes stoppages, and the misuse of kitchen appliances such as grease traps can lead to the accumulation of grease in the

Fig. 4. Floor access point blocked with polythene sheeting

drainage system. Regular cleaning of grease traps is essential, yet only one of the fifteen hospitals visited had arranged for this.

While it is important not to get the incidence of misuse out of perspective, there is no doubt that improved staff training in these matters would help, as would better refuse disposal facilities.

Access to drainage pipework

While much could be done to improve the layout and prevent misuse of drainage systems, some risk of stoppages will always remain. It is important, therefore, and particularly in hospitals, that adequate provision be made for access to the pipework. Many complaints were made of lack of access into ducts—it was either too small, blocked by other services, or in the wrong place. On the pipework itself cleaning eyes often could not be used effectively; some were partly embedded in concrete or turned towards the wall or ceiling, as in Figs. 5 and 6. Access to others was blocked by other services. In many cases it was not possible to use 0.6 m drain rods. Permanent lighting was absent from some ducts; a torch does not give a plumber adequate light to work by.

Some vertical and horizontal pipes with cleaning eyes were encased in brick or hollow pot ducts without access doors. In one example, a stoppage occurred in the enclosed pipe, no drawings were available and holes had to be knocked in the duct to locate the access point; eight hours were needed to clear the blockage which had occurred at a knuckle bend as a result of a build-up of kitchen waste on top of builders' rubbish.

Some horizontal pipes are enclosed within false ceilings, with removable panels providing access. This means that when a stoppage occurs by misuse of an appliance say, on the third floor, it has to be cleared from the second floor since this provides access to the drain above. The routine of two floors is therefore affected, whereas better design might limit the disruption to one floor. Several medical authorities were concerned at the need to clear blockages within the clinical area. In one check test reported by a hospital pathologist, infectious bacteria were found in sewage that had discharged into a clinical area. As well as this risk there is the possible hazard of carrying dirty tools and wearing infected overalls in clinical areas.

Cases were also reported of electrical services being disrupted as a result of sewage leaking round lighting fittings.

Recommendations

While the internal drainage pipework is not normally among the most costly of hospital service systems, the foregoing notes will have indicated the serious consequences of poor design and misuse. Improvements are needed at every stage, from the preliminary sketch plans to the work of the sub-contractor and site supervision, and finally to the use and maintenance of the services. Broad recommendations for each stage can be made as follows:

1. Adequate service space is fundamental to the design of a hospital. The services must be planned within the service area and installed in accordance with the plan. Service inlets and outlets should not be neglected at this stage; the position of sanitary appliances on the sketch plan, and the position and amount of piping needed beneath the floors, must be considered together. The service space should be well lit and capable of being easily cleaned.

2. Bends of large rather than small radius are always to be recommended but are particularly important beneath a ward floor. Similarly, oblique junctions are preferred to T-junctions since they allow loads to flow more readily from branches to main drains.

3. Access doors are advised for all changes of direction except the most gradual. They are also needed at all junctions on horizontal drains. Access doors should of course be easily accessible within

Fig. 5. Access door turned towards wall

Fig. 6. Two access doors turned towards ceiling

28

the service space. Flexible connectors between soil appliances and piping allow their ready removal if this should be necessary for rodding.

4. Planned maintenance is a necessity and thought must be given to the arrangements for maintenance staff to do their work with minimum interruption of ward routines or threat to hygiene. Drawings must be available to them showing the drainage layout and access points. Washing facilities should be provided for them.

5. The education of staff in the properties of sanitary equipment and services is essential.

Single-stack drainage

There are many cases in hospital buildings of various types where a stack serves several appliances close to it on different floors. This is a situation in which single-stack principles are likely to be applicable, and a study of the literature will show the possibilities in this direction.

Further reading

1. AFE Wise, Drainage pipework in dwellings: hydraulic design and performance. HMSO (London, 1957).

2. BRS Digest 80. Soil and waste pipe systems for housing.

3. BRS Digest 115. Soil and waste pipe systems for office buildings.

4. AFE Wise and R Payne. Sanitary services for hospitals: a review of current faults. Architects Journal, 1965, 142 (21st July), pp. 153-6. (Building Research Current Papers, Design Series 39.)

5. Institution of Heating and Ventilating Engineers. Hospital Engineering Services: Present and Future, Symposium held on 12th January, 1966.

Roof drainage: part 1

This digest is published in two parts: Part 1, Sizes for eaves gutters and down-pipes, based on experimental data; Part 2, Sizes for valley and parapet gutters, outlets or receivers, and rainwater pipes, based on experimental data and theoretical studies.

The two parts together replace Digest 107 which is now withdrawn.

Design rate of rainfall

Roof drainage calculations can usually be based on a rate of rainfall of 75 mm/h. Short storms of higher intensity — 150 mm/h or more — do occur and should be assumed for situations where overflowing cannot be tolerated. Regional differences are more significant in relation to total rainfall than to intensities and can usually be ignored for the present purpose.

As an indication of the frequency of these intensities: 75 mm/h may occur for 5 minutes once in 4 years, or for 20 minutes once in 50 years.

150 mm/h may occur for 3 minutes once in 50 years, or for 4 minutes once in 100 years.

Sizes of eaves gutters and downpipes

Flow load

The rate of run-off from a roof is the product of the design rate of rainfall (usually 75 mm/h) and the effective roof area. (1 mm of rainfall on an area of 1 m² is 1 litre of water).

Information on the strength of the wind during periods of intense rainfall is sparse, but it is suggested in CP 308 that the angle of descent of wind-driven rain should be taken as one unit horizontal for every two units of descent (*see* Fig 1). This means that the effective roof area should be taken as the plan area plus half the elevation area — in Fig 1, (b + c/2) × length.

Prepared at Building Research Station, Garston, Watford WD2 7JR
Technical enquiries arising from this Digest should be directed to Building Research Advisory Service at the above address.

Eaves gutters

Each eaves gutter should be designed for the wind direction that will give the maximum rate of flow in the gutter. In practice, it may be convenient to calculate the flow load per metre run of eaves.

A square angle within 4 m of an outlet will reduce the flow capacity of a gutter. This can be allowed for by multiplying the rate of run-off by one of the following factors, instead of adjusting the values for flow capacity of the gutter:

angle within 2 m of the outlet:
— sharp-cornered × 1.2
— round-cornered × 1.1

angle within 2-4 m of the outlet:
— sharp-cornered × 1.1
— round-cornered × 1.05

Table 1 gives flow capacities for level gutters, of half-round, segmental and ogee section, with outlet at one end. The flow capacities of gutters of other profiles and sizes can be calculated from the formula given in Part 2.

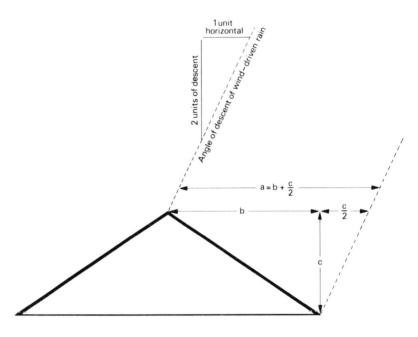

Fig 1 Effective roof area

Table 1 Flow capacities (litres/second) of level gutters with outlet at one end

Nominal gutter size mm	True half-round gutter (1, 2, 5)	Nominal half-round (segmental) gutter (3, 4, 5)	Ogee gutter (2)	(3, 4)
75	0.4	0.3	—	—
100	0.8	0.7	0.9	0.5
115	1.1	0.8	1.4	0.7
125	1.5	1.1	1.7	0.8
150	2.3	1.8	2.6	—

1 Asbestos-cement to BS 569: 1973
2 Pressed steel to BS 1091: 1963
3 Aluminium to BS 2997: 1958
4 Cast iron to BS 460: 1964
5 Unplasticised pvc to BS 4576: Pt 1: 1970

The effective area of roof (in m²) drained by level eaves gutters with outlet at one end, at a rate of rainfall of 75 mm/h, can be found by multiplying the flow capacities given in Table 1 by 48. For other rates of rainfall, the areas will be inversely proportional.

Downpipe sizes appropriate to stated gutter sizes are listed in Table 2. Figures are given both for sharp and round-cornered outlets (see Fig 2). The table allows also for the position of the outlet in relation to the end of the gutter.

Table 2 Minimum downpipe sizes (nominal dia in mm) for various gutter sizes

HR gutter size	Sharp or round-cornered outlet	Outlet at end of gutter	Outlet not at end of gutter
75	sc	50	50
	rc	50	50
100	sc	63	63
	rc	50	50
115	sc	63	75
	rc	50	63
125	sc	75	89
	rc	63	75
150	sc	89	100
	rc	75	100

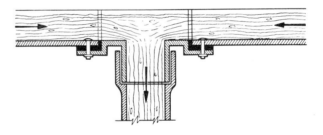

(a) sharp-cornered outlet

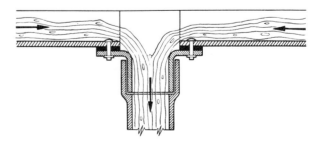

(b) round-cornered outlet

Fig 2 Outlets receiving flow from two directions

Example 1 *Calculate the sizes of eaves gutter and downpipes required to drain a roof 40 m long × 10 m (ridge to eaves) 30° pitch.*

First calculate the effective roof area per metre run of eaves. By calculation, or measurement from drawings, plan length ridge-to-eaves is 8.65 m; the height is 5 m.

Plan length + half the height $= 8.65 + \dfrac{5.0}{2} = 11.15$ m.

The effective roof area is thus 11.15 m² per m of eaves.

With rainfall at 75 mm/h (1.25 1/min m²) the rate of run off will be
$$11.15 \times 1.25 = 13.94 \text{ 1/min}$$
$$= 0.23 \text{ 1/s}$$

From Table 1, a 125 mm (nominal) true half-round gutter could cope with the flow from a length of eaves of $\dfrac{1.5}{0.23} = 6.5$ m

Alternatively, a 150 mm (nominal) true half-round gutter could cope with the flow from a length of eaves of $\dfrac{2.3}{0.23} = 10$ m

Outlets would be needed at not more than double these distances apart, viz 13 m or 20 m for the two sizes of gutter, respectively.

Two possible solutions are shown in Fig 3: the roof could be drained (a) by a 125 mm true half-round gutter and three 75 mm diameter downpipes, or (b) by a 150 mm true half-round gutter and two 100 mm diameter downpipes. The downpipe sizes are read from Table 2 and assume the use of round-cornered outlets.

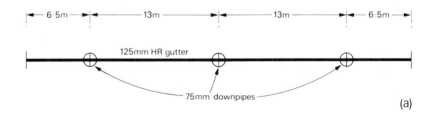

(a)

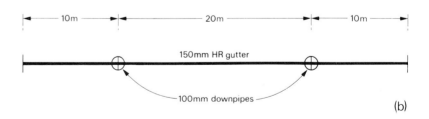

(b)

Fig 3

Roof drainage: part 2

Sizes of valley and parapet gutters, outlets or receivers, and rainwater pipes

For valley gutters, the *plan* area of the double roof should be taken as the effective area to be drained.

The sole of a valley or parapet gutter should be at least 300 mm wide; a convenient profile is obtained by sloping the sides to the same pitch as the roof, up to a height which gives sufficient flow capacity, and then turning a vertical upstand as the freeboard. The depth of the freeboard should be two-fifths of the gutter depth, up to a maximum of 75 mm (see Fig 4).

A gutter that cannot be allowed to overflow should be provided with two or more outlets; a weir overflow at the end of the gutter may also be needed.

Flow capacity for free discharge

Gutters should normally be provided with outlets designed to accept the flow from the gutter without increasing the depth of flow in the gutter ('free discharge') except where the minimum practicable gutter size provides a large surplus capacity (see page 3).

The depth of flow in a straight level gutter discharging freely decreases from the maximum at the far end (the high end) to the outlet. It may be assumed that the depth of flow at the outlet (H_0 in Fig 4) is half the depth at the far end. When the gutter is flowing at its maximum capacity, therefore, the depth at the outlet is half the gutter depth excluding freeboard (D). A fall in the gutter will increase the rate of flow but no figure can be quoted for this; it should be regarded as an increase in the safety margin against overflow.

In a gutter discharging freely, the flow capacity (Q) is approximately

$$\sqrt{\frac{A^3}{B}} \times 10^{-4} \text{ litres/second}$$

where A = area of flow at outlet (mm²) and
B = width of water surface at outlet (mm)

The chart in Fig 5 shows the plan area of roof to be drained for values of B and A. Thus for any rectangular or trapezoidal gutter section and any depth of flow, values of B and A can be calculated or measured from drawings of gutter sections and the equivalent roof area can be obtained; alternatively, the depth of flow at the gutter outlet (required in calculating the size of rainwater pipe) can be determined for a given roof by trial and error.

The charts, Figs 5–7, are based on a rate of rainfall of 75 mm/h. For any other rate of rainfall, the measured plan area must be multiplied by $r/75$ (where r is the design rate of rainfall in mm/h) to obtain the appropriate gutter or rainwater pipe size from the charts. Conversely, the plan area of roof that can be drained by a given gutter obtained from the charts must be multiplied by $75/r$.

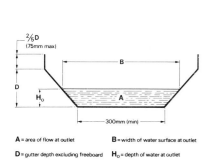

A = area of flow at outlet   B = width of water surface at outlet

D = gutter depth excluding freeboard H_0 = depth of water at outlet

Fig 4   Dimensions at gutter outlet

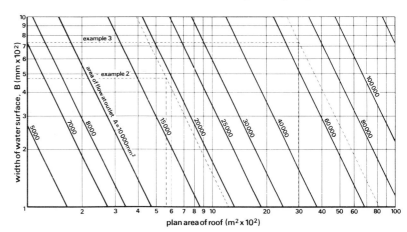

Fig 5   Relationship of roof area and flow dimensions

Prepared at Building Research Station, Garston, Watford WD2 7JR

Technical enquiries arising from this Digest should be directed to Building Research Advisory Service at the above address.

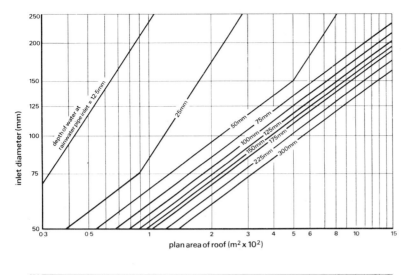

Fig 6 Roof area to be served by rainwater pipe inlet (no swirl)

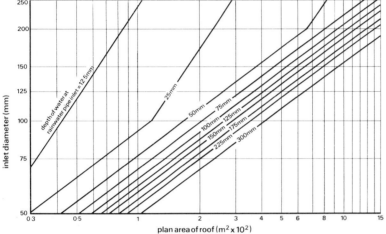

Fig 7 Roof area to be served by rainwater pipe inlet (with swirl)

Box-type receiver

Wherever possible, gutters should discharge into a box-type receiver, the depth of which can be chosen so as to permit the use of a rainwater pipe of convenient size.

The receiver should be at least as wide as the maximum gutter width and it should be long enough to prevent the flow from the gutter over-shooting the box. The top of the box should be level with the top of the gutter except where the box is external to the building, when the outer edge is kept lower to provide an emergency overflow. A suitable shape

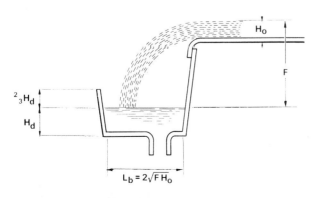

Fig 8 Dimensions of box-type receiver

for an external box-type receiver is shown in Fig 8, in which the recommended depth of the edge is $H_d + 2/3H_d$, where H_d is the depth of water at the inlet to the rainwater pipe required to give the necessary discharge capacity (obtained from Figs 6 or 7 according to whether or not there is swirl, see below). The length L_b of the box should be $2\sqrt{F.H_0}$, where F is the fall between the water surfaces (not less than $H_0 + 2/3H_d$).

Where a box is situated part way along a length of roof, a separate value for L_b should be calculated for each gutter length draining into it and the two lengths added to give the total length of the box.

Rainwater pipes

Where the gutter discharges directly into a rainwater pipe, the rainwater pipe inlet (or gutter outlet) should be designed to allow free discharge from the gutter. The discharge capacity of the rainwater pipe inlet depends on the diameter of the inlet, the depth of water (head) over the inlet and on whether the water swirls about the inlet.

The charts (Figs 6 and 7) give the roof area that can be drained by an inlet of given diameter, at various heads, for inlets with and without swirling, respectively.

Swirl can be neglected if the centre line of the inlet is at a distance less than its diameter from the nearest vertical side of the gutter or box. Swirl can also be suppressed by placing a vertical guide vane along the axis of the gutter, over the rainwater pipe inlet. In the absence of swirl, the inlet acts as a weir for values of the ratio H_i/D up to 1/3 and the discharge capacity of the inlet (Q_i) in litres/second is

$$\frac{D}{5000} \sqrt{H_i^3}$$

where D = diameter of inlet (mm) and

H_i = head of flow over the inlet (mm).

Where H_i/D is greater than 1/3, the inlet acts as an orifice and

$$Q_i = \frac{D^2}{15\,000} \sqrt{H_i}$$

Fig 6 has been prepared from the above expressions. *

Swirl will occur when the centre of the inlet is at a distance greater than its diameter from the nearest vertical side of the gutter or box (unless suppressed by a vane). The inlet then acts as a weir with the same discharge rate as in the absence of swirl for values of H_i/D up to 1/4; above this, the inlet acts as an orifice but swirling will reduce its discharge capacity to

$$\frac{D^2}{20\,000} \sqrt{H_i}.$$

The chart in Fig 7 is applicable to this situation. *

When sizing the inlet to a rainwater pipe leading directly from the gutter, the head H_d should be taken as the depth of flow at the gutter outlet (H_0 in Fig 4). If, however, the gutter discharges into a box-type receiver to which the rainwater pipe connects, the head is the depth of water in the box (H_d in Fig 8).

The diameter of the rainwater pipe can be reduced to two-thirds of the effective inlet diameter provided that the transition is gradual over a length of not less than the diameter of the inlet.

If the inlet is covered by a grating, the effective discharge capacity of the inlet must be calculated from the unobstructed area of the grating.

Gutters with large surplus capacity

The minimum practicable gutter size may be considerably greater than the hydraulic considerations demand. The excess capacity can then be utilised to reduce the required size of rainwater pipe by permitting the depth of flow in the gutter to rise above that for free discharge. This applies only where the rainwater pipe leads directly from the gutter.

The procedure involves some trial and error to discover the smallest size of rainwater pipe inlet

* These formulae use rounded figures but are sufficiently accurate for use in calculations. The exact formulae, before rounding, were used in preparation of the charts, Figs 6 and 7.

that will not cause the gutter to overflow, and is as follows:

Select a convenient size of rainwater pipe inlet and, for the roof area to be drained, read the head H_d (ie the depth of water at the rainwater pipe inlet) from Figs 6 or 7, as appropriate.

From H_d, calculate the width of the water surface B at the gutter outlet, and the area of flow A.

Using these values of B and A, obtain from Fig 5 the equivalent plan area of roof for free discharge.

Calculate $R = \dfrac{\text{plan area of roof to be drained}}{\text{equivalent plan area for free discharge}}$

Using R, from Fig 9 read $X = \dfrac{\text{max. depth in gutter}}{\text{depth at gutter outlet}}$

Calculate $X \times H_d$. If this is greater than the gutter depth (excluding freeboard), the procedure must be repeated using a larger diameter rainwater pipe inlet. If, however, this depth is less than the gutter depth, a smaller inlet can be tried until the smallest which will not result in overflowing of the gutter is arrived at.

Example 2

The plan area of a roof to be drained by a valley gutter is 25 m × 20 m = 500 m²; the roof pitch on either side of the gutter is 30°. The gutter is to be trapezoidal in section, with sides sloping at 30° and a sole width of 300 mm. What depth of gutter and what size of rainwater pipe are required?

Gutter

The required depth is found by trial and error for assumed values of the depth, H_0, at the gutter outlet.

Try $H_0 = 25$ mm:

then, for the given gutter section,

$$B = 386 \text{ mm}; \qquad A = 8580 \text{ mm}^2$$

From Fig 5, the corresponding roof area is under 200 m² which is substantially less than the area to be drained.

Therefore, try $H_0 = 75$ mm:

then, $B = 560$ mm; $A = 32\,250$ mm².

The corresponding roof area (from Fig 5) is over 1000 m², which is on the safe side, but uneconomical.

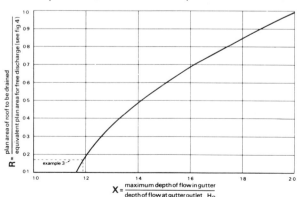

Fig 9 Relationship between depth of flow at outlet and maximum depth of flow for gutter with restricted outlet

36

Therefore, try $H_O = 50$ mm:

then, $B = 473$ mm; $A = 19\ 300$ mm²

and, from Fig 5, the corresponding roof area is 560 m² (approx) which is the approximate plan area of the roof to be drained. (The roof area obtained from the assumed depth of flow will rarely equal exactly the area of roof for which the gutter is required but the exact depth of flow can be obtained by interpolation). For free discharge, the maximum depth of flow in the gutter is twice the depth at the outlet; therefore the required gutter depth is 2×50 mm $= 100$ mm. To this should be added a freeboard of 40 mm.

Rainwater pipe

Where the rainwater pipe is connected directly to the gutter, the depth of flow at the gutter outlet found above, ie $H_O = 50$ mm, is used in calculating the diameter of the rainwater pipe inlet (gutter outlet).

As marked on Fig 7, with $H_O = H_d = 50$ mm, and a plan area of 500 m², the required diameter of the rainwater pipe *inlet* is 180 mm. The *pipe* diameter can, however, be reduced to 120 mm; as explained on page 3, a reduction to not less than two-thirds of the inlet diameter is admissible.

If the gutter discharges into a box-type receiver, it may be possible to use a smaller rainwater pipe inlet by increasing the head over this inlet.

Try a 150 mm rainwater pipe inlet:

from Fig 7, the required head $H_d = 90$ mm, and the difference F between levels of the water surfaces in the gutter and the box should not be less than

$$H_O + 2/3H_d$$

that is, $F \geqslant 50 + (2/3 \times 90)$ mm $= 110$ mm.

The length of the box should not be less than

$$2\sqrt{F}.H_O = 2\sqrt{(110 \times 50)} = 148 \cdot 3 \text{ mm}$$

and the bottom of the box should not be less than

$$(H_d + 2/3H_d) = 150 \text{ mm}$$

below the bottom of the gutter.

Try a 125 mm rainwater pipe inlet:

from Fig 7, the required head $H_d = 175$ mm.

$$F \geqslant 2/3H_d + H_O = (2/3 \times 175) + 50 = 170 \text{ mm}$$
$$L_b \geqslant 2\sqrt{(F.H_O)} = 2\sqrt{(170 \times 50)} = 185 \text{ mm}$$
$$H_d + 2/3H_d = 175 + (2/3 \times 175) = 291 \text{ mm}$$

that is, the use of the smaller diameter inlet requires that the bottom of the box should be at least 291 mm below the bottom of the gutter.

Example 3

A gutter of the same shape as in Example 2 is 200 mm deep, excluding freeboard.

(i) *What is the maximum area of roof that can be drained to it?*

(ii) *If the gutter is to drain a roof 25 m × 20 m = 500 m², what is the smallest rainwater pipe into which it can discharge directly?*

(i) The maximum capacity of the gutter is obtained when it discharges freely. The depth at the outlet is then half the gutter depth, that is,

$$H_O = \frac{200}{2} = 100 \text{ mm};$$

$$B = 646 \cdot 6 \text{ mm};$$

$$A = 47\ 330 \text{ mm}^2.$$

From Fig 5 the maximum roof area which this gutter can drain is 2000 m².

(ii) If the roof area to be drained is only 500 m², the maximum depth of flow in the gutter for free discharge is only 100 mm (*see Example 2*). The excess capacity can be utilised by letting the head at the outlet exceed that for free discharge by reducing the diameter of the rainwater pipe inlet.

Try a 135 mm rainwater pipe inlet:

from Fig 7 the required $H_d = 125$ mm and therefore

$$B = 732 \text{ mm}; \quad A = 64\ 500 \text{ mm}^2.$$

From Fig 5 the equivalent plan area of roof for free discharge is 3000 m².

Therefore,

$$R = \frac{500}{3000} = 0 \cdot 15$$

and using Fig 9 as indicated,

$$X = 1 \cdot 19$$

Therefore the maximum depth in the gutter

$$= 125 \times 1 \cdot 19 = 150 \text{ mm}$$

which is on the safe side. Therefore, try a 120 mm rainwater pipe inlet. From Fig 7 the required H_d is 225 mm, which is greater than the gutter depth. The minimum size of rainwater pipe inlet is therefore 135 mm.

The rainwater pipe diameter can then be reduced to two-thirds of its inlet diameter, ie to 90 mm, as already explained.

Soakaways

A soakaway should have sufficient storage capacity to accommodate a large amount of water from a severe, but comparatively rare, storm and should be able to disperse water at the (lower) average rate of flow into it.

This digest describes a simple test for the permeability of the soil in which a soakaway is to be constructed and how to convert the test results to size of soakaway for a given rate of stormwater flow. Some constructional details are included.

The design rules proposed are applicable to any part of the country, whatever its annual rainfall.

A sewerage system may be 'separate' to accept foul sewage only, all surface water being discharged either to a separate surface-water sewer or direct to watercourses or soakaways; it may be 'partially separate', accepting foul sewage and some surface water with the balance discharged as above; or it may be 'combined' in which a single sewer carries the whole of the foul sewage together with all surface water from roads, footpaths, roofs and yards. Both the partially separate and combined systems will involve spasmodic rushes of storm sewage at the treatment works and this has some practical disadvantages. In general, the present tendency is to favour the adoption of the separate system, with a consequent need for soakaways in some situations.

British Standard Code of Practice CP 301:1971 *Building drainage* suggests that a common method of designing soakaways is to provide water storage capacity equal to at least 13 mm of rainfall over the impermeable area but that on some sites tests of the permeability of the soil through trial boreholes may be needed. This digest indicates a simple method of making such a test and how to interpret the results so as to design a soakaway that will accept all the rainwater from a house or similar small building.

Rainfall

The total amount of rain which falls on a house each year varies with its position in the country, but the amount which falls during a storm of given duration is not so very different in different parts of the country and the most difficult task for a soakaway is to deal with the maximum flow during a rainstorm. The peak rate of rainfall, at the height of a storm, does not last very long, so that the longer the storm lasts the lower the average rate of rainfall is likely to be. This means that a soakaway will have to accept a certain quantity of water in a few minutes but that over several hours, or a day, the average rate of flow of water into it will be less. A successful soakaway is one with sufficient storage capacity to accept the sudden inflow of water and a sufficient rate of dissipation to deal with the average rate of flow.

On the basis of summarised rainfall data in the appropriate form, which is more or less applicable to the whole country, Fig 1 has been prepared. It is very likely that some improvement of this figure could be made for a given locality, by fuller consideration of the actual rainfall data available and applicable for that place. But it is considered that this improvement would not greatly affect the practical recommendations that are advanced here for permeability tests and soakaways.

38

Dispersion

The rate at which water will disperse into the ground depends largely on the permeability of the soil and this can vary tremendously from place to place. Clays such as London Clay, Oxford Clay and Gault are almost impervious (they have been used to line canals and for water cut-offs in dams) and no soakaway formed in them would be able to disperse much water. Sands and gravels, on the other hand, can be very permeable and a soakaway will be able to disperse a great deal of water into this type of soil. The word 'soil' is used here in an engineering rather than an agricultural sense. Between these two extreme types of soil there is a whole range with varying permeabilities.

A soakaway will only be effective when it is wholly above the water table and any available information about the seasonal rise and fall of the water table should be considered in relation to the depth of the soakaway.

Test

An indication of the permeability of a soil is given by the rate at which water will disperse into it from a shallow borehole. By using a trial borehole of the same depth as the proposed soakaway, water from it will soak into the same strata that will have to take water from the soakaway and the measured rate of percolation will be applicable.

The hole should be bored with a 150 mm diameter hand or power auger in the position proposed for the soakaway and initially taken to a depth of 1 m, corresponding to the bottom of the smallest soakaway likely to be required. To perform the test, water should be poured into the hole to a depth of 300 mm and an observation made of the time required for it to soak away. The depth of water can be gauged by marking a stick so that when it is held down the hole its end is 300 mm above the bottom of the hole. If the hole has been cut by the auger to a fairly exact size, $5\frac{1}{2}$ litres of water will give the required depth. The time at which the water is poured into the hole and the time when the water level is seen to reach the bottom of the hole should be noted and the elapsed time expressed in minutes. Where practicable, the test should be repeated once or twice to get an average time.

On completion of this first group of tests, the hole should be bored for a further 1 m, to make it 2 m deep, and a second group of tests made, again using a 300 mm depth of water above the bottom of the hole. If necessary (as discussed below) the hole should be extended in about 1 m steps, and tests made at each depth.

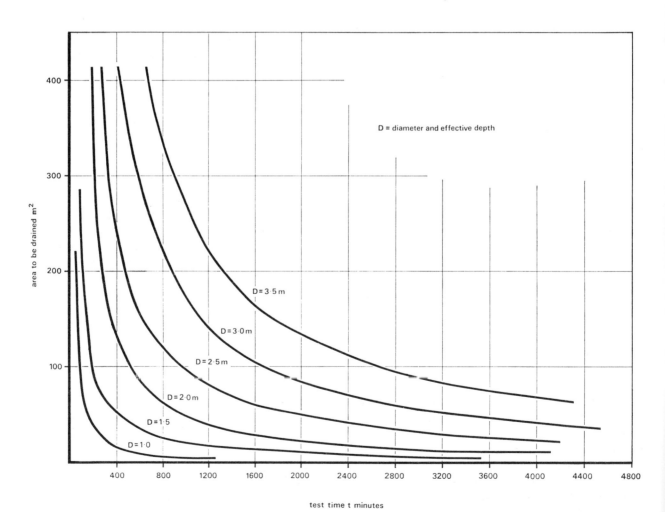

Fig 1   Size of soakaway determined from permeability test time and drained area

Design of soakaway

The rate of rainfall corresponding to a two-hour storm occurring on average not more than once in ten years is 15 mm/hour and this is the basis of the recommendations in this digest. On this basis, over-flowing of a soakaway for as long as two hours is unlikely to occur more than once in ten years and this seems a reasonable risk to take. If a soakaway can accommodate the rate of rainfall in a two-hour storm, it will be able to accommodate the rate of rainfall in longer storms. Also, a soakaway designed for this two-hour storm rate will probably accept the higher rainfall from shorter storms if it is not already full of water; at worst, it might overflow for a period of not more than two hours.

It has been assumed that the soakaway will be cylindrical in shape and will have a diameter about the same as its effective depth, the depth below invert level. The relationship between the diameter (or effective depth) required for the soakaway to suit a give area of roof and/or paved area, and the time t given by the tests, is shown by the curves in Fig 1. To use these curves, a vertical line is drawn upwards from the test time, t minutes, and a horizontal line drawn from the area to be drained. To illustrate the method, if the test time t was 880 minutes and the soakaway is required to take the water from a plan area of 103 m², the diameter (and depth) of the soakaway, from Fig 1, should be 2·5 metres.

If the ground is not very permeable, the size of soakaway can be kept down by splitting up the area to be drained into several parts, with a separate soakaway for each part. In the case of a small house, for example, the two sides of the roof and the paved driveway or yard could be drained to three separate soakaways.

Where the permeability of the ground increases with depth, tests in the deepened holes will give lower values of t, so that it may be cheaper to build a smaller soakaway at a greater depth below the surface.

Construction

Soakaways can be constructed in two main forms, filled and unfilled, depending to some extent on their size.

For small soakaways, an excavated hole can be filled with a coarse granular material such as broken bricks, crushed sound rock, or river gravel with a size range of 150 mm to 10 mm, see Fig 2. The end of the inlet pipe should be surrounded by only the large pieces, to ensure that the rainwater can flow freely into the main mass of granular material. Above the pipe, the size of the pieces should be gradually reduced until the surface of the granular material, at about 0·5 m below ground surface, can be blinded to take a layer of topsoil. This topsoil covering can form a part of a garden or support a lawn.

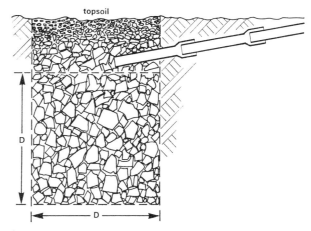

Fig 2 Small filled soakaway

For larger soakaways, the excavated pit may be lined with brickwork laid dry, or with jointed honeycomb brickwork. Alternatively, they can be built with a rigid lining such as perforated precast concrete rings or segments, laid dry, surrounded by granular material to lead the water into the soil, see Fig 3. The top can be covered with a standard reinforced concrete man-hole top to suit the rings or segments, fitted with an access cover so that it is possible to clean out accumulated silt and rubbish.

Caution

Soakaways should be built on land lower than, or sloping away from, buildings and they must be kept at a safe distance from buildings. In land overlying chalk there may be a serious risk of swallow holes and these may be activated by the concentrated discharge from a soakaway. There have been cases of collapse of the corner of a building caused by a soakaway built too close to it, and an example of a swimming pool which lost one end into a swallow hole which developed under a soakaway built to take water from the pool itself. Because of the wide variations in soils and site conditions, it is not possible to give any generally applicable guidance as to the 'safe' distance from a building but the local authority can usually offer advice on this, based on their detailed knowledge and experience of the locality.

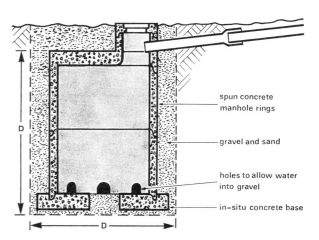

spun concrete manhole rings

gravel and sand

holes to allow water into gravel

in-situ concrete base

Fig 3 Soakaway pit with rigid lining

2 Heating, Thermal Insulation and Ventilation

Standard U-values

This digest provides information which enables U-values for walls and roofs to be calculated on the basis of standard assumptions, in accordance with the IHVE 'Guide, Book A'. This new edition includes a table of thermal conductivities of some building materials but omits the Tables of U-values for wall and roof constructions that were included in the earlier edition.

The calculation of heat losses from ground floors is described in Digest 145; some data relating to windows are included in Digest 140.

Introduction

The thermal transmittance or U-value of a wall, roof or floor of a building is a measure of its ability to conduct heat out of the building; the greater the U-value, the greater the heat loss through the structure. The total heat loss through the building fabric is found by multiplying U-values and areas of the externally exposed parts of the building, and multiplying the result by the temperature difference between inside and outside.

In the past, U-values have been obtained by a variety of methods — by measurement, by adjustment of measured values, or by calculation from thermal resistances of component parts. As a result, different sources often quoted different values for the same construction. In fact, the U-value of a structure does vary to some extent from one situation to another; among other things it depends on the moisture content of the component materials, the wind speed and the internal conditions. The results obtained by measurement depend on conditions during tests and differ from one another as well as from calculated values based on arbitrary assumptions about the conditions of exposure. Although all these methods give values that are accurate enough for heat loss calculation, difficulties arise when regulations require that the U-value should not exceed a stated value.

Standard U-values are needed for comparing different constructions on a common basis or for meeting a stated figure specified by a client or by regulations.

Basis of the standard U-values

Standard U-values are calculated from the resistances of the component parts, which in turn are based on standard assumptions about moisture contents of materials, rates of heat transfer to surfaces by radiation and convection, and airflow rates in ventilated airspaces. The effects of any heat bridging through the structure also have to be taken into account in a standard manner. The standard assumptions are as far as possible typical of practical conditions although they cannot be expected to agree in every case as conditions vary between one situation and another.

Measured U-values cannot be accepted as standard because it is only on rare occasions that the conditions of the test agree precisely with the standard assumptions.

Explanation of terms used

Thermal conductivity (k) is a measure of a material's ability to transmit heat; it is expressed as heat flow in watts per square metre of surface area for a temperature difference of 1°C per metre thickness and may be expressed as: $\frac{Wm}{m^2 {}^{\circ}C}$ but thickness over area $\frac{m}{m^2}$ cancels to $\frac{1}{m}$ and the expression is normally given as $W/m\ {}^{\circ}C$.

Thermal resistivity ($1/k$) is also a property of a material, independent of thickness; it is the reciprocal of conductivity, ie $m\ {}^{\circ}C/W$.

When the thickness of a material is known, its actual **thermal resistance** (R) can be calculated; this is the product of thermal resistivity ($1/k$) and thickness in metres and is expressed as $m^2\ {}^{\circ}C/W$. (If material thickness is quoted in millimetres, for the purpose of making this calculation it must be converted to metres.)

Thermal transmittance (U) is a property of an element of a structure comprised of given thicknesses of material and is a measure of its ability to transmit heat under steady flow conditions. It is defined as the quantity of heat that will flow through unit area, in unit time, per unit difference of temperature between inside and outside environment. It is calculated as the reciprocal of the sum of the resistances of each layer of the construction and the resistances of the inner and outer surfaces and of any air space or cavity. It is given in $W/m^2{}^{\circ}C$.

To summarise the foregoing: a property of any material is its thermal conductivity (k); the reciprocal of this is resistivity ($1/k$). For material of known thickness, the resistance (R) can be calculated (resistivity × thickness) and from the resistances of the various layers comprising a construction and the resistances of cavities, and of inner and outer surfaces, the U-value can be calculated.

For a simple structure without heat bridging, the thermal transmittance coefficient U is expressed as:

$$U = 1/(R_{si} + R_{so} + R_{cav} + R_1 + R_2 \ldots \ldots)$$

where R_{si} = internal surface resistance (see Table 1)

R_{so} = external surface resistance (see Table 2)

R_{cav} = resistance of any cavity within the building element (see Tables 3 and 4)

R_1, R_2 = resistance of slabs of material

Table 1 Internal surface resistance R_{si}

Building element	Surface emissivity*	Heat flow	$m^2°C/W$
Walls	High	Horizontal	0·123
	Low	Horizontal	0·304
Ceilings or roofs, flat or pitched	High	Upward	0·106
Floors	Low	Upward	0·218
Ceilings and floors	High	Downward	0·150
	Low	Downward	0·562

Table 2 Standard outside surface resistance R_{so}

Building element	Surface emissivity*	Surface resistance R_{so} $m^2°C/W$
Wall	High	0·055
	Low	0·067
Roof	High	0·045
	Low	0·053

*Emissivity should be taken as 'high' for all normal building materials including glass, other than unpainted or untreated metallic surfaces such as aluminium or galvanised steel, which should be regarded as 'low'.

Notes

For any part of a building, freely exposed on the underside to an open space, the same values as for roofs should be used.

The values given are applicable to any orientation.

Resistance of cavities (R_{cav}) The thermal resistance of airspaces, such as cavities in hollow wall constructions, depends mainly on the following factors:

1 Thickness of the airspace (its dimension through the thickness of the wall) — resistance increases with the thickness up to a maximum at about 20 mm.

2 Surface emissivity — commonly used building materials have a high emissivity and radiation accounts for about two-thirds of the heat transfer through an airspace with high emissivity surfaces. Lining the airspace with low emissivity material such as aluminium foil increases the thermal resistance by reducing radiation.

3 Direction of heat flow — a horizontal airspace offers higher resistance to downward than to upward heat flow, because downward convection is small.

4 Ventilation — airspace ventilation provides an additional heat flow path but because air movement in such conditions is very variable, estimates for this are necessarily approximate. Ventilation may be either deliberate; for example, ventilated cavity walls, or fortuitous, as in sheeted constructions with gaps between sheets.

Standard values for various airspaces, unventilated and ventilated, are given in Tables 3 and 4. The small amount of ventilation required to avoid condensation in roof spaces does not significantly affect the airspace resistance and, where this is the only ventilation provided, data for unventilated airspace should be used. Similarly, the data in Table 3 are applicable to the airspace in a cavity wall that is ventilated only to normal standards.

Table 3 Standard thermal resistance of unventilated airspaces

Type of airspace		Thermal resistance* $m^2 °C/W$	
Thickness	Surface emissivity	Heat flow— horizontal or upwards	Heat flow downwards
5 mm	High	0·11	0·11
	Low	0·18	0·18
20 mm or more	High	0·18	0·21
	Low	0·35	1·06
High emissivity planes and corrugated sheets in contact		0·09	0·11
Low emissivity multiple foil insulation with airspace on one side		0·62	1·76

*Including internal boundary surface

Materials As explained previously, the resistance (R) of a material is equal to $1/k$ multiplied by thickness. Values of k for insulating materials can be obtained from the IHVE *Guide, Book A*. These materials are intended for use in dry situations and their thermal conductivity in air-dry condition is appropriate.

For the materials commonly used for masonry walling — brick, lightweight concrete, dense concrete — there is a relationship between bulk dry density and thermal conductivity, but the effect of moisture must also be considered. Table 5 sets out for a range of dry densities some average thermal conductivities at moisture contents appropriate to solid brickwork and concrete, protected from rain and exposed to rain as, for example, in the inner and outer leaves respectively of cavity walling.

If available, however, measured k-values should be used for calculating standard U-values. The tests should have been made on specimens at a fairly low moisture content and should be adjusted, using Table 6, to the standard moisture content appropriate to the conditions of use, as indicated by the column headings of Table 5.

Table 4 Standard thermal resistance of ventilated airspaces

(Airspace thickness, 20 mm minimum)	Thermal resistance* $m^2 °C/W$
Airspace between asbestos-cement or black painted metal cladding with unsealed joints, and high emissivity lining	0·16
As above, with low emissivity surface facing airspace	0·30
Loft space between flat ceiling and unsealed asbestos-cement or black metal cladding pitched roof	0·14
As above with aluminium cladding instead of black metal, or with low emissivity upper surface on ceiling	0·25
Loft space between flat ceiling and unsealed tiled pitched roof	0·11
Airspace between tiles and roofing felt or building paper on pitched roof	0·12
Airspace behind tiles on tile-hung wall	0·12

*including internal boundary surface

Table 5 Thermal conductivity of masonry materials

Bulk dry density kg/m³	Thermal conductivity $W/m °C$		
	Brickwork protected from rain: 1%*	Concrete protected from rain: 3%*	Brickwork or concrete exposed to rain; 5%*
200	0·09	0·11	0·12
400	0·12	0·15	0·16
600	0·15	0·19	0·20
800	0·19	0·23	0·26
1000	0·24	0·30	0·33
1200	0·31	0·38	0·42
1400	0·42	0·51	0·57
1600	0·54	0·66	0·73
1800	0·71	0·87	0·96
2000	0·92	1·13	1·24
2200	1·18	1·45	1·60
2400	1·49	1·83	2·00

*Moisture content expressed as a percentage by volume

Table 6 Moisture factors, for use with Table 5

Moisture content (% by volume)	1	3	5	10	15	20	25
Moisture factor		1·3	1·6	1·75	2·1	2·35	2·55	2·75

Table 7 Thermal conductivities (k) of some building materials

Material	Condition (if known)	Bulk density kg/m³	k W/m °C
Asbestos-cement sheet	C	1600	0·40
Asbestos insulating board	C	750	0·12
Asphalt, roofing	dry	1600 —2325	0·43—1·15
Brickwork, *see* Table 5			
Concrete, *see* Table 5			
Cork granules, raw	dry	115	0·052
Corkboard		145	0·042
Fibre insulating board	C	260	0·050
Glass wool, mat or quilt	dry	25	0·04
Hardboard, medium		600	0·08
standard		900	0·13
Metals:			
aluminium alloy, typical		2800	160
copper, commercial		8900	200
steel, carbon		7800	50
Mineral wool, felted	dry	50	0·039
	dry	80	0·038
semi-rigid felted mat	dry	130	0·036
loose, felted slab or mat	dry	180	0·042
Perlite, loose granules	dry	65	0·042
plaster	C	600	0·19
Plasterboard, gypsum		950	0·16
perlite		800	0·18
Plaster, gypsum		1280	0·46
lightweight		400- 960 0·079	0 30
sand/cement	C	1570	0·53
Plastics, cellular			
expanded polystyrene	dry	15	0·037
	dry	25	0·034
polyurethane foam (aged)	dry	30	0·026
pvc flooring	dry		0·40
Plastics, solid			
epoxy glass fibre	dry	1500	0·23
polystyrene	dry	1050	0·17
Stone, *see* Table 5			
Timber, across grain			
softwood	C		0·13
hardwood			0·15
plywood	C	530	0·14
Vermiculite loose granules		100	0·065
Wood chipboard	C	800	0·15
Woodwool slab	C	500	0·085
	C	600	0·093

C = conditioned to constant weight at 20°C and 65% rh

Notes

This table is extracted from Table A3.23 of the IHVE Guide 1970; some of the figures are representative values to be used in the absence of precise information. The values quoted for fibre insulating board and woodwool are obtained from data produced since the publication of the Guide.

Materials commonly used as thin membranes are not included in this Table. The contribution to overall insulation made by a membrane is due largely to the forming of additional airspaces, the resistance of the actual material being too low to be significant.

The thermal resistance of roofing tiles and slates should be neglected because of the airflow through the units; the resistance of this portion of the structure is allowed for in the values for ventilated airspaces given in Table 4.

Heat bridging

A metal or other high conductivity member bridging a structure increases the heat loss. In simple cases, the thermal resistance can be found by calculating separately the thermal transmittances of the different portions of the construction and combining them in proportion to their relative areas.

This may also be done with hollow blocks if the airspaces are not less than 20 mm thick and are wide in proportion to their thickness; it should not be applied to multi-perforated bricks and blocks which do not conform to this condition.

The calculation of U-values

Table 7 sets out the thermal conductivities (from the IHVE *Guide*) of a range of building materials and, in conjunction with Tables 1–6, enables U-values to be be calculated for a wide range of constructions.

Example 1

To find the U-value of an unplastered 'one-brick' solid wall, built of bricks of 1700 kg/m³ density.

From Table 1, $R_{si} = 0.123 \ m^2 \ °C/W$

From Table 2, $R_{so} = 0.055 \ m^2 \ °C/W$

$$R_{brick} = \frac{\text{thickness in metres}}{k\text{-value}} = \frac{0.215}{0.84}$$

$$= 0.256$$

$$U = \frac{1}{0.123+0.055+0.256} = \frac{1}{0.434} = 2.3 \ W/m^2 \ °C$$

Example 2

Calculate U-value of wall shown below.

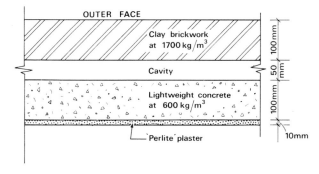

$$U = \frac{1}{R_{si} + R_{so} + R_{cav} + R_1 + R_2 + R_3}$$

From Table 1, R_{si} = 0.123 $m^2 \ °C/W$
From Table 2, R_{so} = 0.055
From Table 3, R_{cav} = = 0.18
From Table 5,
(outer leaf) R_1 = $\dfrac{0.10}{0.84}$ = 0.119

(inner leaf) R_2 = $\dfrac{0.10}{0.19}$ = 0.526

From Table 7,
(plaster) R_3 = $\dfrac{0.01}{0.19}$ = 0.053

$$\overline{1.056}$$

$$U = \frac{1}{1.056} = 0.95 \ W/m^2 \ °C$$

The IHVE *Guide, Book A*(1970)* and *Thermal insulation of buildings* HMSO (1971) lists standard U-values for a selection of external wall and roof constructions. These constructions, or any others for which a U-value is already known, may be varied and the effect of the variation calculated by the following procedures:

1 Find the reciprocal of the U-value (= total resistance of the construction).
2 Deduct from this the resistance of any layers that are to be omitted.
3 Add the resistance of any layers that are to be added.
4 Find the reciprocal of the new total resistance (= the U-value).

Example 3

Calculate the effect of filling with polyurethane foam the cavity in Example 2.

U-value of the original construction = 0.95 $W/m^2 \ °C$

(1)	1/0·95		= 1·056
(2)	Deduct R_{cav}		$\dfrac{0.18}{0.876}$
(3)	Add R_{fill}	$\dfrac{0.050}{0.026}$	$\dfrac{1.923}{2.799}$
(4)	$U = 1/2.799$		= 0.36 $W/m^2 \ °C$

* Section A3 *Thermal and other properties of building structures* is now published separately from the full 'Guide' (by the Institution of Heating and Ventilating Engineers).

Heat losses through ground floors

Digest 109 explained the basis of calculating standardised' U-values and dealt in some detail with wall and roof constructions. For ground floors, either solid or suspended, it is not possible to calculate U-values from first principles but the value of a basic construction can be adjusted according to the nature of the floor finish and any insulation.

Solid ground floors

A solid floor, laid in contact with the ground, with or without a bed of hardcore, is exposed to the air on only one face. The heat flow from inside the building to the outside air is indicated by the flow lines in Fig 1. The greater the distance it has to travel the less is the quantity of heat lost. A U-value for a solid ground floor must, therefore, take account of the size and edge conditions of the slab.

The results of an examination of this problem, published over twenty years ago,[1] still form the basis of the conventional method of dealing with heat loss calculations through ground floors as set out in the IHVE 'Guide'.[2] Table 1 shows the basic U-values for a range of sizes and shapes of solid floor in contact with the ground; the values given are

applicable to dense concrete floors, with or without a bed of hardcore. Because the thermal conductivities of ground and slab are similar, the values may be used for slabs of any thickness. They will not be affected by a hard dense floor finish such as granolithic concrete, terrazzo, clay tiling etc or by a thin finish of little insulation value such as thermoplastic tiles.

In applying these values, the full temperature difference between inside and outside should be used. During the early life of a building, heat flows into the ground to raise it to its final equilibrium temperature; because of the high thermal capacity of the ground, this may take 6–12 months, with an increased demand on the heating during the early period. Steady-state conditions are applicable only to a narrow band round the edge of the floor slab but, nevertheless, the convention is adopted of basing heat loss calculations on a U-value for the whole floor and this will not lead to great error after the ground has been warmed to its equilibrium temperature.

The effect of moisture content on the thermal conductivity of masonry materials is discussed in Digest 108,[3] which describes the standard assumptions to be made for moisture contents of walls in protected and exposed situations. No such assumptions are required for solid ground floors.

Table 1 U-values for solid floors in contact with the earth

Dimensions of floor	U-values			
	Four exposed edges		Two exposed edges at right-angles	
metres	*W/m² °C*		*W/m² °C*	
Very long × 30	0·16*	*6·25*	0·09	*11·11*
× 15	0·28*	*3·57*	0·16	*6·25*
× 7·5	0·48*	*2·08*	0·28	*3·57*
150 × 60	0·11	*9·09*	0·06	*16·67*
× 30	0·18	*5·55*	0·10	*10·0*
60 × 60	0·15	*6·66*	0·08	*12·5*
× 30	0·21	*4·76*	0·12	*8·33*
× 15	0·32	*3·12*	0·18	*5·55*
30 × 30	0·26	*3·84*	0·15	*6·66*
× 15	0·36	*2·77*	0·21	*4·76*
× 7·5	0·55	*1·82*	0·32	*3·12*
15 × 15	0·45	*2·22*	0·26	*3·84*
× 7·5	0·62	*1·61*	0·36	*2·77*
7·5 × 7·5	0·76	*1·32*	0·45	*2·22*
3 × 3	1·47	*0·68*	1·07	*0·93*

* Applies also to any floor of this breadth and losing heat from two parallel edges (breadth then being the distance between the exposed edges)

Figures in italics are reciprocals of the U-values ie the air-to-air resistance

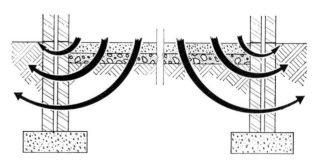

Fig 1 Heat flow through ground slab

Insulation of a solid ground floor

If a floor finish or screed affording some useful degree of thermal insulation is to be used, the U-value of the floor can be calculated accurately enough for most design purposes by taking the reciprocal of the basic U-value, ie the total air-to-air resistance (shown in italics in Table 1), adding the resistance of the additional material and then calculating the reciprocal of the combined resistances to obtain the U-value of the floor.

Assuming, for example, that a concrete floor size 15 m × 7·5 m exposed on all four edges is to be covered with 20 mm softwood flooring:

From Table 1, the basic U-value is 0·62 W/m² °C
and its total resistance, the reciprocal, is   1·61

From Table 5, the thermal resistivity of softwood is 7·7 m °C/W; the thermal resistance of 20 mm

material is therefore $\dfrac{7\cdot7\times20}{1000}$ =0·15

           total air-to-air resistance
           of insulated slab   =1·76
           U-value=1/1·76   =0·57

The effect of an overall layer of insulation (Fig 2) can be calculated in the same manner, by adding the thermal resistance of the insulation to that of the basic floor. The efficiency of the insulation is not, however, constant over the whole area of the floor because the greatest loss through an uninsulated floor is from the edges and the cost of overall insulation is seldom justifiable. An alternative that will give nearly as good results is to treat only the edges

Table 2   Corrections to Table 1 for edge-insulated floors

Dimensions of floor	Percentage reduction in U for edge insulation extending to a depth of:		
metres	0·25 m	0·5 m	1·0 m
Very long × 30	3	7	11
× 15	3	8	13
× 7·5	4	9	15
60 × 60	4	11	17
30 × 30	4	12	18
15 × 15	5	12	20
7·5 × 7·5	6	15	25
3 × 3	10	20	35

of the slab. This can be done in various ways (Fig 3). A vertical layer of insulating material (a) can be used; this should extend from finished floor level down to a depth of not less than 250 mm, but can with advantage be taken down to the top of a strip foundation. Alternatively (b) a horizontal strip about one metre wide can be laid in conjunction with a vertical strip through the full thickness of the floor around all exposed edges. Insulating material used in any of these positions should be of a type that is unaffected by moisture either in its performance or durability, or it should be protected from ground moisture. Data sheets which set out the properties of many insulating materials are included in 'Thermal insulation of buildings'.[4]

Corrections to Table 1 to allow for the effects of edge insulation as in Fig 3 (a) are given in Table 2. The detail shown in Fig 3 (b) will have a performance at least as good as with the same amount of insulation used as in 3 (a).

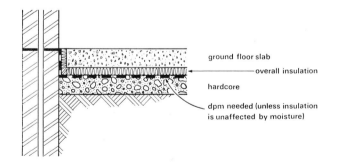

Fig 2 Overall floor insulation

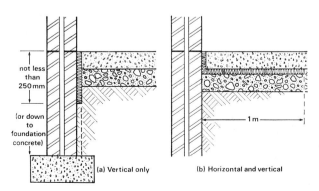

Fig 3 Edge insulation

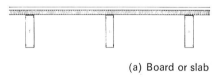

(a) Board or slab

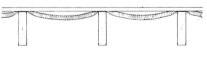

(b) Blanket, quilt or foil

Fig 4 Insulation above joists

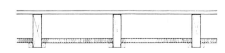

Fig 5 Insulation between joists

Table 3 Suspended floors directly above ground

Dimensions of floor	Basic thermal resistance (any floor structure) $R_{si}+R_a+R_e$	U-values: timber floors		
		Bare or with linoleum, plastics or rubber[1]	With carpet or cork[2]	With any surface finish and 25 mm quilt[3] (as Fig 4b)
metres	$m^2\,°C/W$	$W/m^2\,°C$	$W/m^2\,°C$	$W/m^2\,°C$
Very long × 30	5·35	0·18	0·18	0·16
× 15	2·82	0·33	0·33	0·26
× 7·5	1·67	0·53	0·52	0·37
150 × 60	7·13	0·14	0·14	0·12
× 30	4·58	0·21	0·21	0·18
60 × 60	5·90	0·16	0·16	0·14
× 30	4·02	0·24	0·23	0·20
× 15	2·54	0·37	0·36	0·28
30 × 30	3·42	0·28	0·27	0·22
× 15	2·34	0·39	0·38	0·30
× 7·5	1·55	0·57	0·55	0·39
15 × 15	2·03	0·45	0·44	0·33
× 7·5	1·44	0·61	0·59	0·40
7·5 × 7·5	1·27	0·68	0·65	0·43
3 × 3	0·75	1·05	0·99	0·56

(1) assuming $R_s = 0·20$
(2) assuming $R_s = 0·26$
(3) assuming $R_s = 0·86$

Suspended ground floors

A suspended ground floor above an enclosed air-space is exposed to air on both sides but the air temperature below the floor is higher than the out-side air temperature because the ventilation rate of the underfloor air space is very low. The U-values are therefore lower than was assumed in the past, when the underfloor temperature was taken as the tempera-ture of the outside air.

Table 3 gives the basic thermal resistances of suspended ground floors excluding the resistance of the structure, which must be added in order to cal-culate the U-values. The figures in column 2 are the sums of the inside surface (ie the floor surface) resistance R_{si}, the resistance of an airspace R_a ventilated by 2000 mm² gaps per linear metre of

boundary and the resistance of the earth R_e. To these basic resistances must be added the resistance of the proposed floor structure together with any added insulation R_s. The reciprocal of the sum of these resistances is the U-value.

The results of this calculation for timber floors, either bare or covered with a thin finish of low thermal resistance (R_s assumed to be 0·20), are given in column 3. U-values obtained with a cover-ing of higher thermal resistance (R_s assumed to be 0·26) are given in column 4. U-values obtained with any surface finish and 25 mm quilt, as shown in Fig 4b (R_s assumed to be 0·86), are given in column 5.

Insulation of a suspended ground floor
Additional insulation of a suspended wooden ground floor is commonly provided either in the form of a continuous layer of semi-rigid or flexible material laid over the joists (Fig 4) or semi-rigid material between the joists (Fig 5). With concrete or hollow pot floors it will usually be more convenient to place the insulation above the structural floor.

The thermal resistance of boards or slabs laid over joists (Fig 4a) and of airspaces, or of overall insula-tion above the structural floor must be added to the basic resistance (column 2 of Table 3) and the resistance of the floor structure to calculate the U-value. Standard thermal resistances for airspaces between the various layers of the construction are given in Table 4. Blankets and quilts laid over the joists will be effective only over their uncompressed area between the joists; foils are effective only where they operate in conjunction with an airspace—again, therefore, only between the joists. This is unimport-ant, however, because the resistance of the joists compensates for the absence of insulation.

Table 4 Standard thermal resistance of unventilated airspaces

Type of airspace		Thermal resistance* $m^2\,°C/W$
Thickness	Surface emissivity	
5 mm	High	0·11
	Low	0·18
20 mm or more	High	0·21
	Low	1·06
High emissivity planes and corrugated sheets in contact		0·11
Low emissivity multiple foil insulation with airspace on one side		1·76

* Including internal boundary surface

Vapour barriers

The introduction of an insulating layer on the underside of the floor raises questions as to the incidence of condensation on the colder faces of the construction and, consequently, of the need for a vapour barrier. A vapour barrier could set up a dangerous situation in a wood floor in the event of the space within the floor collecting water either by spillage or by leakage from plumbing or heating installations. The water could not then escape by draining or evaporating to the airspace below the floor and evaporation through the flooring would be so slow that dangerous conditions conducive to fungal attack could persist for a long period. The designer usually has no control over future treatment of the floor, particularly as to the nature of any finish that may be laid later. An impervious or nearly impervious finish such as pvc, rubber or linoleum sheeting would give good protection against spillage but in the event of leakage within the thickness of the floor it would further aggravate the situation. The safest course is not to provide a vapour barrier to a floor that is constructed of or incorporates wood.

References

1 Journal IHVE Nov 51 19 (195) pp 351–372
2 IHVE Guide, Book A : 1970. The Institution of Heating and Ventilating Engineers, 49 Cadogan Square, London SW1. £6·00
3 BRS Digest 108 : Standardised U-values
4 Thermal insulation of buildings; C C Handisyde, D J Melluish : HMSO London 1971. £1·75

Table 5 Thermal properties of some materials

Material	Density kg/m³	Conductivity (k) W/m°C	Resistivity (1/k) m°C/W
Asbestos insulating board	750	0·12	8·3
Carpet		0·055	18·2
—wool felt underlay	160	0·045	22·2
—cellular rubber underlay	400	0·10	10·0
Cork flooring	540	0·085	11·8
Eelgrass blanket	80–215	0·039–0·049	25·6–20·4
Fibreboard	300	0·057	17·5
Glasswool quilt	25	0·04	25·0
Linoleum to BS 810	1150	0·22	4·6
Mineral wool —felted	50	0·039	25·6
	80	0·038	26·3
—semi-rigid felted mat	130	0·036	27·8
—loose, felted slab or mat	180	0·042	23·8
Plastics, cellular —phenolic foam board	30	0·038	26·3
	50	0·036	27·8
—polystyrene, expanded	15	0·037	27·0
	25	0·034	29·4
—polyurethane foam (aged)	30	0·026	38·5
PVC flooring / Rubber flooring		0·40	2·5
Straw slab, compressed	350	0·11	9·1
Wood —hardwood		0·15	6·7
—plywood		0·14	7·1
—softwood		0·13	7·7
Wood chipboard	800	0·15	6·7
Woodwool slab	500	0·10	10·0
	600	0·11	9·1

For a more comprehensive list, see IHVE 'Guide' [2]

Heat losses from dwellings

In the United Kingdom about 30 per cent of the primary energy consumed is used in dwellings. The need to conserve primary energy resources and the cost of energy have led to an increased awareness of the need for thermal insulation; recent amendments to the Building Regulations, introduced to minimise condensation, indirectly limit the amount of energy used for heating new dwellings.

This digest outlines an established method for determining the seasonal heat loss from dwellings and the most suitable measures for restricting it but does not deal with cost benefits. Nevertheless the method may be used to compare heat losses associated with different standards of insulation for ground floors, external walls and roof structures for flats, semi-detached, detached or terraced houses.

In practice, it is frequently found that thermal improvements to existing dwellings partly result in improved comfort conditions and partly in reduced energy consumption, and that it is in only a small percentage of dwellings that the benefits of reduced heat losses are all taken in reduced energy consumption. A survey of energy usage by local authority tenants in Scotland showed that on average only about half of the potential savings from improved thermal insulation were in fact realised as actual energy savings, the other half being used to produce higher temperatures; tenants with the lowest heat usage did not save any energy when the insulation was improved.

The major heat losses occur in dwellings which were built with a low standard of insulation and the examples indicate the savings that can be made by upgrading local authority houses built in the 1940s/1950s. This method of calculating heat losses can be applied to non-standard housing and to other buildings although the wall areas, etc, may be more difficult to determine, and in larger buildings there may be different temperatures in different parts of the building.

Prepared at Building Research Establishment Scottish Laboratory, Kelvin Road, East Kilbride, Glasgow G75 0RZ.
Technical enquiries arising from this Digest should be directed to Building Research Advisory Service at the above address.

Factors influencing the rate of heat loss

In any building where the air temperature inside is higher than that outside, heat loss will occur by conduction through the enclosing structure of the building and by the interchange of the inside and outside air; these two forms of heat loss may be referred to as *Fabric heat loss* and *Ventilation heat loss*. To maintain any required temperature conditions inside the building the rate of heat input must be equal to the rate of heat loss; the heat input in dwellings is provided mainly by the heating system but in addition some heat gain is obtained from the sun, some is due to heat produced by people and some is the result of cooking and the use of domestic hot water.

Many factors influence the rate at which heat will flow from the inside of the heated building to the colder air outside but it may be useful to summarise the major ones briefly.

The thermal transmittance or U-value of the exterior shell of the structure Those external parts of the dwelling which enclose the living space, usually the ground floor, the external walls and the roof and uppermost ceiling have to be considered. The Building Regulations 1975 require that the maximum U-value for the external wall should not exceed 1·0 W/m² °C and for the roof should not exceed 0·6 W/m² °C and assume that the U-values for single and double glazing are 5·7 and 2·8 W/m² °C respectively.

The air change rate This is the rate at which the heated air within the building is displaced by colder air from outside. The air flow will occur through open windows and doors, through any gaps in the structure such as those around closed windows and doors, and through ventilators and flues. The rate of air change varies considerably depending on the magnitude of the two motive forces, wind and the temperature difference between the inside and the outside air.

The difference in temperature between the inside and the outside air An increase in this temperature difference results in a greater heat loss through the exterior shell by conduction in addition to increasing the heat loss due to ventilation. Good thermal insulation is of particular importance in a building where a high standard of heating is maintained.

The area of the external part of the dwelling For a building of any given floor area and construction, the greater the area of the external surfaces the greater will be the rate of heat loss. In this respect a flat generally performs better than a house, and a terraced house better than a detached house of similar shape and size (this is discussed further under *Fabric heat loss*).

The area of windows During sunny periods some solar heat may be gained through south-facing windows; heat loss at other times and on other aspects is increased by large windows although this is influenced to some extent by curtaining. If large windows are desired they should preferably be placed on the south side of the building. The provision of unnecessarily large opening portions may add to the air gaps in the structure and increases heat loss by increasing the rate of air change.

The type of heating system Many modern heating systems are designed without any flues and require no ventilation and this can result in a low ventilation rate in the dwelling. However, if the heating appliance has a flue, appreciable heat loss can occur from an excessive volume of heated air passing up the flue, the influencing factors being the restriction to air flow provided by the appliance and by the cross-sectional area of the flue throat. For example, the volume of air passing up the flue of an open fire with a normal sized throat may be around 170 m³ per hour but if the throat is not greater than 130 cm² in area or if the appliance is a flued gas heater, the air flow is likely to be less than half this volume.

Fabric heat loss

The rate of heat loss by conduction through the fabric can be calculated by multiplying the area of the various components such as the walls, windows, ground floor and roof by their appropriate U-value and by the difference between the inside and outside temperatures. The total obtained by the addition of these various products is the total fabric heat loss.

When considering groups of dwellings, such a procedure is likely to prove tedious as the temperatures in the various parts of the dwelling will vary and the areas of exposed walling, roof and floor structure, and of windows, need to be calculated for each part of the structure according to the variation of inside temperature. There are also likely to be differences in the plans and elevations of the dwellings necessitating the repetition of such calculations for each individual design. For most practical purposes, however, when a comparison of heat losses from different constructions is desired, or the heating load for a group of dwellings is required, certain approximations may reasonably be used in dealing with average dwellings to avoid the need for lengthy calculations.

Heat losses are high in dwellings which were built with a low standard of insulation, and a substantial energy saving could be obtained by upgrading these dwellings. Typical data on the area of windows, walls, etc, and their relation to floor area in dwellings were obtained some years ago for over 200 pre- and post-war local authority dwellings. Little difference was observed between the results for pre- and post-war houses or between different local authorities. Table 1 gives the ratios of exposed wall, ground floor, roof and window areas to the 'floor area' for typical dwellings.

52

Table 1

Floor type	Exposed wall to 'floor area'	Ground floor to 'floor area'	Roof to 'floor area'
Semi-detached (2-storey)	1·00	0·5	0·5
Terraced (middle) (2-storey)	0·75	0·5	0·5
Flat (ground floor)	0·75	1·0	0·0
Flat (top floor)	0·75	0·0	1·0
Detached (2-storey)	1·25	0·5	0·5
Flat (middle floor)	0·75	0·0	0·0

('Floor area' is the total of the floor areas within the external walls. It was found that the ratio of area of windows to floor area was very similar and an average figure of 0·18 can be adopted for typical dwellings)

For most purposes in calculating heat loss from typical local authority dwellings built in the 1940s/1950s, the areas of the various components of the surrounding structure (that is of the exposed walls, ground floor, roof and windows) can be obtained with sufficient accuracy from the above data if the floor area is known.

For these typical houses it is sufficiently accurate to use a mean house temperature as the temperature throughout the whole of the house on which the heat loss calculation can be based. In this way the difference in temperature between the air inside and outside the house may be taken as the same for all parts of the house, avoiding the necessity of calculating the heat loss for each room in turn and enabling the total areas of the exposed walls, ground floor and roof to be used. The temperature difference between the inside and the outside of typical houses with heating methods normally used averages between 5 and 10 °C throughout the heating season.

Example

The following example uses the above data to calculate the conduction heat loss from a typical semi-detached house of 100 m² 'floor area', with U-values for walls, ground floor and roof of 1·7, 0·6 and 1·8 W/m² °C respectively. These U-values are for 275 mm cavity brick walling, concrete ground floor, and a pitched roof of tiles on battens on felt with plasterboard ceiling. The following example shows the conduction heat loss calculated for the various components of the enclosing structure using the ratios of the areas as given in Table 1. The mean temperature difference is assumed to be 7 °C.

To obtain the heat loss for a heating season of 33 weeks, using the 'degree–day' method,* the average heat loss of 2748 W should be multiplied by 5544 (33 weeks × 7 days × 24 hours) to obtain the heat loss in kWh and divided by 278 to convert to Gigajoules (GJ). Thus the conduction heat loss in a heating season is 15,250 kWh, or 55 GJ.

From this calculation the relative value of any desired improvement of the thermal insulation of the walls, ground floor or roof can be assessed. *The average heat loss should not be used to determine the rating of a heating appliance, since the appliance must be able to meet the maximum demand. Calculations of appliance ratings should be based on the design temperature difference between inside and outside.*

Ventilation heat loss

The ventilation rate of dwellings can vary widely but for typical dwellings on sheltered and exposed sites the ventilation rate can be taken as one and two air changes per hour (ach) respectively. In houses without flues (or with appliances with balanced flues), ventilation rates will be lower and rates of ½ ach are not uncommon. To estimate the heat loss due to ventilation, the volume of the dwelling within the external structure should be calculated and a deduction of, say, 10 per cent made for dead space (cupboards, partitions, etc). For a typical dwelling in which the floor area in 100 m² and the ceiling height is 2½ m, assuming a ventilation rate of 2 ach for quite exposed site conditions and an average temperature difference between inside and outside of 7 °C, the ventilation heat loss for a heating season may be calculated as follows:

volume (m³)	× temperature difference (°C)	× 0·33*	× hours in heating season	× no. ach
225	× 7	× 0·33	× 5544	× 2

= 5750 kWh
= 21 GJ

* Where 0·33 = kWh required to raise the temperature of 1 m³ of air through 1 °C.

If the dwelling is in a sheltered position, eg in a built-up area, the typical air change rate is likely to be 1 ach and the ventilation heat loss will then be 10 GJ.

Fabric heat losses for	Area (m²)		U-value (W/m² °C)		Mean temp. difference (°C)		Average rate of heat loss (W)
External walls	(1·00 × 100)	×	1·7	×	7	=	1190
Ground floor	(0·50 × 100)	×	0·6	×	7	=	210
Roof and top ceiling	(0·50 × 100)	×	1·8	×	7	=	630
Windows	(0·18 × 100)	×	5·7	×	7	=	718
							2748

* The 'degree–day' method is an approximate method of calculating seasonal heat losses and does not take account of varying solar heat gains through the heating season.

Total heat loss

It will be seen that the total heat loss from a typical semi-detached house of 100 m² in floor area on a reasonably exposed site is around 76 GJ per heating season, this total being composed of 55 GJ due to conduction loss and 21 GJ due to ventilation loss.

In the absence of any precise data on gain from the sun and from people in the dwelling, an average of 15 GJ per heating season seems a reasonable value to assume. The input required for the heating system in the typical semi-detached house described will, therefore, be around 61 GJ per heating season; this figure includes heat gain from cooking and from the use of domestic hot water.

Reduction of heat loss

The relative value of improving in various ways the structure of a typical house to reduce the heat loss can be obtained from the data given above.

1. Providing a 50 mm layer of insulation material at the ceiling level immediately below the roof will reduce the U-value of the roof to, say, 0·55 W/m² °C, this will result in a reduction in heat loss of about 9 GJ per heating season.

2. Insulating the cavity of a twin-leafed wall will reduce the U-value of the external walling to about 0·5. The reduction in heat loss will be about 17 GJ per heating season.

These two improvements will reduce the heat loss by 26 GJ and the required heat input to maintain the same conditions will be reduced from 61 GJ to 35 GJ per heating season, a reduction of 42 per cent. A 14 per cent reduction is due to the improvement of the roof insulation, an improvement which can be made easily and without undue expense in many existing dwellings.

3. The provision of double-glazing (U-value say 2·8) throughout the house will reduce the heat loss by not more than about 7 GJ per heating season.

4. In addition to any improvement in the thermal insulation of the structure, a reduction of the air change rate will also reduce the heat loss. In some experimental houses where the average ventilation rate was about 2 ach it was found that weather-stripping the two external doors only could reduce the rate by approximately $\frac{1}{2}$ ach on an exposed site. This would reduce the hourly air change rate to $1\frac{1}{2}$

thus making a reduction in the heat loss of about 5 GJ per heating season.

Dwellings built to modern design without flues can have very low ventilation rates and it is suggested that weather-stripping is not undertaken in these cases because of the increase in the risk of condensation.

For well-insulated construction, the grouping of dwellings also assists in limiting the heat loss because of the smaller area per dwelling of exposed surface of the enclosing structure. The heat input for a heating season will be about 37 GJ for a detached house, 35 GJ for a semi-detached house and 33 GJ for an inner house in a terraced block assuming dwellings of 100 m² floor area, having U-values for the external walls, ground floor and roof of 0·5, 0·6 and 0·55 W/m² °C respectively and with single glazing an area 18 per cent of the floor area, a 7 °C difference of temperature between the inside and outside air and a ventilation rate of 2 ach.

The heat loss from a terraced house is, therefore, over nine per cent less than that from a detached house of the same size and construction.

Building Regulations (thermal insulation) give a minimum statutory requirement. From the examples given a saving of about 40 per cent in the heating requirements can be obtained by increasing the insulation above the minimum in walls and roofs and this should be a standard at which to aim.

As any improvement of the wall insulation in existing houses is likely to be expensive, the most reasonable measures to adopt are the provision of roof insulation at the ceiling level.

In providing roof insulation it should be remembered that water tanks or pipes situated in the roof space must also be well insulated and the small part of the ceiling immediately under the tank should be left uninsulated.

It should be emphasised that the calculations give an indication of the fuel savings to be obtained if internal temperatures are not increased after improving the insulation standard. In general this is not the case and part of the benefit of insulation is to provide higher temperatures.

Double glazing and double windows

Two diverse environmental requirements—one, a desire for improved heat insulation, the other, the need for protection against external noise—have stimulated the production and use of various forms of double glazing.

The essential difference between double glazing designs to meet these two needs lies in the width of the air space. For heat insulation, air-space widths as small as 6 mm are of some value; for sound insulation, air-space widths of not less than 100 mm, and preferably more, are essential, but there are other factors which are also discussed in this digest.

Heat insulation

In response to the general desire for improved heat insulation in heated buildings, various double glazing and double window systems are available, ranging from simple 'do it yourself' sealed *in situ* systems to the more sophisticated factory produced hermetically sealed double-glazing units and from double-rebated frames to openable coupled casements and sashes or separate secondary windows (see Figs 1–5). Some of the practical considerations peculiar to the various types are discussed later.

Width of air space

For effective heat insulation, the optimum width of air space for vertical double glazing varies according to the mean temperature of the air space and the temperature difference across it, but it is usually taken as 20 mm. Figure 6 shows that, for all practical purposes, the optimum is maintained down to an air-space width of about 12 mm. Below this width, the heat transmission becomes progressively greater until it approaches the figure for single glazing. For air-space widths greater than 20 mm, the heat transmittance remains practically constant.

Table 1 Thermal transmittance (U) in W/m² °C through glazing (without frames)

Glazing system	Degree of exposure:		
	sheltered	normal	severe
Single	5·0	5·6	6·7
Double			
air space 3 mm wide	3·6	4·0	4·4
6 mm	3·2	3·4	3·8
12 mm	2·8	3·0	3·3
20 mm (or more)	2·8	2·9	3·2
Triple			
each air space 3 mm wide	2·8	3·0	3·3
6 mm	2·3	2·5	2·6
12 mm	2·0	2·1	2·2
20 mm (or more)	1·9	2·0	2·1

Comparative performance of double glazing

Given air-space widths of not less than about 12 mm, the heat insulation of single-glazed windows can be improved appreciably by double glazing; in general terms, an air-space between two layers of glass halves the thermal transmittance. Table 1 shows the comparative thermal transmittance ('U') values for double glazing for four widths of air-space and three grades of exposure. For comparison, the U-values for single glazing and triple glazing are included.

Fig 1 Typical factory-sealed double glazing units

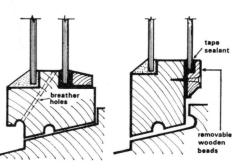

Fig 2 Typical glazed *in situ* double glazing

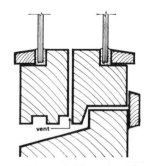

Fig 3 Typical double windows, coupled type

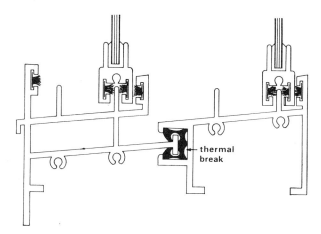

Fig 4   Horizontal sliding type, metal

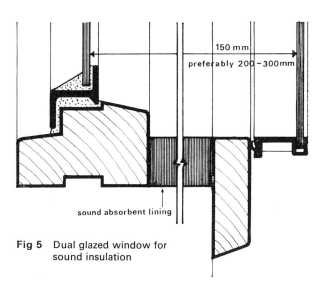

Fig 5   Dual glazed window for sound insulation

The above data relate to transparent and translucent glasses. For glazing systems using heat-reflecting glasses with metallic surface coatings (not laminates) of low heat emissivity, the metallic coating improves the insulation against heat loss. For double-glazing units, the improvement in insulation provided by the unit may be as much as 30–40 per cent.

Effect of window frames

The U-values in Table 1 are for the glass only and do not allow for the effect of the surrounding frame and glazing bars. In wood windows, the thermal resistance of the frame can contribute usefully to the insulation value of the whole window but the reverse is true of metal windows (because of the relatively high conductance of the metal frame) unless thermal breaks are introduced to offset the 'cold bridge' effect. Some proprietary metal double windows now incorporate insulating plastic insets in the frame section to break the thermal continuity (Fig 4). For heat loss calculations, it is often useful to know the average U-value for the window as a whole, and unless the frame section is complex in profile, its effect on the average thermal transmittance can be worked out on a proportional area basis. Table 2 gives comparative U-values for a number of typical wood and metal windows.

Tables 1 and 2 are derived from the IHVE 'Guide'.[1] The three categories of exposure are defined as:

sheltered Up to third floor of buildings in city centres

normal   Most suburban and country premises: fourth to eighth floors of buildings in city centres

severe   Building on the coast or exposed on hill sites: floors above the fifth of buildings in suburban or country districts: floors above the ninth of buildings in city centres.

Loss through windows in relation to the whole

To assess the value of double glazing in terms only of heat saving under steady-state conditions, the heat losses through the glazing must be considered in relation to the overall heat losses. A broad comparison can be made by taking the average thermal transmittances for the different parts of the fabric, multiplying them by the appropriate areas to give the comparative rates of conduction heat loss per degree C difference of temperature between indoors and outdoors and then adding the ventilation heat loss. In a two-storey semi-detached house of 100 m² gross floor area, the component rates of heat loss might be:

Conduction heat loss	Average U-value W/m² °C	Area m²		Rate of heat loss W/°C
Roof	0·50	× 50	=	25
Ground floor	0·76	× 50	=	38
Unglazed walling	1·50	×100	=	150
Windows (single-glazed)	4·30	× 17	=	73

Total rate of conduction heat loss through fabric 286

Ventilation heat loss (assuming 1 air change/hour)

$$= 240 \text{ m}^3 \times \frac{\text{W/m}^3 \text{ °C}}{3} = 80$$

Total 366

Thus the single-glazed windows account for about 20 per cent of the total heat loss including ventilation, or about 50 per cent of the conduction loss through the unglazed external walls. If double-glazed windows with an average U-value of 2·5 W/m² °C are substituted for the single-glazed windows, the rate of heat loss through the windows would be reduced to about 42 W/°C, and the corresponding total rate to about 335 W/°C. The heat loss through the windows would thus be reduced to about 13 per cent of the total.

Table 2 Thermal transmittance (U) in W/m² °C through typical windows (including frames)

Window type	% total window area occupied by frame	Degree of exposure: sheltered	normal	severe
Single-glazed:				
metal casement*	20	5·0	5·6	6·7
wood casement	30	3·8	4·3	4·9
Double-glazed :† metal horizontal sliding window with thermal break	20	3·0	3·2	3·5
wood horizontal pivot window	30	2·3	2·5	2·7

* Metal frame assumed to have a similar thermal transmittance to that of the glass
† With 20 mm air space

Double glazing and comfort

Apart from the reduction in heat loss gained by using double glazing, the increase in surface temperature of the glass facing the room may improve the comfort particularly of persons sitting or working near the windows; the discomfort caused by radiation losses and cold convection currents from the cold surface of single glazing is well known. In general, the surface temperature of the glass facing a heated room in a double glazing system in winter will be some 4–7°C higher than that of single glazing for the same internal and external temperature conditions (Fig 7).

Condensation

An explanation of the conditions which lead to condensation is given in Digest 110.[2] Condensation can occur on the surface of the glass and also on the frame, particularly if it is metal. The condensation on the glass is more troublesome because of the relatively large surface area and because it interferes with the view out. Where condensation on the glazing persists for long periods the run-off of excess moisture can lead to deterioration of the bottom member of the frame and spoil the appearance of the window reveals and areas of wall below the sill. Double

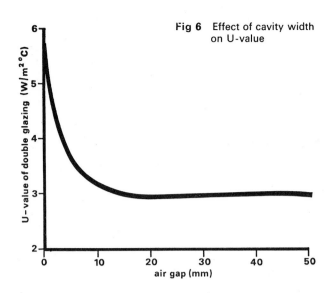

Fig 6 Effect of cavity width on U-value

glazing can reduce the risk of condensation on the glass because the surface exposed to the room is warmer than single glazing and has a better chance of being above the prevailing dew-point temperature. In practice, this will depend on the relative humidity, and in buildings where high humidities occur, coupled with low standards of heating, there is no guarantee that double or even triple glazing will

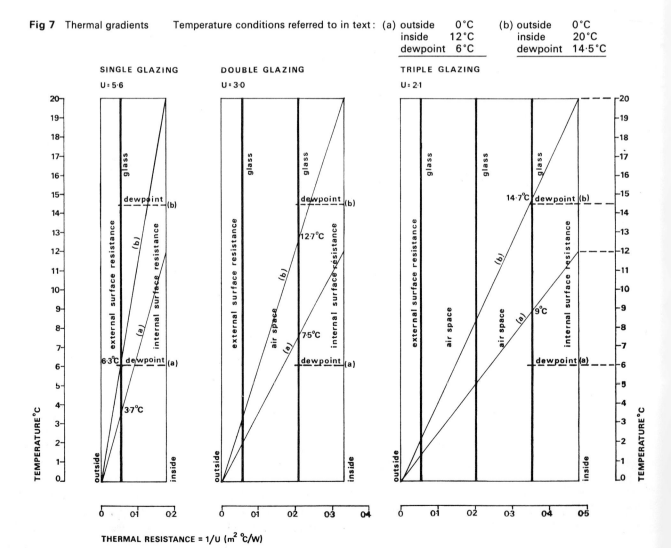

Fig 7 Thermal gradients   Temperature conditions referred to in text: (a) outside 0°C / inside 12°C / dewpoint 6°C   (b) outside 0°C / inside 20°C / dewpoint 14·5°C

THERMAL RESISTANCE = 1/U (m² °C/W)

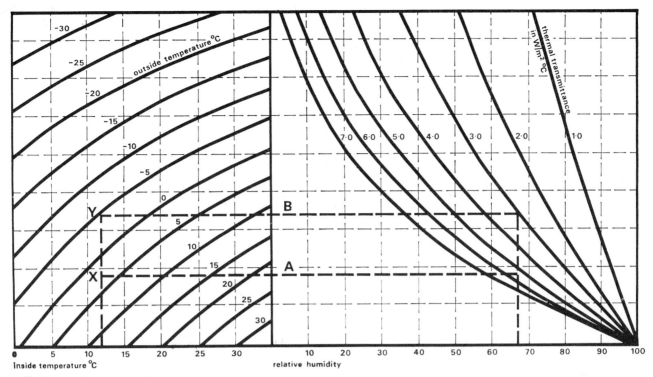

Fig 8 Condensation prediction chart

always avoid condensation on the glass. For instance, assuming an internal temperature of 12°C, an outside temperature of 0°C and a dew-point temperature of 6°C (typical of conditions in a bedroom), condensation would occur on the surface of single glazing, but not on double or triple glazing (see Fig 7). Raising the room temperature to 20°C and hence the glass surface temperature, for the same humidity conditions, there would still be a risk of condensation on single glazing. For conditions of high humidity, such as might occur in a domestic kitchen, assuming an internal temperature of 20°C, an outside temperature of 0°C and a dew-point temperature of 14·5°C, condensation would occur even on double glazing.

If the relative humidity and indoor and outdoor temperatures are known, the possibility of condensation on glazing can be predicted from Fig 8. This gives comparative dew-point curves for a range of U-values which embraces those given in Table 1 for single, double and triple glazing. The chart applies only to steady-state conditions but it can be used for a first check on the likelihood of condensation. As an example, the broken line **A** on the chart indicates the conditions of relative humidity 67 per cent, thermal transmittance 5·6 W/m²°C (for single glazing) and an inside temperature of 12°C; condensation is likely to occur when the outside temperature drops to about 3°C (point 'X' on the chart). The line **B** shows the same temperatures and relative humidity plotted against a thermal transmittance of 3·0 W/m² °C (for double glazing); the outside temperature must now drop to nearly –5°C (point 'Y') to cause condensation on the glass.

Sound insulation

The simplest way of increasing the sound insulation of an element of structure is to increase its mass; in a general sense this is true for windows, but with such lightweight components a substantial improvement can be obtained by double-leaf construction, as shown in Table 3 (taken from Digest 128[3]), provided that the windows are sealed, for example by weather-stripping, and that the air space is wide enough to give the required insulation at the lower frequencies.

Even small gaps in double windows can impair the sound insulation and, therefore, if at all possible, one leaf of a double window should be a fixed light system. If both leaves have openable lights in them, all the openable lights should be weather-stripped. For rooms subject to high humidity, it may be necessary to seal the inner leaf of glazing to prevent condensation in the air space as discussed on page 57.

Table 3 Sound insulation of windows

Description	Sound Reduction (av 100–3150 Hz)
Any type of window when open	about 10 dB
Ordinary single openable window closed but not weather-stripped, any glass	up to 20 dB
Single fixed or openable weather-stripped window, with 6 mm glass	up to 25 dB
Fixed single window with 12 mm glass	up to 30 dB
Fixed single window with 24 mm glass	up to 35 dB
Double window, openable but weather-stripped, 150–200 mm air space, any glass	up to 40 dB
Double window in separate frames, one frame fixed, 300–400 mm air space, 6–10 mm glass sound-absorbent reveals	up to 45 dB

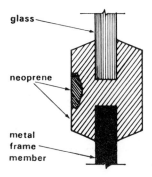

Fig 9 Flexible mounting of glass in neoprene gasket

Width of air space

The optimum air-space width for thermal insulation is about 20 mm, but this is too small to be of any practical advantage for sound insulation. For protection against road traffic noise, in which the low frequency components predominate, a minimum air-space width of 150 mm is recommended and preferably more—say 200–300 mm whenever it is economically obtainable. Unless an air space at least 100 mm wide can be provided, practical insulation against road traffic noise may well be obtained more effectively by heavy single glazing selected to give the same average performance.

Room ventilation

Table 3 shows that to obtain a useful improvement in sound insulation windows must be closed. A room with permanently closed double windows will usually need some form of mechanical ventilation through a duct system. The ducts should be lined with acoustic absorbent material to give sound attenuation comparable with that of the insulation provided by the building enclosure. For the ventilation of individual rooms there are fan-operated ventilating cabinets with sound attenuation designed to match the sound insulation provided by double windows.

Absorbent linings and flexible mountings

The provision of sound-absorbent linings, eg acoustic tiles, on the reveals between double windows can give marginal improvements in sound

Table 4 Effect of comparative area of window on the sound insulation of external walls rated at *40 dB and †50 dB respectively

	Average sound insulation (dB)			
	20 dB window		40 dB window	
Percentage of glazing area to wall area	40 dB wall	50 dB wall	40 dB wall	50 dB wall
0 (windowless wall)	40	50	40	50
10	30	30	40	47
25	26	26	40	45
33	25	25	40	44
50	23	23	40	43
75	21	21	40	41
100 (fully glazed wall)	20	20	40	40

* Equivalent of walls weighing about 120 kg/m², for example, a 50 mm dense concrete composite panel backed with thermal insulation material
† Equivalent of walls weighing about 480 kg/m² for example, a 215 mm solid or a 255 mm cavity brick wall

insulation. Flexible edge-mounting of the glass, for instance in neoprene gaskets, see Fig 9, can also help by promoting resonance damping. Recent experiments have shown that the latter treatment can improve the performance of double windows by at least 5 dB over most of the frequency range 100–3150 Hz, but the extra sound insulation can only be realised if the windows are well sealed; air leakage would tend to mask any benefit from the use of flexible mountings.

Windows in relation to the whole building

The improvement in sound insulation obtainable from double windows depends to some extent on the standard of insulation of the rest of the fabric and on their respective areas, although there is no simple relationship. The net insulation for windows and wall for a range of relative areas is given in Table 4, taken from CP153: Part 3.[4] If the standard of the walling is of the same order as the windows, the percentage area of glazing will have no significant effect on the net sound insulation; if it is appreciably higher, the net insulation of the wall and windows will be less than the wall insulation. The converse is also true as the standard of insulation of the wall or other parts of the building fabric may be lower than that of the windows. For example, in an ordinary house, a tiled roof plus ceiling may have an average sound insulation of about 35 dB as against 40 dB or more for double windows. Thus if it is required to improve the protection against external noise the restricted performance of the roof could inhibit the improvement sought by double glazing.

Effect on light transmission

The transmission of light through glass varies with the angle of incidence and the properties of the glass, but for categorising different types of glazing, a percentage representing the light transmission at normal incidence is usually given by manufacturers. The light transmission through clear glass up to about 6 mm thick could be of the order of 85–90 per cent for single glazing, and about 70 per cent for double glazing. For daylight calculation purposes, it is usually convenient to calculate the daylight factor on the assumption of a single thickness of clear glass and then if necessary to apply a correction factor to allow for the reduced transmission of double glazing or other types of glazing material. Table 5 gives correction factors for a range of typical glazing materials in double glazing, the factor for single clear glass being taken as unity.

For comparison purposes, the transmission of radiant solar heat through various glasses is also given by manufacturers on a percentage basis assuming normal angles of incidence and taking into account the proportion of incident radiation reflected at the outer surface of the glass and the proportion absorbed by the glass and re-radiated.

Table 5 Recommended daylight correction factors and solar gain factors to allow for effect of various types of double glazing

Glasses	Recommended correction factor to apply in daylight calculations	Solar gain factor (S)	Alternating solar gain factor (Sa)	
			Heavy-weight building	Light-weight building
4–6 mm clear single glazing	1·00	0·76	0·42	0·65
4–6 mm clear both panes	0·90	0·64	0·39	0·56
6 mm surface-tinted bronze outer pane 4–6 mm clear inner pane	0·45	0·47	0·32	0·44
6 mm selectively absorbing green outer pane 4–6 mm clear inner pane	0·70	0·42	0·30	0·40
6 mm body-tinted grey outer pane 4–6 mm clear inner pane	0·35	0·39	0·28	0·37
6 mm heat-reflecting laminate (gold) outer pane 4–6 mm clear inner pane	—	0·15	0·12	0·14

When calculating cooling loads in air-conditioned buildings or temperatures in uncooled buildings, it is necessary to know the solar gain through the windows; this varies with the angle of incidence and the properties of the glass. A simplifying concept, adopted in the IHVE Guide[1] allows for these varying characteristics under representative conditions and employs a *solar gain factor* to allow for the effect of a specified glass and any shading devices. In a broad sense, solar gain factors can be used as an approximate measure of the relative effectiveness of different types of glass and sun controls in combating solar gains. For calculating daily mean solar cooling loads or indoor temperatures, the solar gain factor **S** is used; for calculating the fluctuations of solar cooling load or indoor temperature about the mean, the factor S_a is used. Values of **S** and S_a for five types of double glazing and for single clear glazing are given in Table 5.

Practical considerations

Installation

Factory-sealed units Recommended glazing procedures for various types of factory-sealed double-glazing units are given in a booklet[5] issued by the Insulation Glazing Association. The booklet limits its advice to new work; for installing units in existing frames it recommends consultation with the manufacturer.

The glazing instructions relate to minimum depth and width of rebate to accommodate the thickness of the double-glazing unit, fixing of beads, type of bedding compounds, provision of setting blocks and distance pieces to ensure that the unit is correctly positioned in the frame and preparation of the frame. In addition, the booklet gives 'step by step' site glazing procedures appropriate to various kinds of bedding systems, and recommendations for the handling and

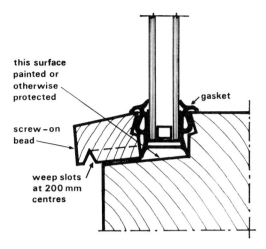

Fig 10 Drained glazing system

storage of units. Some manufacturers provide lists of approved glazing compounds.

Manufacturers' standard instructions are generally in line with the IGA recommendations, supplemented in some cases by additional guidance on the application of the bedding compounds and routine pre-glazing checks.

Failure to follow the glazing instructions of individual manufacturers may invalidate their warranty, which is usually for a period of ten years, but may be longer, and covers failure of the unit to function properly due to deterioration of the seal. If the seal fails, the unit cannot be repaired and can only be replaced by a new one.

Sealed units with fused all-glass edges are being increasingly used and a type using a welded glass-to-metal seal is also available, but there is still considerable use of the form in which flat sheets of glass are bonded to spacing strips and sealed. With the latter, it is important that the edges are kept dry, as continual wetting tends to destroy the bonded joints. For this reason, special care is needed to ensure that the glazing method and compound will protect the edges of the unit. The compound must also be compatible with the bonding materials of the unit. Attention has recently been directed to the value of drained glazing systems because of the difficulty of ensuring, particularly at the bottom edges, that no

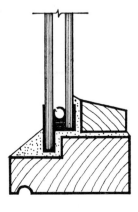

Fig 11 Typical stepped unit

crevices are formed where water may lodge between the unit and the bedding compound; a typical system is shown in Fig 10.

Some firms produce double-glazing units to fit into the rebates of standard sections, wood and metal, without the use of face beads. For frames that are too small, or unsuitable for enlarging to accept the thickness of standard units, stepped units are available (Fig 11).

Sealed double-glazing units with dark coloured, heat-absorbing or heat-reflecting glasses are liable to be heated strongly by the sun. If during a sunny spell the glass is put in partial shade, for example by other buildings or parts of the building, differential thermal movements may cause fracture. There is some evidence also of an increased likelihood of thermal stresses in double-glazed units when tinted glasses are backed by internal blinds of high reflectance. If these conditions are likely to arise the manufacturers should be consulted.

Single-frame double glazing systems sealed *in situ* Many attempts have been made to provide sealed double glazing by glazing to double rebated frames, or by fixing a second line of glazing with wood or plastic face beads, generally to existing frames (see Fig 2). Experience has shown that however well the glazing seals are made, the cavities cannot be expected to remain air-tight indefinitely; under the various movements and shrinkages that occur the seals may break down, water vapour (and dust) then finds its way into the air space and at some times condenses on the inside of the outer glazing.

If the inner panes are bead glazed it may be fairly easy to remove them for cleaning but it is often difficult to reseal the windows effectively and the 'slip-on' or 'press-on' type of added glazing, using proprietary compressible plastic edging strips, may be easier to reseal.

For wood windows, double-glazed *in situ*, the wood exposed to the air space should be painted or varnished to reduce the evaporation of moisture from the timber into the air space and breather holes to the outside should be provided on the basis of one 6 mm diameter hole for 0·5 m² of window. The holes should be plugged with material such as glass fibre insulation or nylon to exclude dust and insects. The frame material of metal or plastic windows will not contribute to a build up of moisture in the air space. The double glazing should be completed under relatively dry conditions (preferably in cold weather) to prevent trapping humid air in the air space and the room-side glazing should be sealed as effectively as possible.

Coupled or sliding double sashes These may be pivoted, with openable coupled sashes, or they may slide, with pairs of sliders operating in the same frame; the former may be wood or metal but the latter

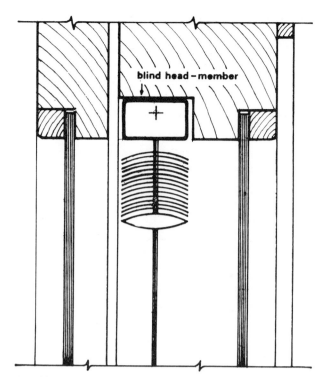

Fig 12 Coupled type double sash with venetian blind

are usually metal (Figs 3 and 4). To reduce the risk of condensation, the inner sashes should be well sealed when the sashes are closed together. For the same reason, in the coupled pivoted type a ventilating slot is commonly left round the periphery of the outer sash to ventilate the air space to outdoors when the sashes are closed. This does not seriously reduce the insulation of the window. In the sliding types, the air space is not usually ventilated because the inner surfaces are readily accessible for cleaning. The air space in the double-sash types is usually sufficient to allow retractable blinds to be installed between the inner and outer glazing (Fig 12). The blinds, usually venetian or pleated, are controlled by cords or rod action, but fully automatic control of the venetian types is available.

Double windows with separate frames When wide cavities are required for sound insulation, or it is not practicable to double glaze single-frame windows, secondary windows can be fixed to the existing frames or to ancillary frames. With the latter, a space can be left between the two panes, depending on the depth of the reveals, for fixing acoustic absorbent material to give additional sound insulation. There are proprietary secondary windows available in aluminium or PVC. They may be either hinged or sliding, to give access for cleaning; fixed lights can also be incorporated. Units can be supplied for screw fixing to the original window by householders, but specialist firms will supply and fix. The casement types usually have a compressible strip to provide a seal against the existing window or the ancillary frame; the sliding types usually have a wool pile seal in the track. Both types can be removed for cleaning or storage but large glazed units are awkward to handle. Where the secondary window has to provide access to the opening lights of the outer frame for cleaning it is necessary to ensure that the opening lights in both frames work conveniently together. In this respect, difficulties can arise with pivoted windows because of the unavoidable projection of their opening lights across the cavity.

If tinted glass is used to reduce solar heat gains, blinds of high reflectance in the inter-space may give rise to increased thermal stresses as already mentioned.

Maintenance

General recommendations on the durability and maintenance of windows are given in CP 153 : Pt 2.[6]

The glazing joints of factory-sealed double-glazing units should be checked at intervals recommended by the manufacturer and when repainting the window frames. For low-cost schemes using self-hardening compounds and without beads, a maintenance check every three years may be required; for silicone acrylic-based or polysulphide sealants a check every ten or twelve years may be sufficient. Where there are surface cracks, extrusion or retraction of the original compound, it should be raked out to a depth of 3 mm and replaced with a compound preferably of the same type. The joint should be shaped to direct water away from the glass on the outside and room side. Glazing beads should be re-secured as necessary and drainage holes in drained joint and gasket systems should be cleared of any dirt or other obstructions. For gasket glazing, maintenance checks at one-year intervals are recommended by specialist firms.

With sealed in situ systems it may be necessary to remove the glazing on the room side, clean the glass and reseal at intervals of two years or more depending on the effectiveness of the seal. The appearance of the inner surface of the outer glazing will show whether it is necessary to remove the inner glazing. The joints of wood frames, specially the bottom joints, should be checked for cracks, and when the glass is removed the opportunity should be taken to re-paint and re-varnish the wood exposed to the air space. The glass should be replaced under dry conditions, preferably a cold dry day, with the room amply ventilated to avoid entrapping warm moist air from the room.

For coupled or sliding double sashes or double windows with separate frames, maintenance will be as required for single windows but the fit of the inner opening lights should be checked to ensure that the seal on the room side is maintained as effectively as possible and if necessary weather-strip or flexible seals replaced by new.

Tests

There are currently no British Standard tests that are specific to sealed double-glazing units or double windows, but the IGA booklet [5] includes performance tests for glazing compounds for use with double-glazing and multiple-glazing units. BS 4254[7] includes methods of test for two-part polysulphide-based sealing compounds.

In general, for glazed in situ double-glazed single-frame windows and coupled or sliding dual sashes, tests related to the performance requirements for single-glazed windows would apply. The provisional recommendations published by the British Standards Institution in DD4[8] for resistance to wind loads, air infiltration and water penetration are relevant.

The thermal transmission of double windows is usually calculated, taking into account the width of the air space, the type of frame and the proportional areas of frame and glazing. A description of the method of calculation and the concept of standardised U-values is given in the IHVE Guide.[1]

For sound insulation measurements of windows, the standard methods given in BS 2750[9] apply.

References

1 IHVE Guide: Book A, Design Data (1971); Institution of Heating and Ventilating Engineers, London

2 BRE Digest 110 Condensation

3 BRE Digest 128 Insulation against external noise—1

4 British Standard Code of Practice CP 153 Part 3: 1971 Windows and roof lights—Sound Insulation

5 Glazing requirements and procedures for double glazing units; Insulation Glazing Association, London 1968

6 British Standard Code of Practice CP 153 Part 2: 1970 Windows and roof lights—Durability and maintenance

7 British Standard BS 4254: 1967 Two-part polysulphide-based sealing compounds for the building industry

8 British Standard Draft for Development DD4: 1971 Recommendations for the grading of windows

9 British Standard BS 2750: 1956 Recommendations for field and laboratory measurement of airborne and impact sound transmission in buildings

Further Reading

Beckett, H. E. and Godfrey, J. A.—Windows, Crosby Lockwood Staples: London (1974)

Ventilation of internal bathrooms and WC's in dwellings

The ventilation of internal bathrooms and WC's requires effective substitutes for openable windows. Natural and mechanical systems of ventilation can be designed for single dwellings or a number of dwellings, and are discussed here under the headings of individual and common-duct systems. The major part of the digest is concerned with mechanical extract ventilation by vertical common-duct systems. This digest replaces No 78, which is now withdrawn.

Natural ventilation

On the windward side of a building, the outside air pressure is normally higher than the inside pressure; on the leeward side it is normally lower. Air will move from the higher to the lower pressure zones through any apertures in the walls. The higher the wind speed, the greater the pressure difference and therefore the rate of air change.

Air is also moved by temperature difference. Air warmed within a building will tend to rise and to escape through high-level outlets; it is replaced by cooler air entering at low level. The greater the temperature difference, the faster will be the rate of air flow. This type of movement is described as 'stack effect'.

Natural ventilation in buildings depends on wind pressure, on the stack effect, or on both, and duct systems can ventilate internal rooms by these natural forces without a fan. If properly designed, for some limited applications such as individual systems serving low-rise buildings, they can perform adequately. Although such systems may contravene British regulations and byelaws, they have been

allowed and are discussed here because they could be the subject of waivers.

Individual systems

Examples of systems of natural ventilation for individual dwellings employing one or two separate horizontal or vertical ducts, or a combination of horizontal with vertical ducts, are shown in Figs 1a, b, c. Because there is a risk of these systems operating in reverse, replacement air should be drawn from the outside by an inlet duct (Fig 1a) rather than from an adjoining room or lobby (Fig 1b).

There is no connection *between* dwellings but in buildings of more than four floors, the space requirements for individual vertical ducting become excessive and wind speeds are often too high for trouble-free operation, especially of horizontal systems.

Common-duct systems

Common-duct systems for natural ventilation could effect a saving in the space needed for ducting but are not recommended because of the risk of air flow between rooms in different dwellings.

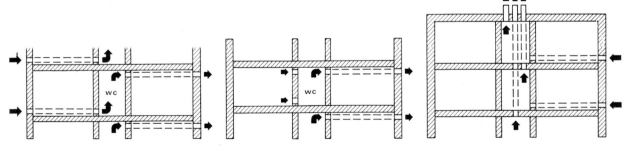

Fig 1a (Section) Horizontal system with inlet and outlet ducts

Fig 1b (Section) Horizontal system with outlet duct but no inlet duct (replacement air enters from adjoining lobby)

Fig 1c (Section) System with combination of vertical outlet ducts and horizontal inlet ducts

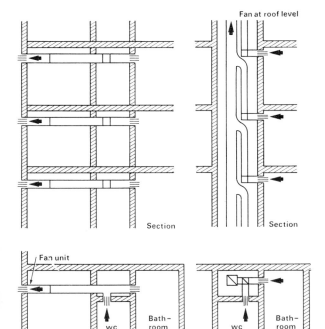

Fig 2 (Sections above, plans below) (a) individual horizontal ducts (b) common-duct with vertical shunts

Mechanical ventilation

The risks of variable or reverse air flow associated with natural ventilation systems, and their unsuitability for tall buildings, have led to the general use of mechanical ventilation in the UK. Basic types of mechanical extract system are illustrated in Fig 2.

Individual systems

Individual mechanical systems require a single duct, either vertical or horizontal, connecting the internal room to the outside air; an extract fan installed in the duct operates continuously or intermittently. There is no connection between dwellings. Such systems are specially suited to converted buildings and to owner-occupied dwellings in which the responsibility for running and maintaining the system lies with the occupier. Most local authorities and landlords prefer one or two large fans external to the dwellings, as in a common-duct system, whereas individual systems in a large building need numerous small fans and outlet points, often on external façades.

Intermittent operation may be allowed and the following notes are a guide to reasonable design. The fan should operate while the compartment is occupied and for at least 20 minutes thereafter. It should extract during any single operation at least 20 m³ of air from a WC or from a bathroom with no WC, or 40 m³ from a bathroom with WC. The air speed in the duct should not exceed 3·5 m/s. Adequate provision must be made for the entry of replacement air. A spare fan or motor, to take over in the event of breakdown, is *not* necessary.

Individual systems should also conform to the recommendations given under the following headings on pages 2 and 4:

Ventilation requirements	Section 1(b)
Noise	Section 2(f)
Flow measurement and testing	Section 2(g)
Entrance lobby and replacement air	Section 4

Common-duct systems

Most buildings with internal bathrooms have common-duct mechanical extract systems operating continuously. Horizontal branches or vertical shunts link the internal rooms of each dwelling with the common duct. Air is extracted through a grille from each room and replacement air flows into the rooms through any gaps round the doors or through grilles in corridor walls.

The notes that follow conform generally, though not in every detail, with:

The Building Regulations 1972
The Building Standards (Scotland) (Consolidation) Regulations 1971
Greater London Council Byelaws
BS CP 352:1958 *Mechanical ventilation and air-conditioning in buildings*
BS CP 3: Code of basic data for the design of buildings: Chapter IV *Precautions against fire*
BS CP 413:1973 *Ducts for building services*

1 Ventilation requirements

(a) The minimum air extract rate from a WC or from a bathroom with no WC is 20 m³/h; from a bathroom with WC, it is 40 m³/h.

(b) Extracted air should discharge to the outside air and the system should be separate from any other ventilating plant.

The ventilation requirement is expressed in terms of volume of air to be extracted rather than air changes per hour, because this can deal with rooms of different sizes. Consideration of flow in these terms simplifies testing and the specification of standard duct sizes; it meets statutory requirements for all practical purposes.

2 The system

(a) **Design air extract rates** should exceed the minimum rates stated in 1(a) by about 20 per cent, to allow for variation after balancing.

(b) **Shunt and common main ducts.** Inlets from rooms should be connected to the common main duct through a vertical shunt duct at least 1 m long; this should terminate at its lower end with an inlet piece inserted through the service duct wall from the internal room. The air speed in a shunt duct with smooth sides should not exceed 3·5 m/s, or in a main duct 7·5 m/s; higher velocities may lead to noise and cause a nuisance.

A shunt duct system in which air extracted from an internal room is passed vertically upwards (Fig 2b) is preferred to a system with horizontal branches because it is more compact in layout, offers better sound attenuation between dwellings and is less likely to allow carry-over of smoke between dwellings in the event of a fire. Because of noise transmission, flow variation and fire risk, shunt ducts from dwellings on the same floor level should not be connected to the same vertical main duct.

(c) **Extract grilles, balancing devices and dampers.** An extract grille should be fitted to each ductwork inlet piece; it should be sub-divided into no more than six equal parts. The pressure loss through a grille should be at least 25 N/m² at the minimum allowable air flow rate, so as to limit the flow variation due to external pressure changes. As an adjustable grille is closed, the pressure drop increases, air speed increases, and at some point noise will occur. The designer should ensure that the system can be balanced and will operate satisfactorily before this point is reached. 'Balancing' means adjusting the various resistances to air flow in the system so as to obtain the required flow conditions throughout. A system may be prebalanced, when resistances are set by the contractor before the ductwork is installed, or site-balanced after installation.

The air-flows into a system from various floor levels should all be approximately equal. The balancing of air-flows into a common main duct, if necessary, is best done by using an extract grille adjustable at the face, or a fixed circular orifice plate within the shunt duct. The area of the orifice should be at least one quarter that of the shunt. Main duct flows can be balanced by dampers at the heads of the vertical main ducts.

There is no statutory upper limit to the air extract rate but it is suggested that the recommended design rate should not be exceeded by more than 20 per cent; too high a rate can waste heat during the heating season. Air extract rates should, therefore, be within the following limits:

WC, or bathroom with no WC   20–28 m³/h
Bathroom with WC   40–56 m³/h

In the past, blockage of balancing devices within shunt ducts by accumulated dust and dirt has been a major cause of deteriorating performance. There should be no adjustable damper, balancing device or other adjustable flow restricter within the vertical duct-work, nor should any such device be fitted immediately behind a fixed extract grille. If blockage occurs at the grille face, it can be seen and easily cleared away. Access points should be provided to allow ducts to be swept. Dampers at the tops of main ducts are a potential source of noise because of the high air velocities through them. They are best positioned away from the uppermost dwelling, eg at points A in Fig 3.

(d) **Fans.** Fans should be capable of extracting the design total flow against the resistance imposed by the duct system and extract grilles. The fan static pressure should be at least 125 N/m² for buildings up to 18 m high and at least 185 N/m² for higher buildings, but it may be necessary to increase this to counteract wind pressures. If the loss of pressure through the ducting is less than the recommended fan static pressure, the fan will extract more than the design total air flow. This may be reduced, if desired, by fitting a damper at the fan outlet or inlet.

With sufficient fan power and in-built resistance to air flow, an extract system can be made largely immune to external changes in air pressure—mainly the effects of wind and of doors opening—which would otherwise cause fluctuations in flow.

Duplication of fans and motors to provide alternative power in the event of failure is a requirement of the GLC byelaws; the Building Standards (Scotland) (Consolidation) Regulations 1971 require duplication of the motor if the system serves more than one house. If duplicate fans are permanently fitted to a system, the change-over damper or dampers, whether manual or automatic, should be so designed that the air pressure within the system assists in holding them in the correct position. The fan unit should be regularly serviced to the manufacturer's specification.

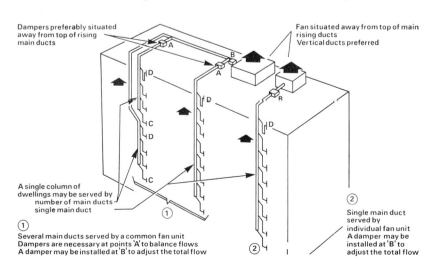

Dampers preferably situated away from top of rising main ducts

Fan situated away from top of main rising ducts
Vertical ducts preferred

A single column of dwellings may be served by number of main ducts single main duct

① Several main ducts served by a common fan unit
Dampers are necessary at points 'A' to balance flows
A damper may be installed at 'B' to adjust the total flow

② Single main duct served by individual fan unit
A damper may be installed at 'B' to adjust the total flow

Fig 3 Typical vertical common-duct systems

(e) **Ducting.** Primary factors in selecting a material for ducting are the expected life, fire risk, cost, availability and weight. The system should be capable of easy assembly and have permanently air-tight joints.

Ductwork is usually enclosed in service ducts and, therefore, inaccessible; it should last without maintenance for as long as the building itself. The traditional ducting material, galvanised steel, is suspect from this standpoint; alternatives with an acceptable life expectancy are asbestos-cement, aluminium, and unplasticised PVC.

Vertical ductwork is a potential route for fire spread between dwellings or compartments of a building and whilst it is usually recommended that ventilation ductwork should be non-combustible, unplasticised pvc systems, carefully designed, have been used successfully. In any case, the materials for both ductwork and insulation should be carefully selected to ensure that fire and smoke spread through the system is reduced as far as practicable.

To avoid air leakage into the ductwork requires careful jointing. Circular ducting with spigot and socket joints is preferred and the materials mentioned above are available in this form. Spirally-wound aluminium pipes require care in installation and probably specialist skill. The jointing of asbestos-cement should prove satisfactory if done to the manufacturer's specification but in small service ducts it might be difficult to make the joints. A leak-tight system is easily constructed with PVC soil pipes and fittings.

(f) **Noise.** Noise from the system should not exceed 35 dB(A) in any room intended for habitation, or 40 dB(A) in any other room.

The fan is a potential source of nuisance from noise and attention must be paid to its inherent noise level, its siting and mounting. Noise may be transmitted through the ductwork or through the building structure to habitable rooms. Structure-borne noise can be reduced by carefully siting the fan and by using anti-vibration mountings. Transmission through the ductwork is usually less troublesome but fans should be located as far from the heads of rising ducts as is practicable (*see* Fig 3).

(g) **Flow measurement and testing.** The system should be checked, after commissioning, for air flow, using a hooded vane anemometer (Fig 4) of diameter not less than 100 mm; the local authority should do this. Testing with smoke pellets is not satisfactory. As a basis for testing, a diagram of the ductwork layout (similar to Fig 3) is required; air-flow rates should be measured at extract points at the top and bottom of each vertical main (points C and D on Fig 3) and at several other points.

There is no simple, reliable test for duct leakage that can be recommended.

3 Fire precautions
Ventilation ductwork must comply with the relevant requirements of Part E of the Building Regulations and should, in most cases, be contained within a service duct, the walls of which have a fire resistance similar to that required of the surrounding structural elements. Inspection or access doors and frames should have similar fire resistance. Recommended practices are described in CP 413 *Ducts for building services.*

Fire and smoke spread via the protected shaft should be limited by the provision of fire dampers and fire stopping. The latter has ancillary advantages —providing a means of fixing the ductwork and helping to increase the sound insulation between dwellings. Digest 158 describes honeycomb fire dampers which are suitable for use in some ventilation ducts.

4 Entrance lobby and replacement air
Internal WC's and bathrooms should be entered from an entrance lobby. Replacement air should be drawn from this lobby, not from external air.

Fig 4 Hooded vane anemometer

Domestic chimneys for independent boilers

Because modern boilers are highly efficient, they require properly designed chimneys if they are to perform satisfactorily without hazard to health and safety or risk of structural damage from the effects of condensation.

Chimney design should aim to cut down heat losses in the flue, thereby increasing the buoyancy of the flue gases, raising the temperature of the flue surfaces and minimising the risk of condensation; the chimney fabric must be protected against the effects of condensation should it occur.

To meet these aims, this Digest recommends flue sizes in relation to boiler input ratings and standards of thermal insulation for domestic fuels. The characteristics of various lining materials are discussed.

The recommendations apply equally to new work and to the repair or conversion of older chimneys to serve modern appliances.

Fig 1   Water-vapour plume from chimney terminal

Functions of a chimney

The functions of a chimney are to help induce the flow of air needed for complete combustion of fuel and to vent the products of combustion safely to the outside atmosphere. A badly designed chimney will fail in either or both of these duties, especially when the outside air temperature is only marginally different from that inside and there is little or no wind.

Insufficient draught to burn fuel completely will result in the formation of carbon monoxide and other dangerous substances that endanger health and safety if they are allowed to accumulate. The hazard may take the form of a flue explosion, with risk of damage to the chimney structure, or the leakage of gases into living areas with possible fatal consequences. If the flue surface temperatures are too low, combustion products will condense on the flue surfaces; in the absence of a protective lining, this may damage the chimney fabric and cause staining inside the dwelling.

Both types of malfunctioning can be prevented by designing to make the best use of the small quantities of heat entering the flue with boilers which have a high thermal efficiency. Account must be taken of differences in the mode of operation of domestic boilers with various types of fuel.

Modes of operation

Solid-fuel boilers operate continuously or occasionally, depending on the user's needs. The rate of burning is controlled by opening or closing a primary air inlet, either by hand or by a thermostat in the water-jacket. The fire-bed is highly resistant to the flow of air through it and so, unless the appliance is fan assisted, the flue draught is needed both to maintain the required burning rate and to vent the combustion products to outside air. A small fixed opening for the admission of air above the fire-bed is a common design feature. This secondary air assists combustion and dilutes the products of combustion. A damper for manual control against wind effects is sometimes fitted at the appliance outlet.

Oil-fired boilers operate under thermostatic control either at two levels (high/low flame) or intermittently at one level only (on/off). Burning rate depends on the rate at which oil is fed to the burner which may be a pressure jet or a vaporiser. Both types of burner can operate under natural or forced draught. Where the combustion process is fan-assisted, flue draught is only needed to vent the products of combustion to the outside air; under natural draught conditions, however, part is required to assist the combustion process. Draught in excess of that required for these duties is spoiled by allowing air from the room to enter the flue through an opening in the appliance flue-pipe. This opening may be pre-set manually or may include a draught stabiliser which adjusts automatically to compensate for variable wind conditions.

Gas-fired boilers operate intermittently in response to a pre-set thermostat coupled to a relay valve. When the valve opens, a pilot flame ignites the main burner which then operates at full output. Flue draught is not required for the combustion process but is needed to vent the products to the outside air. A fixed opening at the appliance outlet admits air into the flue to spoil draught in excess of that required; this opening is termed a draught-diverter and provides protection against flame failure should down-draught occur.

Combustion

Calorific value and air requirement

The heat given off when unit weight of fuel is completely burned is termed its heating or calorific value. There is a theoretical requirement for the amount of oxygen, and therefore of air, required for the chemical process of combustion but, in practice, an excess of 25–50 per cent over this amount is supplied to ensure good mixing and to assist combustion. Because of differences in the chemical composition of fuels, there are small variations in calorific value and therefore in theoretical air requirements, but the ratio of air required/unit of heat does not vary greatly. Table 1 lists some characteristics of typical domestic fuels.

Products of combustion

Water vapour and carbon dioxide are the chief combustion products. Depending on the type of fuel burned, sulphur oxides, nitrous oxides, ammonia, tar acids, free water and unburned particles of fuel may also be present and the flue gas stream will contain excess air.

Condensation and chemical attack

Unlined chimneys of traditional construction were prone to chemical attack. The temperature of the flue, particularly if serving slow combustion appliances, was insufficient to prevent condensation and the condensate could act on sulphur compounds in the flue deposits, supplemented sometimes by sulphates contained in the bricks, to set up sulphate attack of the mortar. This caused the mortar to expand, and because the attack was usually uneven the stack would distort. Attack was progressive because many of the salts were hygroscopic and took up moisture from the air when the chimneys were not in use.

Building Regulations now require all new chimneys to be lined. Problems arising from condensation are consequently less acute but only if the flue linings are of suitable material and correctly installed. They should be resistant to acids and impermeable to liquids or vapours.

Table 1 Combustion characteristics of domestic fuels

Fuel (kg unit)	Gross calorific value (A) 10^3 kJ	Air requirement (B)		Ratio B/A	
		Theory kg	Practice kg	Theory	Practice
Coke	29·2	10·0	15·0	0·34	0·50
Anthracite	30·3	10·1	15·1	0·33	0·50
Kerosene	46·5	14·7	19·1	0·32	0·41
Gas oil	45·4	14·8	19·2	0·33	0·42
Town gas	32·6	9·4	14·1	0·29	0·43
North Sea 'B'	54·4	16·9	25·4	0·31	0·47

Condensation occurs when the temperature of a flue surface is lower than the dewpoint of the combustion products. Table 2 lists the water vapour concentrations and dewpoint temperatures of domestic fuels. The values are based on the air quantities which in practice give maximum concentration of condensable products and allow for a background moisture content of 10 g/kg in the air supplied. The risk of condensation is greater with a higher concentration of water vapour and the standards of thermal insulation recommended in Table 4 are therefore matched to the fuel used.

Air dilution and thermal insulation

Dilution with air reduces the concentration of combustion products but it also lowers the flue gas and surface temperatures. If the temperature of the flue surface is depressed below the dewpoint of the combustion products, the risk of condensation is increased. Dilution can also reduce the buoyancy of the flue gases to the point where down-draught results and therefore has limitations as a complete remedy for condensation. Sizing and insulating the flue to reduce the heat loss from flue gases and to keep the flue surfaces warm is preferred.

Gas- and oil-fired appliances have spoil-draught features to limit flue draught to the required operating level. These are openings into the flue-pipe, allowing a dilution air-flow about equal to that through the combustion zone of the appliance with no wind acting. For solid-fuel appliances, small amounts of dilution air may be advantageous and can be provided by an opening of about 13 cm² at the bend in the smoke pipe.

Design aims

The main objectives in chimney design are: to cut down heat losses in the flue, thereby increasing the buoyancy of the flue gases and raising the temperature of the flue surfaces; to minimise the risk of condensation but to protect the chimney fabric from its effects should it occur.

One way of cutting down the heat lost by the waste gases as they pass along the flue is by insulation.

A better and more economic method is to reduce the flue size and thus the area through which the gases can lose heat. Reducing the size of the flue has two opposing effects: one is to raise the average temperature of the flue-gas stream and thus the buoyancy force; the other is to increase the frictional drag at the flue surface as the gases speed more rapidly through the flue. Initially buoyancy increases at a faster rate than frictional drag so that there is a net gain in flue draught. Ultimately the increase in buoyancy is offset by the loss from frictional drag so that there is a lower size limit below which the flue draught diminishes. There is therefore an optimum size of flue for a known flue-gas flow rate. Since the flow rate depends on the rate at which fuel is burned, the flue sizes recommended later are related to boiler input rating, that is, to the burning rate at full-load operation. The greatest single advantage with this method of flue sizing is the gain in stability in varying wind conditions. With aiding wind, friction losses increase very rapidly for only a marginal increase in flow rate; the additional head due to wind is largely offset by the opposing frictional drag. With adverse wind, the flow-rate decreases slightly, because part of the buoyancy force is now used to counter down-draught, but frictional drag is reduced and this compensates for the loss in buoyancy. A nearly constant pressure difference is thus maintained across the system so that the flow rate remains stable.

Because the flue area is reduced with deposition from boilers which burn oil or solid fuel, and this reduction is more critical the smaller the flue, a lower limit of 100 mm is proposed for the diameter of a flue liner to serve these appliances. Apart from this limitation, the size of flue need never be greater than is necessary to allow the free passage of waste gases when the boiler is run at its maximum output. Surrounding the flue lining with thermal insulating material reduces the risk of condensation on the flue surfaces, especially during periods when the boiler is operated at less than its peak load. The flue sizes and thermal insulation standards recommended in Tables 3 and 4 take account of variations in appliance load and outside temperature. The design

Table 2 Water-vapour concentration in flue gas for domestic fuels

Fuel (kg unit)	Water produced (free and from combustion) kg	Total water content of flue gases kg	Water vapour concentration* in flue gases g/kg	Water vapour dewpoint °C
Coke	0·224	0·374	25·0	28·5
Anthracite	0·098	0·248	16·6	21·9
Kerosene	1·260	1·451	76·0	47·8
Gas oil	1·210	1·402	73·2	47·0
Town gas	0·957	1·098	78·0	48·1
North Sea 'B'	2·203	2·457	96·8	52·0

* Includes a background moisture content of 10 g/kg in the air supplied.

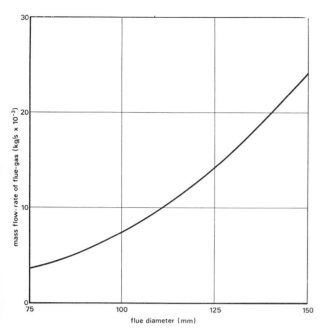

Fig 2 Relationship between mass flow-rate of flue gas and
flue diameter for optimum performance

combination of flue size and thermal insulation
should thus ensure that condensation is only likely
to occur when the appliance is first lit and the
surfaces are relatively cold. Condensate produced
at such times is prevented from coming into contact
with the chimney fabric by the use of an impervious
flue lining and, if present in quantity, can be collected
and drained away. Condensate remaining on the
liner surface will subsequently be re-evaporated
and vented to atmosphere as the flue temperatures
rise.

Design procedure for new construction

1 Select the size of flue

Studies of the heat-transfer and fluid-flow processes
in boiler flues at the Building Research Station
showed that flue sizing is the key to stable chimney
performance. Fig 2 shows the relationship between
the mass flow rate of flue gas and flue diameter to
provide optimum performance and is the basis of the
recommendations for flue sizes in Table 3. A boiler
heating efficiency of 80 per cent is assumed; if the
efficiency is less than this, the sensible flue loss is
greater and the problems of flue draught and condensation are correspondingly smaller.

Flue-sizing requirements for gas-fired boilers are
dealt with more fully in BS CP 337 : 1963.

2 Determine the required standard of thermal insulation

Although condensation is reduced by the use of
smaller flues, additional thermal insulation is necessary to minimise the risk when appliances are
operated at low burning rate. The standard of thermal
insulation required depends very much on the type
of fuel in use, as illustrated by the water vapour
concentrations and corresponding dewpoint temperatures given in Table 2. Requirements will differ
for those parts of a chimney exposed to the outside
climate compared with the parts that are entirely
enclosed within the building fabric. In recommending
the standards of insulation given in Table 4, a thermal
efficiency of 80 per cent is assumed and allowance
is made for the latent heat of the water vapour in the
flue gases. The effect of dilution of the flue-gas stream
inherent in appliance design has also been taken into
account. The recommended insulation standards are
sufficient to maintain flue surfaces at a temperature
equal to the water vapour dewpoint when the boiler
is operated at low burning rate.

Table 5 suggests some methods of construction that
will provide the levels of thermal insulation recommended in Table 4. The constructions listed are in
general based on the traditional brick chimney as an
outer shell, with a range of liner sizes and various
insulating materials as infills. The thermal insulation
value of an unlined brick chimney with traditional
flue is included for comparison. The table also shows
that the heat loss from a bare pipe reduces as the
size is reduced and shows the increased effectiveness
of a given thickness of insulation with reduction in
flue size. The construction selected should provide
a thermal insulation standard equal to or better than
that recommended in Table 4 for the size of flue.

3 Select the type of construction

Structural chimneys

Suitable lining materials for new construction are:
1 Clay flue linings complying with BS 1181 : 1971.
2 Non-perforated and acid-resistant clay pipes and
   fittings, with socketed joints, complying with
   BS 65 & 540 : Part 1 : 1971.

Table 3 Flue sizing for optimum performance at full-load operation

	Upper limit of boiler input rating (kW)			
Coke	—	14	28	47
Anthracite	—	16	31	53
Kerosene	—	21	39	65
Gas oil	—	20	38	63
Town gas	10	20	38	62
North Sea 'B'	9	19	35	58
Flue diameter (mm)	75	100	125	150

Table 4 Thermal insulation standards for minimum condensation risk at low burning rate

Boiler input rating kW	Coke	Anthra-cite	Kero-sene	Gas oil	Town gas	North Sea 'B'
				m°C/W		
7·5	1·17	1·04	2·40	2·34	2·60	3·00
	0·80	0·55	1·60	1·56	1·80	2·10
10·0	0·90	0·72	1·35	1·30	1·45	2·20
	0·64	0·44	0·97	0·94	1·03	1·59
15·0	0·69	0·52	0·91	0·89	0·96	1·47
	0·48	0·29	0·64	0·63	0·68	1·04
20·0	0·53	0·40	0·68	0·67	0·74	1·15
	0·39	0·22	0·50	0·50	0·52	0·78
25·0	0·42	0·29	0·55	0·54	0·59	0·94
	0·31	0·15	0·42	0·41	0·43	0·62
30·0	0·36	0·26	0·45	0·44	0·48	0·77
	0·25	0·13	0·35	0·34	0·36	0·51
35·0	0·31	0·23	0·38	0·37	0·41	0·64
	0·20	0·12	0·30	0·29	0·30	0·43
40·0	0·29	0·20	0·33	0·33	0·36	0·55
	0·17	0·11	0·25	0·24	0·26	0·38
45·0	0·26	0·17	0·30	0·29	0·32	0·49
	0·15	0·10	0·22	0·21	0·22	0·34
50·0	0·24	0·16	0·27	0·26	0·30	0·44
	0·13	0·09	0·20	0·19	0·20	0·30

Note: For external chimneys, use the higher set of values.
For internal chimneys, use the lower set of values.

Table 5 Thermal insulation levels provided by various constructions

Chimney	Lining	Infilling	75	100	125	150
Constructional details:			Flue diameter (mm)			
				m°C/W		
Asbestos-cement flue-pipe (heavy quality)	—	—	0·70 / 0·26	0·54 / 0·20	0·45 / 0·17	0·38 / 0·14
Traditional brick; 225×225 mm flue with half-brick surround*	Asbestos-cement pipe (heavy quality)	Air space	1·19 / 1·13	1·04 / 0·96	0·93 / 0·86	0·86 / 0·79
	Clay pipe	Air space	1·24 / 1·17	1·07 / 1·00	0·96 / 0·89	0·89 / 0·82
*Thermal insulation value (unlined): Internal 0·20 External 0·13		Crushed brick aggregate concrete	1·10 / 1·02	0·87 / 0·81	0·72 / 0·67	0·58 / 0·54
		Loose exfoliated vermiculite	2·47 / 2·41	1·98 / 1·91	1·60 / 1·53	1·27 / 1·21
	Flexible metal	Air space	1·38 / 1·30	1·14 / 1·07	1·00 / 0·93	0·91 / 0·84
		25 mm wrapping of glass-silk or mineral wool, plus air space	3·31 / 3·24	2·70 / 2·63	2·27 / 2·21	2·00 / 1·94
	Combined lining and infilling of vermiculite concrete of density 40 kg/m³		2·20 / 2·13	1·77 / 1·70	1·43 / 1·36	1·16 / 1·09

Note: Upper/lower values refer to internal/external chimneys, respectively.

3 Rebated or socketed linings made from kiln-burnt aggregate and high alumina cement.

In addition to the materials listed above, the following may be used to line existing chimneys:

4 Heavy quality asbestos-cement pipes with spigot and socket joints.

5 Flexible metal linings.

6 Lightweight insulating concretes, cast *in situ.*

Non-structural chimneys

1 Asbestos-cement flue pipe (heavy quality). For gas appliances, CP 337:1963 specifies the conditions in which the internal surfaces of these pipes should be coated with an acid-resisting compound, eg one prepared from a vinyl acetate polymer or a rubber derivative base.

2 Proprietary insulated precast concrete flue blocks.

3 Factory-made insulated metal chimneys which conform with BS 4543:1970.

Other design features

Terminal The effectiveness of any additional thermal insulating material will be reduced, particularly with external chimneys, if rain is allowed to enter the space between liner and brickwork.

When the exposed part of the stack is more than about 1 metre high, good weather protection, eg suitable flashings and damp-proof course, is needed where the stack emerges from the roof. Good protection is needed at the top and a precast concrete cap is suitable; it should be in one piece, with sloping top, ample overhang and properly throated. Cappings of brickwork and tile creasing, even though flaunched with mortar, cannot be relied on to keep out moisture indefinitely and require a damp-proof course immediately below them.

A solid cap, to prevent the direct entry of rain into the flue, should be raised sufficiently to provide a free area not less than the cross-sectional area of the flue lining, so as not to restrict the escape of flue gases.

Soot box The gases are hottest at the bottom of the flue and to minimise heat loss at this point the area of a soot box should be no greater than is needed for access for sweeping when the boiler is fired by solid fuel or oil. Because condensation can occur when the boiler is started up, the chimney may be attacked at the soot box unless contact with condensate is avoided. It is therefore recommended that the liner should extend below the junction with the smoke pipe and be provided with a drip so that any condensate can drain into a collecting vessel and not down the sides of the soot box.

Bends Flues should be as straight as possible. Bends and offsets increase heat losses from the gases and reduce the available draught, with a consequent increase in the risk of condensation. If bends are necessary, eg where the boiler outlet pipe is connected to an external chimney, the angle the pipe makes with the horizontal should never be less than 45°.

Lining existing chimneys

The recommendations for sizing and insulating new flues apply equally to existing ones and should be followed as closely as possible when chimneys are repaired, so as to avoid a recurrence of troubles, or when new appliances are to be fitted.

Asbestos-cement flue pipe complying with BS 835 is sometimes used in straight chimneys because the long lengths available require few joints. Several lengths are arranged socket uppermost and jointed with high alumina cement before lowering into the chimney. The lining is centred within the brick flue and supported at the base by a plate inserted just above the soot box. (Asbestos-cement flue pipe is not permitted within 1·8 m of the junction of the flue with a solid-fuel or oil-fired appliance.) An air space is sometimes left between the lining and the brickwork but it is more usual to fill the space with dry exfoliated vermiculite or a weak mix of lightweight concrete. To prevent the formation of voids, the infill should be well tamped; it should be protected at the top by a cement flaunching to prevent rain entering the space and reducing the insulating properties of the material.

Clay and refractory concrete flue linings require temporary access holes to be cut through the brickwork for jointing. Both are available in various diameters and lengths and with rebated joints to keep any condensate inside the linings. A range of standard bends is available. Thermal insulation can be improved by filling the cavity between the lining and the brickwork with suitable materials, as the work progresses.

Flexible metal linings are intended for use with gas- or oil-fired boilers. The linings are corrugated and available in a range of diameters. They can be cut to any length and, being flexible, are particularly useful where bends in an existing flue have to be negotiated. The lining may be of stainless steel (316S16) or aluminium alloy, or made in sandwich form with an outer leaf of aluminium separated from an inner leaf of aluminium or lead foil by corrugated paper. To provide additional thermal insulation, they may be wrapped with lightweight material such as glass-silk or mineral wool before insertion into the chimney; alternatively, the space between lining and brickwork may be filled with a lightweight insulating material such as exfoliated vermiculite.

Care is needed during installation to avoid the liner being damaged by abrasion as it is pushed down the chimney. The lining is supported at the top of the chimney by a metal plate which acts also as a cover to prevent rain entering the space around the lining.

Lightweight concrete infills An inflatable rubber tube is inserted into the chimney and centred in the flue-way by metal spacers attached to it at points where bends occur in the chimney. The tube is then inflated to the required diameter of the finished flue and a metal sealing plate fixed at the bottom of the chimney, just above the soot box. Lightweight aggregate concrete (perlite or vermiculite) is then pumped through a plastics hose which is lowered into the space around the rubber former. When the concrete is sufficiently set, usually after a day or more, the rubber core is deflated and removed.

This method gives a reduced size of flue with additional thermal insulation in one operation.

3 Acoustics, Vibration and Noise Control

The acoustics of rooms for speech

Although many rooms are used for both music and speech, the acoustic requirements are quite different. The acoustic problems of multi-purpose auditoria are too complicated to cover adequately in a digest, as are the acoustic problems of auditoria used mainly for music, ie concert-halls and opera houses. This digest therefore deals only with the simplest case, that of rooms used mainly for speech. It replaces Digest 82 which is now withdrawn. The next digest discusses speech reinforcement systems.

Why are some rooms acoustically bad ?

There are three main reasons why rooms may be acoustically bad for speech : they may be too reverberant for the speech sounds produced in them; they may be noisy because of the feeble resistance they offer to the penetration of noise from outside or from nearby rooms; or they may be shaped so that the speakers are more or less screened from their audience or part of it (this may be particularly serious in rooms intended for discussions).

Effect of reverberation on speech

Part of the sound from a speaker, the *direct* sound, passes directly to the ears of his audience. Another part travels to the room surfaces and is reflected, eventually reaching the listeners' ears from many directions at close intervals; this is called *reverberant* sound (see Fig 1). If there is too much reverberant sound, the listener is still receiving the reflected sounds of previous syllables when he is trying to cope with the direct sound of later syllables. The two then interfere, and speech is difficult to understand. The amount of reverberation is usually expressed as the reverberation time, which is the time taken for the sound to die away after its source has stopped. It is determined by the volume of the enclosure and its capacity for sound absorption and is defined as the time taken for the sound to

Fig 1 Direct and reverberant (multi-reflected) sound

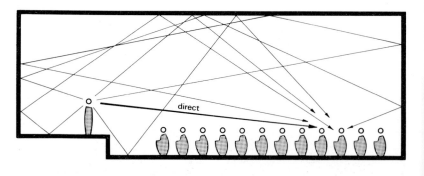

Prepared at Building Research Station, Garston, Watford WD2 7JR
Technical enquiries arising from this Digest should be directed to Building Research Advisory Service at the above address.

decay by 60 decibels. It may vary from half a second in an ordinary living room to 12 seconds or more in a large cathedral. If this time is one second or less, the reverberant sound will die away quickly enough not to interfere with the direct sound of speech. If it is much longer than this the reverberant sound will persist long enough to interfere and possibly become a nuisance. This simple picture is complicated by the fact that the interference of reverberant sound in understanding speech depends also on the distance from the speaker to the listener. It is common experience that if we are close to performers on stage the volume of direct sound reaching us is such that it hardly matters how much reverberant sound there is; the direct sound is so strong. But at 5 to 10 m the direct sound has begun to fall away appreciably and the effect of reverberant sound is more important, though not intolerable if the reverberation time is no more than two seconds. At 30 m the intensity of the direct sound will have fallen off seriously, while the reverberant sound, which stays at much the same level throughout the room, may now have become an equal if not dominant rival. Speech then becomes difficult or even impossible to follow if the reverberation time is about two seconds. If it is down to one second, a distance of 30 m between speaker and listener may be just tolerable, provided the speech is loud and clear and not too fast. A poor speaker will not be heard properly at this distance.

The optimum reverberation time for speech is not particularly critical: somewhere between 1·0 and 1·2 seconds at the frequency of 500 Hz will be suitable or even as high as 1·4 seconds. With times less than one second the room tends to sound 'dead' and is slightly unpleasant to speak in.

Calculating the reverberation time

The reverberation time T (in sec) is given by the formula $T = 0·16V/A$ where V is the volume of the room (in m³) and A is the total sound absorption (in m²) of the room inclusive of furniture and occupants. The total absorption (A) at any frequency is obtained by multiplying the area of each different type of surface finish by its absorption coefficient (the proportion of sound energy it absorbs) at the given frequency and adding these products together, then adding on the absorption of occupants furniture, etc. The reverberation time should be calculated for at least three different frequencies, first at 500 Hz to get within the optimum range for this frequency, and then at 125 Hz and at 2000 Hz to make sure that the reverberation times at these frequencies are between 75 and 125 per cent of the value at 500 Hz. At 2000 Hz some contribution to the absorption is made by the air in the room and this should be added to the previous total.

Fig 2 Placing of absorbent materials

The sound absorption (A) is the main factor for controlling the acoustics because the volume of the room (V) is usually determined mainly by other factors. If the preliminary calculations show that it will be necessary deliberately to introduce some sound-absorbent materials, then it might be advisable at this stage to employ an acoustic consultant if money is not to be wasted on ill-sited or ineffective materials. However, as a general rule, the absorbent should go first on the rear wall, then on the ceiling margins and side walls, and last of all on the centre of the ceiling (see Fig 2).

Fig 3 High platform or raked floor—direct sound and 'first reflections'

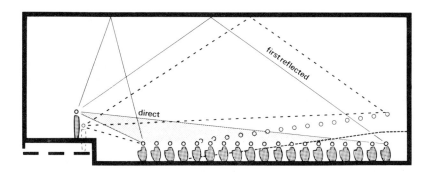

Loudness of direct sound

Ideally, the direct sound should be loud enough for everyone with normal hearing to hear it, whatever their position in the room. There should be a clear view of the speaker; this means raising the speaker relative to the audience, raking the seats of the audience, or using a combination of both methods (see Fig 3). Another help is to augment direct sound with 'first reflected' sound, ie sound reaching the listener after a single reflection at a ceiling or surface placed for maximum effect. If the reflector is suitably positioned, the direct and the first reflected sound reach the listener with only a short time-interval between them, and interference is at a minimum; the reflected sound merges with the direct sound and strengthens it. The difference in path lengths for the two sounds should not exceed 9 m; it may be better if it is rather less. Some rooms are of such a height that the centre part of the ceiling is an effective reflector, provided it has not been covered with an acoustically absorbent material to reduce the reverberation time. To act as a good reflector, the surface of the ceiling should be hard and smooth. If the ceiling is too high for first reflections to strengthen the direct sound effectively, or if it has an absorbent surface overall, a reflector can be installed over the speaker. A plane-surfaced reflector is usually adequate, and it can be so placed that the difference in path lengths is reduced to a minimum (see Fig 4). It should not be too light in weight; for example, a 25 mm thick plasterboard reflector painted on the underside would be suitable. Such a reflector is particularly useful when, as frequently

Fig 4 Difference in path direction of reflected sound from ceiling and reflector

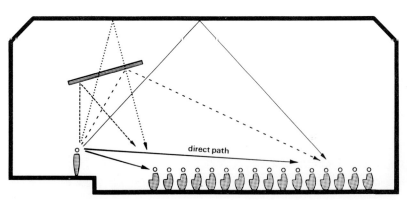

happens, the audience floor is horizontal and the speaker is raised only slightly above it. In these circumstances the speaker's remarks are addressed almost at grazing incidence to the heads of his audience, and the direct sound is rapidly absorbed. First reflected sound, not being at grazing incidence, is more effective in carrying sound to the back of the audience.

Complaints may be made of poor hearing in rooms with low ceilings and deep transverse beams. This is because the beams cut off the ceiling reflections to certain areas and often reflect the sound back to the speaker (see Fig 5). Remedies are a flat suspended ceiling under the beams or a tilted reflector above the front of the platform. To keep up the strength of the direct sound the listeners should be as close as possible to the speaker. Every effort should be made in sparsely filled rooms to persuade audiences to sit nearer the platform. Having done that, it is up to the speaker to ensure that his own delivery is acoustically beyond reproach. This means speaking slowly and, if necessary, loudly, with syllables spaced out as evenly as possible.

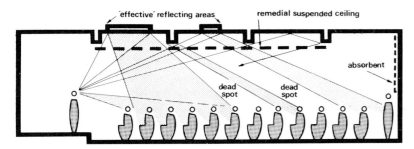

Fig 5 Patchy ceiling reflections (dead spots) due to transverse beams

If an existing room is too reverberant for speech, temporary measures can sometimes help. These include using absorbent materials such as carpets and curtains to blank off reflective surfaces and, if possible, the introduction of upholstered chairs, particularly if audiences are sparse. Even overcoats spread over empty chairs can have a useful effect. More permanent steps to reduce reverberation times involve fixing sound-absorbent materials on appropriate surfaces, as already discussed.

Further reading
Parkin, P. H. and Humphreys, H. R., Acoustics, noise and buildings. Faber and Faber. (London, 1958, metricated 1969.)
Purkis, H. J., Building physics: Acoustics. The Commonwealth and International Library. Pergamon Press (Oxford, 1966.)
British Standard Code of Practice CP 3, Chapter III: 1972; Sound insulation and noise reduction. (British Standards Institution, London.)

BRE Digests
102   Sound insulation of traditional dwellings: part 1
103   Sound insulation of traditional dwellings: part 2
128   Insulation against external noise: part 1
129   Insulation against external noise: part 2
143   Sound insulation: basic principles
187   Sound insulation of lightweight dwellings

Loudspeaker systems for speech

Loudspeaker systems tend to have a bad reputation although they are capable of functioning perfectly satisfactorily. Shortcomings can be avoided if installations are well designed, well installed, well maintained and properly used. This digest reviews the general principles that should be followed to achieve satisfactory performances. It also outlines the types of system available to help those generally responsible for the design and use of rooms but who are not directly concerned with the technical details; these will be the concern of the equipment supplier. It does not discuss loudspeaker requirements for music: any auditorium where music reproduction is of importance will usually warrant an acoustical consultant and his brief would include the loudspeaker system.

In this digest, the word 'speaker' means the person speaking as opposed to the loudspeaker; a 'speech reinforcement system' is defined as one that has the speaker and the loudspeaker in the same room whilst a 'public address system' has them in separate rooms. The design principles are the same for both, except that there is no feed-back problem with a public address system and there is no need for the sound to appear to be coming from the person speaking: this is desirable with some speech reinforcement systems. Especially in a large room, a loudspeaker system helps speech in one or both of two ways: it makes the speech louder so that it can be heard above the background noise (eg traffic noise) and it can mitigate the effect of the reverberant sound on the intelligibility of speech.

As described in Digest 192, the direct sound from any source will fall off with distance according to the inverse square law, ie a drop of 6 dB for every doubling of the distance. The intelligibility of speech depends on this direct sound being louder than both the background noise and the reverberant sound, both of which are roughly at the same level throughout the room. Feed-back limits the loudness of speech reinforcement systems: if the limit is exceeded, sound from the loudspeakers is picked up by the microphones, amplified and re-emitted from the loudspeakers; the sound gets louder each time round the circuit and results in howl. The effect can be minimised by using directional microphones and loudspeakers and, as far as practical, positioning them so that their least sensitive sides are towards

Prepared at Building Research Station, Garston, Watford WD2 7JR
Technical enquiries arising from this Digest should be directed to Building Research Advisory Service at the above address.

each other. The loudness of public address systems is limited by the size of the amplifiers and the loudspeakers. Keeping the direct sound above the reverberant sound (and limiting feed-back in speech reinforcement systems) depends mainly on the design and position of the loudspeakers; this is the most difficult part of designing a system for speech.

Any repetition of the sound which arrives at the listener not more than about 35 milliseconds after the arrival of the direct sound blends with the direct sound and is a help to the intelligibility. Any repetition coming later than this hinders the intelligibility, and the louder the repetition is relative to the direct sound, and the further behind it is, the more it hinders. When speech is being reinforced by one loudspeaker, this should be placed so that the sound from it arrives at the listener within 35 milliseconds of the sound from the speaker. This means that the distances from speaker to listener and from loudspeaker to listener should not differ by more than about 10 m.

If the listener is further from the loudspeaker than from the speaker, the sound from the loudspeaker will reach the listener later than the sound from the speaker and even if the loudspeaker sound is as much as twice as loud as the sound from the speaker, all the sound will appear to be coming from the speaker. This is known as the 'precedence' or 'Haas' effect and is important where it is desirable to maintain some illusion of reality. To sum up: for intelligibility the various sounds should arrive within 35 milliseconds of each other; for realism the sound from the speaker should arrive first.

There are two main classes of speech loudspeaker systems, low-level and high-level (level refers to loudness of the output from each loudspeaker and not to their height above the floor).

Low-level systems consist of a large number of smallish loudspeakers distributed throughout the room usually just above head-height. No listener is very far from a loudspeaker and all get adequate direct sound from the one nearest. If it is physically possible to have loudspeakers close enough to everyone, the acoustics of the room hardly affect the intelligibility of the speech. There have been systems where every seat has had its own loudspeaker but this is not always possible or convenient. When loudspeakers have to be, say, 10 m or more away from the listeners (if they must be fitted to the walls for example) this type of system has limitations partly because the nearest loudspeaker may now be too far away to be louder than the reverberant sound and partly because the sounds arriving from different loudspeakers may be of much the same loudness but different enough in arrival times to hinder intelligibility. In all low-level systems the sound will be coming from the loudspeaker nearest to the listener and not from the speaker; there is no realism. This may not matter but the overall sound of a lot of small loudspeakers can be rather unpleasant.

With low-level systems, and for speech from fixed microphone positions, feed-back is usually not much of a problem because, although the loudspeakers are not usually directional, they are operating at a low volume and can be kept far enough away from the microphone. If microphones are to be used from anywhere in the room, eg at a conference, feed-back may be more of a problem and it may be necessary to switch off the loudspeakers close to the microphone in use.

High-level systems use only a few loudspeakers,* perhaps only one, operating at a higher sound output. The main advantages over low-level systems are that they may be easier to install (fewer electrical connections are necessary) and it is much easier to maintain realism. Greater care in design is needed than with low-level systems for adequate loudness and intelligibility over the whole audience area.

* Ordinary loudspeakers are rarely used in high-level systems because they are directional only at the higher frequencies. Their use, therefore, introduces feed-back problems and they are only a little help in keeping the direct sound above the reverberant sound. This help comes from placing the loudspeakers higher than the speaker; the direct sound from the speaker is at grazing incidence over the heads of the audience and gets attenuated at a greater rate than the inverse square law; the sound from the loudspeaker is not so much attenuated.

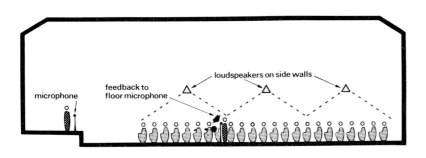

Fig 1 Low-level loudspeaker system

Fig 2   High-level loudspeaker system

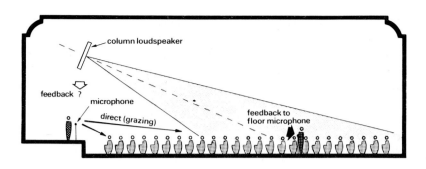

The majority of high-level systems use directional loudspeakers and, although there are several types, the most common is the loudspeaker column. This consists of several ordinary loudspeakers mounted vertically one above the other which has the effect of concentrating most of the sound in the vertical plane into a beam at right angles to the column whilst having no effect in the horizontal plane (some columns are designed to be directional in the horizontal plane as well as in the vertical to reduce feedback). The result is something like a fog lamp beam: narrow in the vertical plane and broad in the horizontal plane.

This directionality benefits in two ways. Firstly, feed-back can be minimised by positioning the loudspeaker so that the microphone is out of the main beam. Secondly, the main beam of sound is directed at the audience and is there largely absorbed before it can become reverberant sound. Thus, the direct sound level is kept well above the reverberant sound level. A minor advantage is that because of this concentration of sound, less power is required from the amplifiers and the loudspeakers.

Obviously the columns must be carefully positioned so that the sound is directed most usefully. This usually means that the base of the column is slightly above audience head height (by not more than about a metre) and tilted forward about 5 to 10 degrees. The centre of the beam of sound should hit the audience at about two-thirds of the way back. This ensures that most of the audience is in the beam and that there is some compensation for the inverse square law reduction.

Depending on the design of the room it may be necessary to move the column higher and tilt it more forward. This will tend to reduce feed-back but the column must be less directional or the whole of the audience will not be in the beam. The directionality of a column depends on its length relative to the wave length of sound: the longer the column the more directional it is for a given frequency. This means, of course, that a column gets more directional the higher the frequency and this is a disadvantage: the column may not be long enough to be sufficiently directional at low frequencies and be too directional at higher frequencies. Columns are available which overcome this disadvantage but they are more expensive; more sophisticated columns are becoming available which do not need to be tilted forward.

The number of loudspeaker columns necessary will obviously depend on the size of the room and how reverberant it is. For speech from one end of the room only, one loudspeaker column to one side of the room at that end will normally be enough, but listeners near the front on the other side of the room will not hear much loudspeaker sound. A column may be needed on each side in wider rooms. If one column is used, it should for realism be positioned slightly further back (1·5 m) than the speaker, at the risk of increasing the feed-back problem because the column will be only slightly directional in the horizontal plane, ie radiate less to the sides and back than to the front.

If a column is used on each side realism can be achieved only by introducing delays in the electronics.

If two columns are used, they must be in phase but there will always be a slightly odd effect for listeners on the longitudinal centre line of the room because, due to the Haas effect, they will hear all the sound apparently coming from one side or other of the room, depending on which way they lean their heads.

In a fairly small room, seating, say, up to about 200 people, and with a reasonable reverberation time (not more than 1·5 seconds) a column about 1·5 m long will be suitable. For larger rooms and longer reverberation times, a column 2 m or even 3 m long may be necessary. It is impossible to be more specific than this, because of the many types of column available.

With any type of system, the microphone should be as close to the speaker as possible to reduce the amplification necessary and the chance of feed-back. For speech from a fixed position the microphone can obviously be on a stand. A halter microphone has the advantage of keeping a constant distance between the mouth and the microphone, (although there is a noticeable difference in quality when moving the head down to look at notes and up to look at the audience or turning the head to one side). However, the speaker is free to walk around or to use chalk boards etc. With the microphone on a stand, the speaker cannot move away from the stand without losing practically all aid from the loudspeaker system. Even moving the head about so that the distance from the microphone varies between say, 100 mm and 300 mm will result in the amplified sound varying by a distracting factor of two in loudness.

If there has to be speech from the floor, microphones are unavoidably in the main beam and it is necessary to rely on the directionality of the microphone to reduce feed-back, ie to use a microphone which has the least sensitivity at its back.

Any system must be well maintained. Microphones are vulnerable to mechanical damage; amplifiers can develop intermittent faults; loudspeakers are probably the most reliable part. For a large system it is probably wise to have a maintenance contract with the supplier, although with the simple systems described in this digest there should be little trouble

with good quality equipment. Ideally, there should be a trained operator to adjust the volume of the system to suit different speakers and to guard against feed-back if, for example, the microphone is moved.

Finally, no system will turn mumbling speech into clear speech and the speaker must speak up, speak clearly, keep a fairly constant distance from the microphone, and not vary much in loudness from word to word, or even from syllable to syllable. The more reverberant and noisy a building, the more important it is that the speaker should remember these points. Notes to guide preachers in reverberant churches were written some years ago by Mr Hope Bagenal.

His advice, quoted here in part, cannot be improved upon: *'A good microphone technique is one that proceeds at a slow pace; maintains an even distance from the microphone; does not greatly vary the loudness. Of these, going slow is very important indeed: no amplification system, in a church having a long reverberation, can enable a preacher to be heard if he goes too fast. Some slight variation of loudness is permissible in order to avoid monotony, but the microphone cannot deal with the shout or the dramatic whisper: the technique is more a reading than an oratory . . . it is the syllable, not the word which is the unit of sound: hence in reverberant conditions a slowness of separate syllables is required rather than long pauses between words, or between sentences . . ..'*

Vibrations in buildings—1

Part 1 : *An explanation of some of the terms used in vibration studies and their relationships; human sensitivity to vibration—the Reiher-Meister and Dieckmann scales; vibration and damage to buildings.*

Part 2 : *(Digest 118, see page 90) Some ways in which nuisance from vibration can be avoided or reduced to a tolerable level—by treatment at source and by providing protection against vibration from an external source.*

Developments in machinery and transport tend to increase vibration and noise to an extent that may be not only objectionable to people but may interfere with laboratory work and some trade processes. Fears are also expressed that buildings may be damaged by vibration.

During the past twenty-five years, the Building Research Station has carried out a considerable number of investigations of ground and structural vibrations from sources which included forge hammers, compressors, machinery, blasting, pile-driving, road traffic and aircraft, church bell ringing and domestic activities. In no instance was any observed damage found to be directly attributable to the effects of vibration alone. The risk of damage to normal buildings is extremely small, even when the level of vibration is considered objectionable or intolerable by the occupants of the building.

The control of noise is not a primary concern of this digest. But the frequencies of vibrations encountered in buildings lie mostly within the range 5–50 Hz and when the frequency exceeds about 20–30 Hz it passes into the audible range. If its energy is sufficient, i.e. above the threshold of audibility of sound in air, the vibration will then be perceived by the ear as sound. Furthermore, even sub-audio frequencies generated in a structure may induce higher (audio) frequencies in some parts of the structure; this possibility is an important reason for isolating vibrating sources so as to reduce the energy transmitted to the structure.

Terms used in vibration studies

Vibration is repeated movement about a position of rest and many vibrations, particularly those produced by machinery, are of periodic (simple harmonic) nature, like that of a vibrating string. Specific terms are used to describe the extent of the movement, the number of times it is repeated in unit time and the nature of the vibration— whether it is free or forced.

Cycle : a completed sequence of repeated events.

Period (T) : the time taken, usually a fraction of a second, for one complete cycle of movement.

Frequency (f) : the inverse of *period*; the number of times the cycle is repeated in unit time, usually one second in which case it is expressed in hertz (1 Hz=1 c/s).

Amplitude of displacement (A) : the maximum displacement from the mean position (position of rest), or half the total, peak-to-peak, displacement. A convenient unit is the micron, μ; $1\mu = 10^{-6}$ m. In this digest, the simpler term *amplitude* is used to denote amplitude of displacement.

Velocity : the rate of change of displacement. For a vibration of sinusoidal form, the peak velocity for maximum amplitude A and frequency f is given as $2\pi A.f$. The peak velocity for an amplitude of 400μ and frequency 20 Hz would thus be approximately 50 mm/sec.

Acceleration : the rate of change of velocity. The maximum acceleration in a vibration of sinusoidal form is given as $4\pi^2 A.f^2$. It is sometimes convenient to express acceleration in terms of the gravitational value g; a maximum acceleration of 0·01 g is given, for example, by a sinusoidal vibration of maximum amplitude 25 μ and frequency 10 Hz.

Free vibration takes place when an elastic body or system is disturbed from its position of rest by a force or impulse that then ceases to act on it; for example, when a weight resting on a spring is displaced and then released. The frequency of the free vibration is called the 'natural frequency' and its particular value depends on factors such as the elasticity, or stiffness, and the weight or load. The amplitude of free vibration will diminish with time, at a rate dependent on the amount of damping present in the system.

There is a relationship between the theoretical natural frequency of vibration of a system and its deflection under static, point, loading; this is given as $f_n = 15\cdot8/\sqrt{y}$, where f_n is the natural frequency in Hz and y is the static deflection in millimetres. The relationship is useful in selecting suitable types of antivibration mountings for machinery, apparatus and structures.

If, for example, a spring-mounting system deflects statically 7 mm under the load it supports, the theoretical value for natural frequency of the system is $\dfrac{15\cdot8}{\sqrt{7}}$ Hz=6 Hz. If the static deflection were four times the previous value, the natural frequency would be halved.

Simple formulae are available (see item 2 of **Further reading**) for the determination of fundamental natural frequencies of beams for various end-fixing conditions and types of loading. For a simply supported beam centrally loaded, the expression given in the previous paragraph is applicable; thus, if the deflection at the centre is 7 mm, $f_n = 6$ Hz. If the same deflection were given for uniform loading, however, the natural frequency would be higher—about 6·7 Hz in this case.

Forced vibration is produced in an elastic system when it is affected by some external source of vibrational energy. Then the system will vibrate at the same frequency as that of the source of vibration (the 'forcing frequency') and not at its natural frequency. For example, a beam may have a natural frequency of 6 Hz. If it is affected by an electric motor producing vibration of frequency 50 Hz then it will be caused to vibrate at a frequency of 50 Hz.

The movement (or response) of a system affected by vibration will depend on the relationship between the forcing frequency and the natural frequency and also on the degree of damping present. In the special case where the frequencies coincide, resonance occurs and a significant increase in amplitude and force can result. At this critical stage even a small change in the frequency of the excitation will usually result in a noticeable diminution of response. Although structural materials do not generally possess a great degree of intrinsic damping, there is usually sufficient to keep amplification at resonance within reasonable bounds. The small amount of damping does not greatly affect response other than at, or near, the resonance condition however.

Human sensitivity to vibration

Under certain conditions the human body can detect amplitudes as small as one micron, whilst amplitudes of the order $0.05\ \mu$ can be detected with the finger-tips. The basic data concerning 'whole-body' sensitivity to vibration are provided by the Reiher-Meister scale. Although this was developed over thirty years ago, its validity is

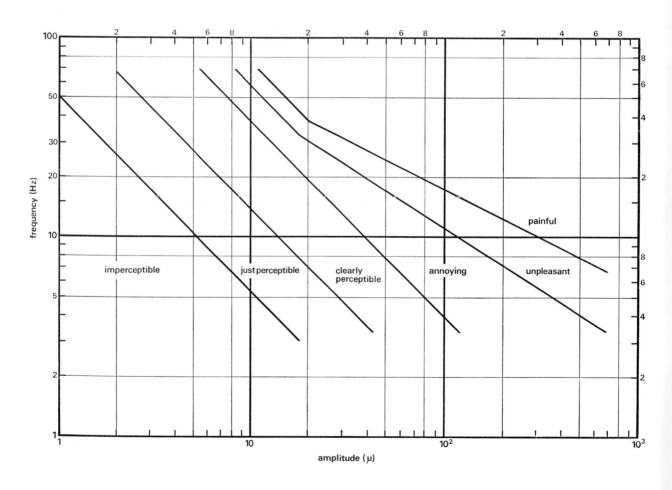

Fig. 1   Human sensitivity : Reiher-Meister

still accepted for steady-state vibrations, but for transient vibrations, e.g. floor vibrations produced by people walking, there is recent evidence that amplitudes much greater than those given by the scale are necessary to produce a given sensation at a given frequency.

In the Reiher-Meister investigation it was noted that vertical vibration was most readily detected when people were standing, whereas horizontal vibration was more noticeable when they were lying down. The sensation produced depends on frequency and amplitude. An amplitude of 100 μ constitutes an annoying vibration if the frequency is 5 Hz and a painful vibration if the frequency exceeds 20 Hz. An amplitude of 10 μ is just perceptible at 5 Hz, but would be annoying at 50 Hz. Expressed in terms of peak velocity, the threshold of perception corresponds to a velocity of 0·3 mm/sec and a vibration is annoying if the velocity exceeds 2·5 mm/sec. The data for vertical vibration (persons standing) are given in **Fig. 1.**

More recent data are provided by the Dieckmann scale, which extends into lower ranges of frequency and, hence, may be useful also in determining the effect on people of building sway in high winds. In this scale, zones of equal intensity are defined and are given K-values ranging from 0·1 to 100, the effects of various intensities being given as follows:

K=0·1 ; lower limit of perception

K=1 ; allowable in industry for any period of time

K=10 ; allowable only for a short time

K=100 ; upper limit of strain for the average man.

The charts for vertical vibrations and horizontal vibrations are given in **Figs. 2** and **3**. The Dieckmann K-values for vibrations of given amplitudes and frequencies can be simply calculated, as shown in **Table 1.**

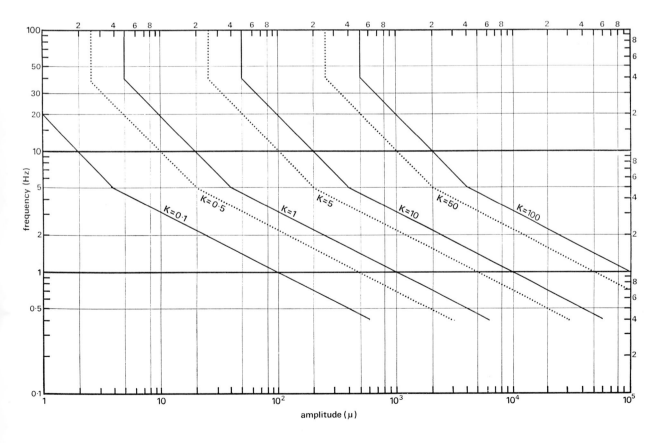

Fig. 2   Dieckmann values: vertical vibrations

Table 1   Calculation of Dieckmann K-values

Vertical vibrations	Horizontal vibrations
Below 5 Hz: $K=0.001 A.f^2$	Below 2 Hz: $K=0.002 A.f^2$
5–40 Hz: $\quad K=0.005 A.f$	2–25 Hz: $\quad K=0.004 A.f$
Above 40 Hz: $K=0.2 A$	Above 25 Hz: $K=0.1 A$

(A being amplitude in microns and f the frequency in Hz)

Two examples illustrate the calculation of vibration intensity and the effect on people. In the case of a very severe building vibration, due to the operation of reciprocating machines at high level, the amplitude of horizontal vibration was 750 μ and the frequency 4 Hz. The Dieckmann K-value is 12 and the Reiher-Meister classification is *annoying to unpleasant*. A typical result for vertical vibration for street traffic on the other hand is an amplitude of 10 μ at frequency 20 Hz. This gives a strength of K=1 on the Dieckmann scale and is graded as *clearly perceptible* on the Reiher-Meister scale.

Vibration and damage to buildings

It is not easy to define, with absolute certainty, what constitutes a damaging vibration, nor does any universally accepted criterion exist that may be used to compare the effects of vibrations of differing amplitudes and frequencies. In assessing the possibility of damage many factors should be taken into account, for example, the additional stresses set up by the vibration, the size and type of building, the fatigue properties of the materials, and the possibility of resonance.

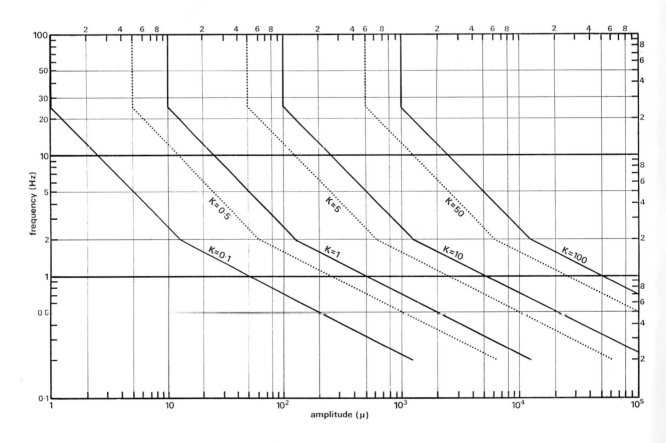

Fig. 3   Dieckmann values: horizontal vibrations

It is often difficult enough to perform a static stress analysis for a complete structure; analysis of the dynamic stresses set up by vibration in a masonry structure is virtually impossible.

The criteria that assist in determining the effects of vibration include those of maximum amplitude, velocity, acceleration and energy, as well as the use of the human sensitivity scales. A comparison with cases where no damage has been produced by vibrations of known characteristics may often be useful and information available concerning vibrations and shock from blasting, pile driving and earthquakes is also relevant. The various scales of earthquake intensity and associated acceleration values should not however be applied, without modification, to vibrations of buildings due to more usual sources of excitation.

Much information is available on the effects of blasting vibration. Some early work in the USA indicated that conventional types of property, either wood-framed or of brick construction, can withstand amplitudes of about 400 μ. From the results of such tests, working rules were adopted enabling the operator to determine the maximum charge that could safely be fired at certain distances, the maximum amplitudes of vibration permitted ranging from 125 μ to 400 μ. In these earlier tests, and in similar tests more recently carried out in Sweden, the UK and elsewhere, damage was caused deliberately by firing charges far in excess of those normally permitted. Large amplitudes, ranging from 250 μ to 5000 μ, were involved.

Current opinion is that the maximum velocity ('peak particle velocity') is a more realistic criterion for damage than is the simpler criterion of maximum amplitude and this is in general accord with an

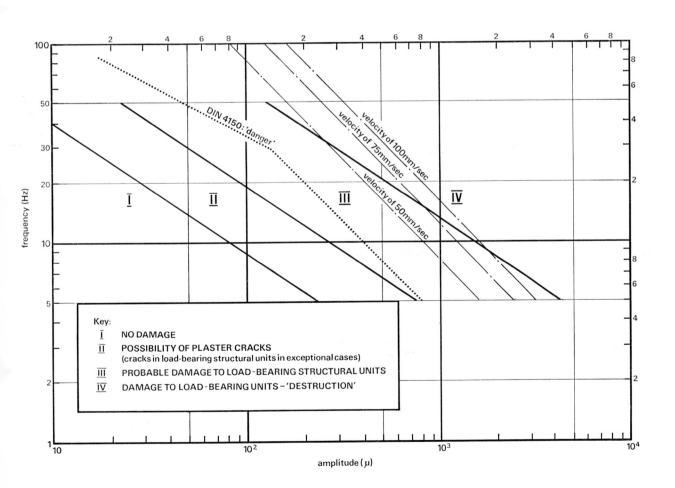

Key:
I   NO DAMAGE
II   POSSIBILITY OF PLASTER CRACKS
     (cracks in load-bearing structural units in exceptional cases)
III PROBABLE DAMAGE TO LOAD-BEARING STRUCTURAL UNITS
IV   DAMAGE TO LOAD-BEARING UNITS – 'DESTRUCTION'

Fig. 4 Possible damage to buildings

'energy ratio' criterion suggested in 1949. The latter indicates that structures are safe if the peak particle velocity does not exceed 80 mm/sec and that there is danger if the value exceeds 120 mm/sec. Swedish information is that there is minimal damage at maximum velocity of 70 mm/sec, slight damage at 110 mm/sec, moderate damage at 160 mm/sec and serious damage at 230 mm/sec. A Canadian Building Digest* issued in 1965 gives the threshold of damage due to blasting for a horizontal or vertical velocity of about 75 mm/sec, as recorded in the foundation wall nearest to the blast.

For vertical vibrations of frequencies 5–40 Hz, a maximum velocity of 75 mm/sec corresponds to a Dieckmann K-value of 60, and such a vibration would normally be deemed extremely unpleasant. Vibration from blasting is usually of short duration and it would appear that higher amplitudes or velocities can be tolerated than would be the case for sustained, steady-state, vibrations, e.g. from machinery.

The data given in **Fig. 4** for frequencies from 5–50 Hz, the range most commonly encountered in buildings, is the most reliable information at present available for the estimation of structural damage from vibrations and is based on a review carried out in 1961. So far as is known, no case of damage has ever been reported for vibrations in Zone I of the chart. For Zone II, cracking in load-bearing structural units is likely only when reserves of strength have been used up prior to the dynamic loading, For purposes of comparison, the danger-zone according to DIN 4150 (1939) and maximum velocities of 50, 75 and 100 mm/sec have also been indicated on the chart. There is probably little risk of damage if the quantity A^2f^3 is less than 50 mm^2/sec^3.

Cases where vibration has caused definite damage to buildings have been extremely rare and the generalisation can be made that vibration must become unpleasant or painful to the occupier long before there is any possibility of damage to the building itself. Nevertheless, complaints are often received that minor damage such as the cracking of plaster ceilings, brickwork and window glass, as well as the loosening of roofing tiles, has been caused by some particular source of vibration. That such complaints are so common may be due to the extreme sensitivity of the human body. When a new source of heavy vibration arises and is such as to disturb the occupier of a building he may become concerned about the structural effects of such vibration. He may then assume that cracks which have escaped notice for some time have been caused by the vibration. There are many causes of cracking in buildings, see for example Digest 75 'Cracking in buildings'; cracks should not be attributed to the effects of vibration until other possible causes have been eliminated. For example, much of the cracking in plaster ceilings must be attributed to movements of the plaster itself or of the timber supports due to changes in moisture content and such cracking often occurs in areas known to be free from external sources of vibration. The earlier work in the USA in connection with blasting vibrations included tests on floors and ceilings where deliberate attempts were made to cause damage by excessive blasting or by mechanical agitation, even under sustained resonant conditions. Sustained amplitudes of vibration of the order 2500 µ (2·5 mm) were required to produce severe cracking of plaster and these are far in excess of the amplitudes produced by traffic, factory machinery, compressors, or other common sources of

---

* **Blasting and Building Damage,** Canadian Building Digest 63(1965), NRC, Canada.

vibration. Door slamming or walking about heavily are likely to produce greater vibration and shock than will most external sources of vibration.

In building vibration problems there is no evidence that fatigue is significant. For steel beams it can be shown that there is usually a large factor of safety against fatigue failure even when the vibration is such as to be unpleasant to a person. The additional stresses set up by vibration, even in bridges, are usually very small when compared with normal working stresses. Information obtained in recent years has shown that the resistance to fatigue of reinforced and prestressed concrete is also sufficient to ensure that fracture or damage is unlikely to result if the level of vibration can be tolerated personally.

Further reading

1   **Vibration in Civil Engineering**   B O Skipp (*Editor*); Butterworth London 1966

2   Current Paper Engineering Series 37   **Some aspects of structural vibration**   R J Steffens BRS 1966

3   Library Bibliography No. 199   **A bibliography on vibrations 1955–1965** BRS 1966

4   Library Bibliography No. 217   **A bibliography on vibrations in laboratories**   BRS 1969

5   Current Paper 14/70 **The problem of vibrations in laboratories** R J Steffens BRS 1970

6   International Series of Monographs in Civil Engineering, Vol. 2   **Building on springs**   R A Waller Pergamon Press 1969

Vibrations in buildings—2

The first of this pair of digests (Digest 117, see page 82) explained some of the terms used in vibration studies and their relationships; it discussed human sensitivity to vibration and the effects of vibration on buildings.

This second part discusses some of the ways in which nuisance from vibration can be avoided or reduced to a tolerable level—by treatment at the source, by reducing the transmission of vibration and by providing protection against vibration from an external source.

The numbering of figures is continued from the earlier Digest.

The isolation and reduction of vibration

Vibration can usually be isolated or reduced to a tolerable level by fairly simple and relatively inexpensive means if the precautions are incorporated in the initial design. It is likely to be more difficult and expensive to provide a remedy in existing buildings and installations.

The ways in which nuisance from vibrations can be treated include:

a reduction of the vibration at its source;

b reduction of the vibration transmitted to the immediate surroundings of the source:

c placing machines that cause vibration as far as possible from people and equipment likely to be affected;

d protecting sensitive apparatus against external vibration;

e protecting a whole building or a part of it against external vibration.

It may be necessary to apply two or more of these measures to any one problem. For example, if *a* alone is unlikely to give the required improvement, then *a* and *b*, *a*, *b* and *c*, or other combinations may be needed.

Reducing vibration at source

This can be done by the maker of the machine by dynamic balancing of the moving parts so far as this is possible. An alteration in the speed of the machine may be effective in some cases, particularly if a state of resonance would otherwise exist between the machine and the structure.

Reducing vibration transmitted from machines

The transmission of vibration from machines may be reduced by various means:

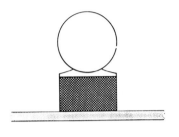

By mounting the machine on a heavy base, of weight comparable to that of the machine, resting directly on the ground; the vibrations are then reduced by the increase in mass.

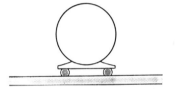

By placing an insulating mounting between the machine and a slab of concrete resting on the ground; this is particularly suitable for low or medium power machines that are not likely to generate very low frequency vibrations.

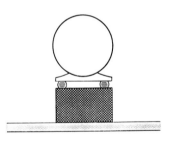

By the combined use of a heavy base and an insulating mounting.

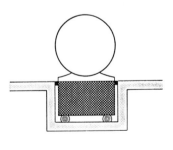

By the combined use of a heavy base and an insulating mounting sunk into a pit. The insulation may be either above or below the machine base. An air space should be left all round the block but the gaps may be sealed at the top by a strip of flexible material to prevent the ingress of debris.

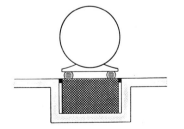

The insulating material may be of cork, felt, rubber or composite materials; steel springs are sometimes used. Each material has its advantages and disadvantages for any given conditions and the choice is a matter for specialist advice. Some firms deal only with a particular material or type of unit but others will advise on the type most suitable for a given installation. In many cases a compromise solution may have to be accepted. For example, bolting a machine tightly to a rigid foundation may prevent vibration of the machine but may transmit harmful vibrations elsewhere. On the other hand, a very flexible mounting, while it may greatly reduce the transmitted vibration, may result in a large increase in amplitude of movement of the machine which may be inconvenient to the operator, or even harmful to the machine.

A satisfactory result is likely when the natural frequency of the spring-supported mass is no higher than about one-third of the forcing frequency, the latter depending on the rate of operation of the machine. If the required natural frequency can be determined the required static deflection can be found and, hence, the stiffness of the spring or mounting. The suggested frequency ratio of 3 to 1 would, theoretically, reduce the transmitted vibration amplitude by more than 80 per cent; a higher value of frequency ratio would achieve a still greater reduction. The efficiency of a mounting system will, in practice, be less than the theoretical value calculated from the static deflection alone because the dynamic stiffnesses of isolating materials are much greater than the static values. Fig 5 illustrates the effect on energy reduction of resilient machine mountings and shows that if the frequency ratio is less than $\sqrt{2}:1$ there is no benefit to be derived from the mountings.

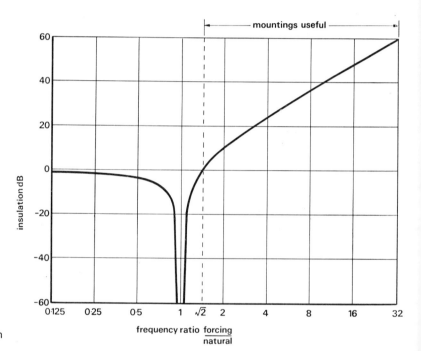

Fig 5 Effect of mountings on energy reduction

The relationships given in Fig 1 (Digest 117) between human sensitivity and the amplitudes and frequencies of vibrations can be used in choosing a suitable mounting. If a machine is producing vibrations that constitute some degree of nuisance, the information given in Fig 1 would indicate by how much the amplitude should be reduced to bring the vibration down to a tolerable level and from this a suitable mounting could be selected. For example, if a machine was producing vertical amplitudes in nearby property of 20μ at a frequency of 30 Hz, the vibration intensity would be deemed *annoying* (Dieckmann K-value = 3). If the machine was then fitted with a mounting having a natural frequency of 10 Hz for which the required static deflection under load would be about 2·5 mm, the amplitude would, theoretically, be reduced to about 2·5 μ and the vibration would then be *just perceptible* (K-value = 0·4).

Makeshift remedies such as placing a pad of some sort under the machine are all too often useless, generally because the mounting is too stiff and the low frequency structural vibrations have been virtually unchanged; they might even make matters worse.

Siting of machinery

The amplitude of vibration will usually diminish with distance from the source. Machinery or plant causing vibrations should, therefore, be kept as far as possible from buildings likely to be affected. No precise relationship between amplitude and distance can be stated since both the nature of the ground and the type of vibration or ground shock are involved, but in many cases doubling the distance will reduce the amplitude to about one-third.

Protection of apparatus and equipment

The vibration problem should be considered at an early stage by the designer in collaboration with the client (or user) to ensure that vibration levels will be acceptable for the type of apparatus to be used.

Sensitive apparatus should, preferably, not be placed on brackets attached to walls. Benches should be robust and kept from direct contact with walls. Excessive vibration of floors due to people walking is a common source of trouble, particularly in a suspended timber floor, or one of light construction. Well-designed floors, of adequate stiffness, reduce the risk. Floor vibrations can sometimes be reduced by fitting dynamic vibration absorbers. The low natural damping can also be improved by the use of 'sandwich layers'—a comparatively new technique. Where floor vibration is likely to be important, sensitive equipment should be located away from the middle of a room. In general, very delicate work should not be carried out on upper floors. The disturbing effects of door-slamming should also be considered; swing doors or special door closers may have to be provided.

After all other precautions have been taken it may still be necessary to protect individual pieces of sensitive equipment. One method is to build an isolated pier upon which the apparatus can stand. The pier, of brick, stone, or concrete passes through the floor into the foundation or basement of the building. The gap between floor and pier should be left open if possible, but if material is inserted for dust-sealing or waterproofing it should be of a suitable flexible type. Care should be taken to ensure that the base of the pier does not pick up vibrations from other sources; if there is a possibility of this, suitable anti-vibration material can be inserted either at the base or at the top of the pier.

In many cases, the use of commercially available instrument mounts will avoid the need for more complicated measures. The mountings are usually of rubber, loaded in shear, and are similar to some of the mountings used for machinery but of smaller size and load capacity. A very wide range is available, the appropriate type depending on the frequency of vibration affecting the apparatus, the degree of isolation required and the weight to be supported. Some types of apparatus are affected by amplitudes of $0.1\ \mu$ or less, or by very low values of acceleration, and need a very high degree of isolation.

Special mountings, developed particularly for use in aircraft, are also available for the protection of racks of electronic and electrical equipment. Sensitive chemical balances may require special mounting systems that incorporate damping arrangements, and a self-contained mounting unit with viscous damping has been marketed. Where isolating systems of very low natural frequency are required, special design may have to be undertaken, and steel springs, rather than rubber unit mountings, used. Air-bellows mountings are an alternative and convenient method of providing a high degree of isolation at low frequencies.

The isolation of buildings

Existing buildings The isolation of a complete existing building is a difficult matter and in many cases the work required would be elaborate and costly. The isolation of limited areas, however, has been successfully achieved on many occasions, usually by mounting floors or small rooms on rubber carpet mountings or on special rubber or steel spring mountings.

Trenches have been used as a means of preventing external vibrations affecting a building and can be effective in the protection of small buildings, for example, seismological observation posts. The method, however, is not usually applicable to a large building, particularly if the ground vibrations are deep-seated rather than close to the surface. Modern theory is that the depth of trench should be about one-third of the wavelength of the vibration. Thus if the velocity of wave propagation in a particular soil were 300 m/sec and the frequency of vibration 20 Hz, the wavelength would be 15 m and the trench would have to be at least 5 m deep; it would also have to be kept open or, possibly, filled with thixotropic fluid.

New buildings Provision can be made at the design stage for a new building to be specially mounted to protect it from the effects of vibration that cannot otherwise be controlled.

For many years, buildings in the USA have been mounted on sandwich pads of lead and asbestos, inserted under girders and columns. The original purpose of this treatment was to protect the buildings from the effects of noise from road and rail traffic but it has been claimed that the low-frequency vibration problem is also adequately dealt with. This view is, however, not universally supported.

In London there are now several buildings in which special measures have been taken to combat vibration and low-frequency noise.

In the Barbican Development, the underground railway runs just below the area and the heavy deck of the track itself is supported on rubber units giving a natural frequency of about 6 Hz. The possibility of resonance of the prestressed beams of the deck has also been taken into account and the intrinsic damping improved by the use of a sandwich damping layer of specially selected material.

In the new Marble Arch Odeon scheme the whole auditorium, weighing about 4,600 tonnes, has been mounted on sandwich pads to protect it from vibration and noise from nearby railway lines.

A block of flats, Albany Court, near Victoria, has been mounted on a low-frequency isolating system to protect it from vibrations from the underground railway running directly below the building. A large number of sandwich pads of rubber and steel were used, the individual loads carried ranging from 60 tonnes to over 200 tonnes and the natural frequencies obtained were 7 Hz vertically and 2·5 Hz horizontally. The possibility of wind-excitation of the spring-mounted building was also taken into account. The cost of mounting the building, including the structural modifications to accommodate the springs, is likely to be more than offset by the amenity value that would otherwise be lost.

The pathological laboratories of St Mary's Hospital Medical School (Fig 6) are located near to three railway lines, one on the surface and two underground, which interfered seriously with work in the old building. Measurements showed that vibrations due to trains were mainly in the range 25–40 Hz. The new building of five storeys was built on a reinforced concrete platform supported by four columns, each about $1 \cdot 2 \times 1 \cdot 5$ m capable of carrying a load of 660 tonnes and resting on a rubber mounting. The V-shaped bases (Fig 7) take care of the vertical component of vibration and of the horizontal component at right-angles to the Vs; the horizontal component in the other direction is dealt with by a 50 mm thick rubber pad, clamped to the parallel sides of the column feet (Fig 8). An airspace of about 50 mm was left between the new building and its neighbours on each side and sealed with a rubber strip. The effectiveness of the design is shown by the two record traces (Figs 9 and 10) of train vibrations measured in the old building and in the electron microscope room of the new.

Fig 6

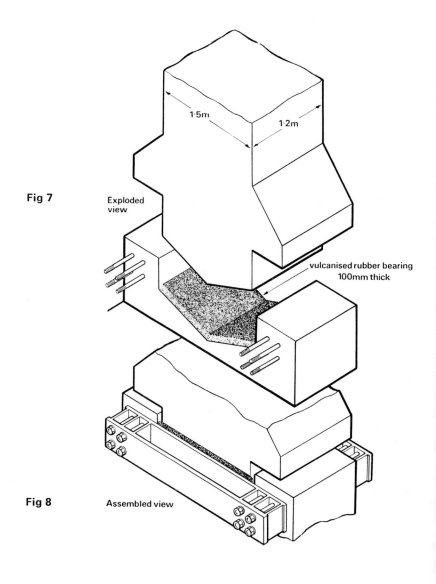

Fig 7

Exploded view

1·5m   1·2m

vulcanised rubber bearing
100mm thick

Fig 8   Assembled view

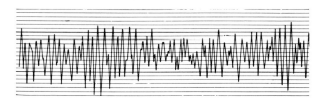

Fig 9 Vibration in the old building

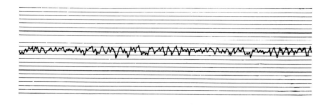

Fig 10 Vibration in the new building

Figs 6–10 reproduced by permission of Oscar Faber and Partners.

Sound insulation: basic principles

The use of novel designs and the recent amendment of the Building Regulations 1965 to include a performance standard in the deemed-to-satisfy provisions demand that increasing attention be paid to sound insulation between dwellings. This digest outlines the basic physical principles involved and complements other digests which contain more specific recommendations.

Types of sound transmission

Insulation is required against sound generated in two different ways. On the one hand, a source such as a radio may produce sound waves in air which in their turn produce vibrations in a party wall or floor (Fig 1a). On the other hand, a wall or floor separating two dwellings may be excited by the direct impact of a solid object, eg footsteps (Fig 1b). The two kinds of insulation have to be measured separately. Only party floors are required to meet a prescribed standard of impact insulation, but both party walls and party floors are required to provide a standard amount of insulation against airborne sound. This digest is concerned mainly with airborne sound.

Measurements to test the airborne sound insulation between two rooms are concerned with the total amount of energy transmitted, by whatever means, but, to understand what occurs, a further distinction

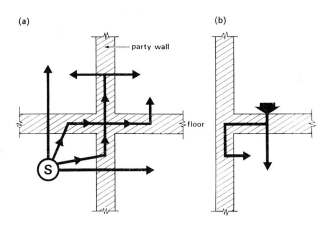

Fig 1 Vertical sections illustrating paths of sound transmission
(a) airborne
(b) impact

must be made. It is a familiar fact that a radio or television in a ground floor room of one house can often be heard in the bedroom of a neighbouring house. In this instance vibrations set up in the party wall at ground floor level are able to travel up the party wall and then radiate into the upper floor rooms. This is known as *flanking transmission*. It can still occur when the rooms adjoin, either horizontally or vertically, and often provides a serious addition to sound passing by the obvious path through the common wall or floor.

Specification of airborne sound insulation

In standard tests (BS 2750) the insulation is measured in each of sixteen one-third-octave bands, the centre frequencies of which range from 100 to 3,150 Hz. Measurements to find whether a construction can be deemed to satisfy the Regulations by way of the performance standard must be carried out in actual buildings and the results are expressed in terms of *normalised level difference*; this is the difference in decibels (dB) between the energy levels in the rooms corrected to allow for a standard amount of absorption representative of normal furnished conditions. Laboratory measurements are usually expressed in terms of the *sound reduction index*, which is the difference in decibel measure between the amount of energy flowing towards the wall in the source room and the total amount of energy entering the receiving room. The numerical value of the term which converts from normalised level difference to sound reduction index is usually small and either quantity can be used in considering principles; for this purpose it is therefore convenient to use the general term 'insulation'.

Prepared at Building Research Station, Garston, Watford WD2 7JR
Technical enquiries arising from this Digest should be directed to the Building Research Advisory Service at the above address.

The required values of normalised level difference range from 36 dB (for party floors) or 40 dB (for party walls) at 100 Hz to 56 dB at 1,600 Hz and above. The value of a quantity in decibels is equal to ten times the common logarithm of its value, relative to an arbitrary standard, in ordinary measure (joule/m³ for energy level or watt/m² for energy flow). In the usual situation where the conversion term between sound reduction index and normalised level difference is small, this means that the fraction of the energy falling on a party wall which can be allowed to pass through it is only 1 part in 10,000 at 100 Hz and 2·5 parts in a million at the highest frequencies. A common misconception is that the problem of sound insulation is one of absorbing the sound energy which falls on a wall. In fact, to approach the high levels of absorption that would be required all walls would have to be made like those of the anechoic chambers used in acoustic laboratories and this is not practicable for dwellings. Thus the problem of sound insulation is almost entirely one of reflecting energy back into the source room; the role of absorption is limited to supplementing reflection at high frequencies in some types of wall or floor.

'Mass Law'

Most of the principles involved in direct transmission are revealed by considering a single-leaf wall. The most widely known term in relation to sound insulation is the *mass law*. A strict mathematical derivation is outside the scope of this digest but Fig 2 shows in broad physical terms how the so-called 'mass law' arises. It seems obvious that the amplitude of sound waves radiated into the receiving room must depend only on the amplitude of vibrations in the wall, and not on any other properties of the wall; the amplitude of wall vibrations depends on the amplitude of the oscillatory part of the pressure in the source room which acts on the wall. According to Newton's law of motion,

$$\text{force} = \text{mass} \times \text{acceleration}$$

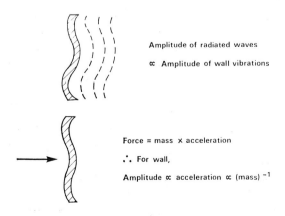

Amplitude of radiated waves
∝ Amplitude of wall vibrations

Force = mass × acceleration
∴ For wall,
Amplitude ∝ acceleration ∝ (mass)⁻¹

Fig 2 Physical ideas behind the 'mass law'

and, in an oscillatory motion, amplitude and acceleration are proportional to one another. Thus the amplitude of wall vibrations is inversely proportional to the mass of the wall. It follows that the amplitude of sound waves radiated into the receiving room is inversely proportional to the mass of the wall. Transmission is considered in terms of energy. Since this is proportional to the square of velocity, and thus to the square of amplitude, the transmitted energy is inversely proportional to the square of the mass of the wall. This means that by doubling the mass of the wall transmission is reduced to a quarter. In decibel measure, insulation is increased by

$$10 \log_{10} 4 = 6 \text{ dB}$$

The argument can be modified to deduce that a doubling in frequency should also produce a 6 dB increase in insulation. Broadly, these two statements about the effects of changes in mass and frequency constitute the 'mass law'.

When due account is taken of other properties of a wall, theoretical treatment becomes very complicated. However, whatever refinements are introduced, it is nearly always found that, as between single-leaf walls of a similar type, a doubling in mass produces an increase of about 6 dB in insulation, and that, except for the rapid change which is discussed shortly, there is an increase of the same order when the frequency is doubled.

'Coincidence'

There are two important features of the transmission process to which it was unnecessary to draw attention while sketching the basis of the mass law.

First, insulation is always measured for a 'diffuse' incident field which may be regarded as composed of plane waves approaching the wall from all possible directions. The more oblique the incident wave, the better it is transmitted; if the wall were large enough, waves approaching at a glancing angle would always be perfectly transmitted.

As part of the transmission process, a plane wave approaching at an oblique angle produces a forced motion in the wall (Fig 3) which has a greater wavelength (the 'trace' wavelength) than that of the incident waves in air. Because walls possess stiffness as well as mass, it is possible for free bending waves to propagate in a wall. The amplitude of wall vibrations is greatly magnified when the wavelength of the free bending waves is near to that of the wave impressed on the wall by the incident sound wave. This in its turn greatly increases transmission. The effect is known as *coincidence*.

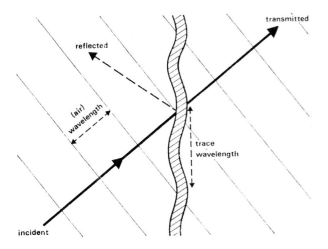

Fig 3 Forced waves in a wall have a longer wavelength than incident sound waves

The velocity of free bending waves increases with frequency. Since wavelength = velocity/frequency, the wavelength of free bending waves is less than that of sound waves in air at the lowest frequencies and greater than that of waves in air at the highest frequencies. It has been seen that the trace wavelength is greater than that of waves in air. Consequently coincidence can only occur when the wavelength of free bending waves is greater than that of waves in air; the frequency where these two wavelengths are equal is known as the *critical frequency*. Above this frequency transmission is dominated by coincidence. Near the critical frequency, coincidence occurs for glancing waves. Thus in this region two effects which lead to high transmission act together, and there is a large decrease in insulation.

The result is that, apart from some modification at low frequencies due to the relatively small area of practical walls, all single-leaf walls have a curve of insulation against frequency approximating to that shown in Fig 4. No origin is shown on either scale, since both the general level of insulation and the position of the critical frequency depend on the particular wall considered; it may be necessary to take measurements above or below the standard range of 100 to 3,150 Hz to detect the critical frequency dip. As 'coincidence' depends on a kind of resonance, the depth of the dip depends on the amount of damping present. For common amounts of damping, insulation at the critical frequency is often as much as 10–15 dB below the level suggested by the trend at lower frequencies. Insulation usually remains 5–10 dB below the low frequency trend for an octave above the critical frequency.

With sufficient mass, adequate insulation can be obtained even at the critical frequency. However, the depth of the dip suggests that it is desirable that a wall should have a critical frequency either below 100 Hz or above 3,150 Hz.

For reasons touched on later it is not too difficult to cope with a critical frequency dip at frequencies somewhat below 3,150 Hz. However, the general principle that critical frequencies between just over 100 Hz and 1,000 Hz should be avoided whenever possible is one of the most basic in sound insulation.

The critical frequency is given by

$$f_c = \frac{c_o^2 (1-\sigma)}{1 \cdot 8 c_c h \sqrt{(1-2\sigma)}}$$

where

f_c = critical frequency,

c_o = sound velocity in air,

$$c_c = \sqrt{\left\{ \frac{E(1-\sigma)}{\rho(1+\sigma)(1-2\sigma)} \right\}}$$

= compressional wave velocity in wall material

E = Young's modulus of wall material,

σ = Poisson's ratio of wall material,

ρ = density of wall material,

h = thickness of wall.

Of the parameters involved, c_o, the sound velocity in air, cannot be altered. With a sufficiently wide choice of materials the wave velocity, c_c, and Poisson's ratio, σ can vary widely, but between most practicable materials the variation is limited. The most important parameter is usually the thickness, which can vary over a range of 50:1 between conceivable walls.

Practical implications for single-leaf walls

Some typical values of critical frequency are shown in Table 1. The critical frequency of a brick wall is not easily determined and the properties of brick walls are variable, but it appears that the critical frequency of a one-brick (215 mm) wall is usually in the neighbourhood of 100 Hz. A one-brick wall thus satisfies the condition of having a critical frequency near or below the lower limit of the practical range. In practice the insulation given by a brick wall near 100 Hz is better than might be expected from the idealised curve of Fig 4 because of effects related to

Table 1 Approximate critical frequencies

Material	Thickness mm	Surface mass kg/m²	Critical frequency Hz
Brick	215	400	100
Brick	102·5	200	200
Lead	18	200	15,000
Plasterboard	10	9	2,900–4,500
Plasterboard	20	18	1,400–2,300
Glassfibre reinforced gypsum	10	18	2,000
Plywood	10	7·5	1,300
Steel	3	25	4,000

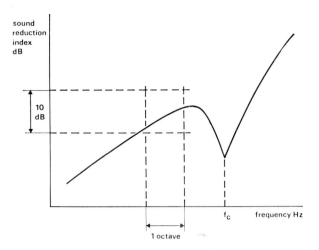

Fig 4 Typical form of insulation curve for a single-leaf wall

the limited area of practical walls. These arise because near the critical frequency, 100 Hz, the sound wavelength in air and the bending wavelength in the wall are both equal to about 3·4 m, and are therefore comparable with the linear dimensions of a normal room. Another departure from simple theory is that the insulation given by a one-brick wall levels off near 1,000 Hz instead of continuing to increase, because new modes of transmission arise in a thick wall. In practice this is not serious because the insulation is adequate.

Concrete walls—including those made of lightweight concrete—fall into the same category as a one-brick wall, having a critical frequency near 100 Hz if they are thick enough to provide the mass required.

With thinner brick or concrete walls insulation is poorer not only because the mass is reduced but because the critical frequency goes up to a frequency at which its effects are more serious.

Since brick and concrete walls operate mainly above the critical frequency, the insulation they provide is affected by the amount of damping present. It is therefore theoretically possible to improve the insulation such walls provide by modifying the materials used to increase their internal damping. There has in fact been research in University and in Industrial Laboratories on the use of polymers to increase the internal damping of concrete.

It is logical to consider the possibility of going to the other extreme, ie using walls with a critical frequency near or above the upper end of the frequency range at 3,150 Hz. Table 1 contains examples of panels with a critical frequency above 1,000 Hz. However, the mass required for a single-leaf wall with a high critical frequency to meet the party-wall grade can be calculated, and is just over 200 kg/m². This is about half the mass of a one-brick wall. It does not appear feasible to combine the required mass with a high

critical frequency in a practicable panel. Materials which would provide the required combination in a homogeneous panel exist, but they are expensive; the most suitable is lead, which (Table 1) would have a critical frequency of 15,000 Hz with a surface mass of 200 kg/m². With stiffer materials, a sophisticated sandwich construction would be necessary. For practical purposes, single-leaf walls with a high critical frequency can be ruled out; to obtain practical walls with a high critical frequency it is necessary to use two or more leaves separated by a wide gap.

Flexible double-leaf walls

The basic idea behind a double-leaf wall is shown in Fig 5. If three rooms were arranged in a row with successive pairs separated by walls giving 25 dB insulation, then the insulation between the end rooms would be 50 dB. It is reasonable to hope that insulation of the same order can still be obtained if the middle room is reduced to the sort of air gap that is practicable within a party wall. In contrast, if the two original walls could be joined together into a single wall without lowering the critical frequency, then, according to the mass law, the total insulation would be only

$$25+6=31 \text{ dB}$$

Practical experience has confirmed that a double-leaf wall with flexible (high critical frequency) leaves can provide satisfactory insulation with much less mass than is needed with a single-leaf wall. Walls already in use provide satisfactory insulation with a total mass of 60 kg/m²—only 15% of the mass of a one-brick wall. However, a double-leaf wall performs better than a single-leaf wall of the same total mass only above a certain frequency which depends on the cavity width. This frequency is the 'mass-spring-mass' vibration frequency f_o of two rigid bodies of

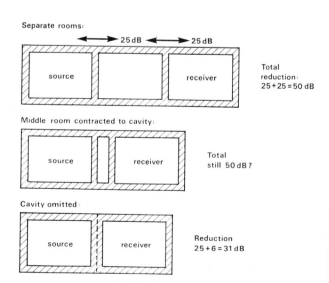

Fig 5 Basic idea behind a double-leaf wall

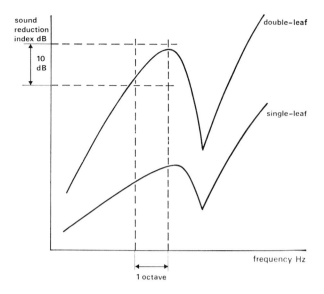

sound reduction index dB

10 dB

double-leaf

single-leaf

frequency Hz

1 octave

Fig 6 Comparison of shapes of typical insulation curves for single and double-leaf walls

the same mass as the leaves connected by a spring with the stiffness of the air in the cavity:

$$f_0 = \frac{1}{2\pi} \sqrt{\frac{\gamma P}{d}\left(\frac{1}{m_1} + \frac{1}{m_2}\right)}$$

where P = air pressure (approx 10^5 N/m^2)
  γ = ratio of specific heats (approx 1·4)
  m_1, m_2 = masses of leaves per unit area
  d = cavity width

To gain advantage from double-leaf construction, it is essential that the cavity is wide enough for the 'mass-spring-mass' frequency to be below 100 Hz, and desirable that it is as wide as possible. Thermal double glazing with a narrow cavity does not provide good sound insulation because the 'mass-spring-mass' frequency is well above 100 Hz. A sharp dip in the insulation curve usually occurs near the 'mass-spring-mass' frequency. A similar effect can occur in more complicated arrangements such as dry-linings on masonry walls.

It is sometimes said that a double-leaf wall 'beats the mass law' Whether this is so is essentially a matter of definition and it is more convenient to regard single and double-leaf walls as obeying different kinds of mass law. A double-leaf wall provides nearly twice the insulation which each leaf would provide on its own and the effect of changing the mass of both leaves is twice the effect of a similar change in the mass of a single-leaf wall; thus if the total mass is doubled the insulation is increased by approximately

2 × 6 = 12 dB.

It will be recalled that for single-leaf walls insulation increases by approximately 6 dB per octave except where it dips near the critical frequency. For double-

leaf walls the corresponding slope is 12 dB per octave. Consequently the curve of insulation against frequency for a double-leaf wall has the typical shape shown in Fig 6.

Because of the steep slope, it is usually found that while it is difficult to obtain sufficient insulation with lightweight double-leaf walls at the lowest frequencies, there is then a wide band of frequencies in which the insulation is well above what is required. Above this region, there are again problems because the dip near the critical frequency is even more pronounced than with a single-leaf wall. At low frequencies, the only solution, after choosing the largest practicable value for the cavity width, is to have sufficient mass, but having dissimilar leaves and using an absorbent curtain or filling of a material like glass fibre helps to overcome problems arising from the critical frequency dip. If the leaves have different critical frequencies, the dip, although broader, is not so deep. Absorbent material in the cavity is useful because the thickness needed to make an effective contribution decreases with frequency and reaches a practicable value at the higher frequencies. However, even in the critical frequency region absorption serves only as a supplement to reflection, which still makes the major contribution.

Although double-leaf lightweight walls can provide satisfactory insulation, it is not yet possible to predict with much confidence whether a given design will prove satisfactory. In particular, more needs to be known about the factors which can spoil insulation. There is a large difference between the vibration levels in the two leaves when a wall is behaving satisfactorily. As a result, insulation can easily be spoilt by solid bridges. With lightweight panels, studding is usually needed to increase static stiffness. It is almost certain that insulation will be inadequate unless the two leaves are attached to different studs. With low-rise housing, it is possible to build structurally detached houses with a gap of 300 mm or less between them, but in high-rise buildings some solid connection at the edges of a party wall is unavoidable. Study of a similar problem for windows has shown that mounting the panes in a neoprene gasket goes a long way towards providing the desired isolation between the leaves. There is good reason to expect that similar mountings can be useful in double-leaf walls.

Stiff double-leaf walls
The use of cavity constructions has been discussed at some length for flexible lightweight walls, because with such walls a cavity is essential. With stiff, heavy walls the situation is different. In practice, the choice is between a single-leaf wall and a double-leaf wall

of similar total mass, eg a one-brick wall or two half-brick leaves. As was seen earlier, the loss in insulation which results from using a thinner wall is more than would be expected from considering the mass alone, because the critical frequency goes up into a region where its effects are more serious. This shift in critical frequency tends to offset the advantages of having a cavity. For practical arrangements, the two effects roughly cancel out and there is little to choose between single and double-leaf constructions. As with lightweight walls, ties between the leaves are liable to spoil the insulation, and it is advisable to keep to designs which experience has shown to be acceptable.

Floors

A heavy concrete party floor insulates against airborne sound on the same principle as a single brick or concrete wall: its effectiveness depends on the critical frequency being near the lower end of the practical frequency range, and the mass required for a given level of insulation is the same as with a wall. To obtain satisfactory insulation with less weight, it is again necessary to use a double-leaf construction. The obvious practical difference as compared with walls is that the floor must be stiff enough not to deform too much under loading. This requirement rules out having a critical frequency near 3,150 Hz and consequently the aim must be to make the floor as stiff as possible in order to have a critical frequency near or below 100 Hz. For ceiling panels which have

(a) Party wall vibrationally isolated from external wall

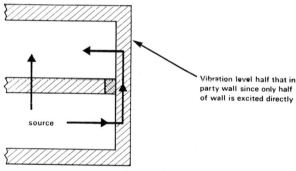

Vibration level half that in party wall since only half of wall is excited directly

source

(b) Party wall built into external wall

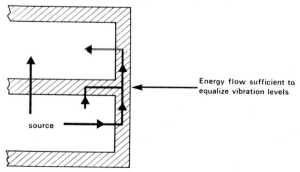

Energy flow sufficient to equalize vibration levels

source

Fig 7 Flanking transmission with heavy single-leaf party and external walls

only their own weight to support, a critical frequency near 3,150 Hz should be acceptable. The best known lightweight construction is a timber-joist floor combined with a pugged ceiling. Although the details are complicated, this conforms roughly to the basic principle of combining a low critical frequency floor and high critical frequency ceiling.

Flanking transmission

The subject of flanking transmission is a complex one, because this type of transmission does not depend solely on the properties of the flanking wall, but also on the properties of the party wall, on whether the flanking wall is continuous, and on how firmly it is bonded to the party wall. The principles are most fully understood for heavy walls which operate almost entirely above the critical frequency.

The simplest case is that of a single-leaf external brick wall. It is convenient to consider first a hypothetical arrangement in which the external wall is vibrationally isolated from the party wall (Fig 7a). In considering direct transmission through a wall operating above the critical frequency, it was seen that most of the vibrational energy of the wall is associated with the 'coincidence' effect and is contained in free bending waves. In the arrangement of Fig 7a these waves can travel up and down the external wall without hindrance. Consequently the vibration level is the same in the parts of the wall adjacent to the receiving room as in the parts adjacent to the source room. As compared with the same construction used as a party wall, there is an inflow of energy over only half the wall instead of the whole, but energy is dissipated over the whole area. Thus the vibration level over the whole wall is half what it would be in a similar party wall. The radiation into the receiving room for a given vibration level is the same as if the wall were acting as a party wall. Consequently, for approximately square rooms, transmission via the external wall is about half (in decibel measure 3 dB lower) that which the same wall would transmit if acting as a party wall. With similar party and external walls, total transmission is then nearly 2 dB more than direct transmission.

If, as is more usual, the party wall is bonded to the external wall (Fig 7b) there is extra restraint at the junction and energy is less freely transmitted between the two parts of the external wall. However, the energy levels in the two parts of the external wall will still be virtually the same if a packet of energy traverses a section of the wall a fairly large number of times before being dissipated: this situation is approached for practical values of damping. A complicating feature is that energy flow between the party wall and the external wall is of the same order of magnitude as that between the two parts of the external wall. If the walls are of the same thickness,

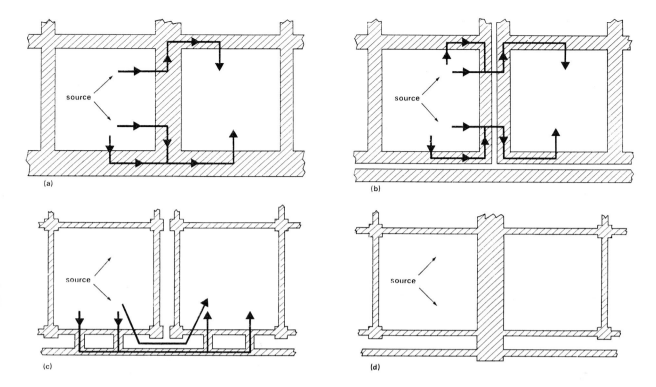

Fig 8 Principal flanking paths: (a) heavy single-leaf party and external walls; (b) heavy double-leaf party and external walls; (c) lightweight double-leaf party and external walls; (d) heavy single-leaf party wall, lightweight double-leaf external wall (flanking negligible)

the result is that the energy levels in the party wall and the two parts of the external wall become nearly equal. To produce this result, there must be a net flow of energy from the party wall to the external wall. As far as the party wall is concerned, this loss of energy has the same effect as additional damping in the wall, so that direct transmission is less than if the two walls were vibrationally isolated. But because of the additional energy received by the external wall, flanking transmission becomes nearly the same as direct transmission instead of being 3 dB less.

This result holds if the external wall has the same thickness as the party wall. Clearly flanking transmission depends on the properties of the wall in a very similar fashion to direct transmission. Thus if the external wall is much thinner than the party wall, flanking transmission is likely to exceed direct transmission.

Flanking transmission also occurs via internal partition walls; indeed, the fact that these are usually thin suggests that they are likely to contribute more than external walls. However, the position is more complicated because resistance to transmission across the junction with the party wall is usually high enough to produce a significant lowering in the vibration level of the partition wall. At present,

only actual measurements can determine the precise importance of internal partition walls.

Possible flanking transmission paths for a number of types of construction are sketched in Fig. 8. Two paths are indicated for a wholly lightweight construction (Fig 8c): one via the air in the cavity, and one via ties across the external wall. The danger of a serious amount of transmission by the latter path is reduced by the fact that, except over short distances (as across the ties), structure-borne transmission depends on resonant vibrations. In discussing direct transmission, the only resonant vibrations mentioned were free bending waves above the critical frequency. Actually natural vibrations of panels below the critical frequency can increase direct transmission and contribute flanking transmission, but these vibrations radiate poorly, and it should be possible to produce designs with which their effects are unimportant.

Impact insulation

In several respects, the requirements for impact insulation are the same as for insulation against airborne sound. The importance of mass may be deduced in the same way as for airborne sound (Fig 2): the energy radiated by a floor depends on the vibration amplitude, and Newton's law of motion indicates that the vibration amplitude is inversely proportional to the mass of the floor whether the exciting force is air pressure or the impact of a solid object. It is also clear that, as for airborne sound, a useful part can be played by a cavity construction which ensures that the vibration level in the radiating leaf is much lower than in the leaf directly excited. For best results, transmission through solid connections must be less than the unavoidable transmission through the air in the cavity. This is clearly not practicable when the upper leaf is supported on the lower. Instead, to raise impact insulation to a satisfactory level, use is commonly made of the lesser degree of isolation between the leaves provided by a floating floor. In this, the leaves are separated by a resilient layer, usually glass wool. A complicated theory is needed to give a satisfactory explanation of the behaviour of floating floors, but there is a fairly simple criterion for the minimum thickness of a resilient layer of given stiffness: it needs to be thick enough for the 'mass-spring-mass' frequency of the system floating layer/resilient layer/structural flow to be below 100 Hz.

Further reading
BS 2750:1956 Recommendations for field and laboratory measurements of airborne and impact sound insulation in buildings (with 1963 amendment) British Standards Institution.
National Building Studies Research Paper No. 33 Field measurements of sound insulation between dwellings: P H Parkin, H J Purkis, W E Scholes HMSO 1960

Digests:
102 Sound insulation of traditional dwellings—1
103 Sound insulation of traditional dwellings—2
128 Insulation against external noise—1.
129 Insulation against external noise—2
187 Sound insulation of lightweight dwellings

Sound insulation of traditional dwellings—1

This Digest and the next are mainly concerned with the design of new dwellings—houses and flats—of traditional construction but the principles discussed are equally applicable to the conversion of existing buildings.

Part 1 deals with: requirements of building regulations; explanation of the grading system; sound insulation of party walls in traditional construction; sound transmission between rooms in the same dwelling; improvement of existing dwellings.

Part 2 deals with: floor constructions, concrete and wood joists.

Requirements of building regulations

The building regulations for Scotland (1963) and for England and Wales (1972) incorporate mandatory requirements controlling sound transmission between dwellings. There are some differences in the standards of performance required in Scotland and in England and Wales, but in no case is the required performance below Grade I. Lower grades are, however, still of interest, for example, in the improvement of existing dwellings or in defining performance in other types of building for which there are no mandatory requirements for sound insulation.

The grading system

Party-wall grade. This grade is based on the performance of the one-brick party wall. It reduces the noise from neighbours to a level that is acceptable to the majority; a lower standard certainly could not be justified on present evidence. A higher standard is not yet practicable, mainly because at this level of insulation flanking transmission is usually about equal to direct transmission and there is little to be gained from improving only direct transmission.

Grade I. This is the highest insulation that is practicable at the present time *vertically* between flats. It is based on the performance of a concrete floor construction with a floating floor, which gives the best floor insulation obtainable by normal structural methods. Noise from the neighbours causes only minor disturbance; it is no more of a nuisance than other disadvantages which tenants may associate with living in flats.

Grade II. With this degree of insulation the neighbours' noise is considered by many of the tenants to be the worst thing about living in flats, but even so at least half the tenants are not seriously disturbed.

Worse than Grade II. If the insulation between flats is as low as 8 dB worse than Grade II, then noise from the neighbours is often found to be intolerable and is very likely to lead to serious complaints. With better insulation than '8 dB worse than Grade II' the likelihood of complaint decreases gradually, but when there are also other reasons for dissatisfaction serious complaints about noise may occur if the insulation is worse than Grade II.

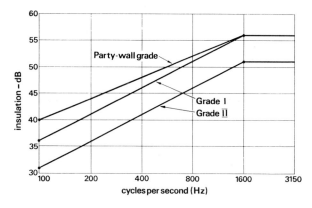

Fig 1 Grade curves for airborne sound insulation

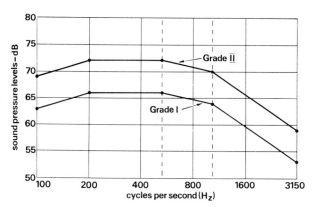

Fig 2 Grade curves for impact sound insulation

The grade curves

The levels of airborne and impact sound insulation that satisfy the party-wall grade and Grades I and II are given in Figs 1 and 2. To qualify for a particular airborne sound grading, the insulation should be *not less than* the value shown at each frequency in Fig 1. To meet an impact grading, the measured noise levels produced underneath a floor by a standard impact machine operated on the floor should not *exceed* the value shown at each frequency in Fig 2.

Wall constructions are required to meet the grade for airborne sound only; floor constructions, in order to be classified under a particular grade, must satisfy that grade for both airborne and impact sound insulation.

Measured insulation curves seldom follow the grade curves exactly and measurements satisfying grade requirements over most of the frequency range may very likely fall short at one or two frequencies; it is not intended that constructions should be condemned or graded down on account of minor faults that may have little significance, and in practical grading assessments a suitable tolerance is allowed. For strict grading purposes, such as conformity to Building Regulations, tolerances must be accurately defined, and it is also necessary to correct, or normalise, measurements to standardised conditions; without this correction a construction would, for example, vary in its insulation value depending on whether it was measured in an occupied or an empty dwelling, because of the differing amounts of absorption present.

All the gradings in this Digest and in Digest 103 are based on measurements made in accordance with BS 2750:1956, normalised to 0·5 sec reverberation time in the receiving room. The permitted tolerance for compliance with a particular grade is a total adverse deviation from the grade curve at all $\frac{1}{3}$-octave frequency bands between 100 and 3150 Hz of not more than 23 dB. The performance of party wall and party floor constructions is given in terms of their sound insulation grading relative to the grade curves.

Wall constructions

The sound insulation gradings of various forms of construction for party walls are shown in Table 1. The values given are based on the assumption that the remainder of the construction is traditional, with all structural elements firmly bonded together.

Another assumption is that the planning is traditional; that is, the size and layout of the rooms is normal or average for dwellings. Large rooms separated by small areas of party wall create favourable conditions for sound insulation; conversely, small rooms, with large areas of separating wall, are unfavourable. At present, no practical guidance on the effects of planning on sound insulation is considered feasible because of the complications of flanking transmission.

Party walls in houses
Solid walls

The basis of the party-wall grade is the one-brick wall, plastered on both sides, weighing not less than 415 kg/m² (including plaster*), but any solid walling material of this weight, plastered on both sides, is likely to meet the grade.

On concrete of open texture, eg no-fines concrete and many lightweight concretes, the plastering must not be omitted from any part of the surface; flues from fireplaces, etc, should be lined or rendered so as to seal off all air-paths through the porous material.

Attempts to seal porous concrete walls with linings of wallboard on battens, plaster strips or dabs, instead of plastering, have usually proved to be unsatisfactory for sound insulation. Non-porous walls lined with plasterboard on battens should be satisfactory if the overall weight is not less than 415 kg/m², but there is some doubt about the performance of plasterboard attached by plaster dabs; this requires further investigation. Dense concrete cast in permanent shuttering of wood-wool slabs, with the surface of the wood-wool plastered, has been found to provide only poor insulation at an important part of the frequency range.

* Throughout this Digest, 'plaster' implies dense, two-coat work at least 12 mm thick weighing not less than 24 kg/m²

Table 1: Gradings of party walls in traditional dwellings

SOLID WALLS		Weight incl any plaster	Grade
		kg/m²	
One-brick wall	pbs	415	P–w
In situ concrete or concrete panels with joints solidly grouted	plaster optional	415	P–w
175 mm concrete at 2320 kg/m³	without plaster	—	P–w
175 mm concrete at 2080 kg/m³	pbs	—	P–w
Lightweight concrete or other material	pbs	415	P–w
300 mm lightweight concrete 1200 kg/m³	pbs	—	P–w
225 mm no-fines concrete at 1600 kg/m³	pbs	—	P–w
200 mm ditto	pbs	365	I
Lightweight concrete, or other material	pbs	220	II
Half-brick wall	pbs	220	II

CAVITY WALLS, with wire ties of butterfly pattern and plastered on both sides	Cavity width	Weight incl any plaster	Grade
	mm	kg/m²	
Two leaves each consisting of: 100 mm brick, block or dense concrete	50	415	P–w
Lightweight aggregate concrete—sound absorbent surfaces to cavity (see text)	50	300	P–w
Ditto	75	250	P–w
Ditto	50	250	I
50 mm lightweight concrete at 1280 kg/m³	25	—	II
100 mm hollow concrete blocks	50	—	II

Notes: pbs = plastered on both sides

P–w = Party-wall grade

All weights, thicknesses and widths are minima.

Cavity walls

Cavity walling of two half-brick leaves, separated by a 50 mm cavity, was formerly recommended for party walls on the basis of the higher single-figure insulation. All available evidence now shows that this has no sensible advantage over the one-brick solid wall; indeed, unless the cavity width is maintained as a minimum and wire ties of butterfly pattern are used, the party-wall grade will not be attained.

Materials other than brick (eg concrete or stone) used for the leaves of a cavity wall will attain party-wall grade if the wall is plastered and its overall weight is not less than 415 kg/m².

If light-weight aggregate concrete is used, presenting sound-absorbent surfaces to the cavity, the weight of the wall can be reduced to 300 kg/m² with a 50 mm (minimum) cavity. If the width of the cavity is increased to 75 mm or more, the weight of the wall can be further reduced to 250 kg/m².

In all cases, the requirements already stated regarding plastering and wall ties must be met.

Wall linings

Wall linings on battens or studs can improve sound insulation but their performance is variable because of many factors. Although firm directions cannot be given, some guidance based on present knowledge might be helpful.

No improvement of the wall is of any value if flanking transmission already equals or exceeds direct transmission.

On a solid wall weighing at least 250 kg/m², wall linings can sometimes bring the insulation up to party-wall grade. The weight of the lining may need to be as much as 25 kg/m² but this could depend on the width of the air-space between the lining and the wall. The air-space should always be as wide as possible because of the benefit at low frequencies, where the needs of sound insulation are usually most difficult to meet; an air-space of 150 mm may well be necessary. Because isolation of the lining from the wall is desirable, a lining on independent studding is more effective than one on battens, but lining on battens may occasionally give the small improvement necessary to up-grade a wall that otherwise fails by only a small margin. A moderate amount of sound absorption in the cavity behind the lining (eg a suspended quilt of glass wool or mineral wool) is nearly always beneficial.

Linings over shallow air-spaces (eg on battens) are of little value for insulation of the low frequencies.

Party walls in flats

Although flanking conditions vary between two-storey houses and multi-storey flats, the effect of this on sound transmission in traditional construction is not usually significant. Any of the constructions mentioned in Table I should therefore attain the grading attributed to them if they are built as panels within the members of normal heavy steel or reinforced concrete framed multi-storey flats, provided that the connection between the frame member and the wall panel is made air-tight and rigid by sealing with mortar. The practice of inserting strips of non-rigid material, such as cork or felt, round the edges of the panel to separate it from the frame is not recommended; in certain special constructions such strips can be designed to improve the insulation but in general they are likely to reduce it. In particular, horizontal resilient membranes alone inserted in the walls at floor levels—usually with the object of reducing the flanking transmission up and down the walls—are of no value for the purpose and are best omitted.

A solid wall, weighing at least 365 kg/m², is likely to meet Grade I requirements.

Attention to detail

To achieve the grade that a construction is capable of giving, care must be taken to ensure that the wall construction does not fall below the standard in any area, such as in under-building or roof space, within the thickness of floors and ceilings or behind heating appliances. Eaves and dormer windows may also need special care.

Internal partitions

Although there are no mandatory requirements for sound insulation of partitions, it is desirable to provide a reasonable degree of privacy between rooms. Many dwellings have partitions either of plaster-board-faced panels with a honeycomb core or of plasterboard on each side of 50 mm or 75 mm stud framing; the sound reduction* is about 28–30 dB; this is a low standard of privacy between habitable rooms. Although no systematic investigation has been made of the amount of complaint it provokes, a reduction of about 35–40 dB between most rooms seems acceptable. This can be achieved by a solid partition of brick- or block-work, plastered both sides, weighing 75–150 kg/m², or by a sealed double-leaf dry partition of half the weight with a cavity not less than 50 mm wide. However, this cannot be a universal recommendation because the pattern of living within the home varies widely and it is difficult to determine whether the demand justifies imposing a standard between internal rooms.

Existing houses

It is rarely economical or even practicable to improve the insulation of a completed dwelling. This is because sound insulation is a function of the whole construction, not of the party wall alone and still less of any surface treatment of the party wall. Between well-built houses with one-brick solid party walls

* The performance of internal partitions is expressed according to common practice as a single-figure average over the usual frequency range.

and external (flanking) walls, with all brick joints and frogs filled with mortar, the sound insulation should be better than the party-wall grade. This is close to the maximum normally obtainable with ordinary methods of construction. Reducing the direct transmission alone (by insulating the party wall) is pointless because by-passing by flanking transmission leaves the net insulation almost unchanged. It is not often possible to reduce the flanking transmission.

Where the direct transmission is appreciably greater than the flanking transmission, and treatment of the party wall could be beneficial, it will be worth while to engage an experienced consultant. If there is leakage via air-paths, perhaps through underfloor spaces that are linked by gaps around the ends of joists built-in to the party wall, these paths should be sealed before any further treatment is given. Suitable treatment may sometimes take the form of wall linings, though a wide cavity between the lining and the wall behind may be needed to give adequate improvement at low frequencies, as already discussed.

Sometimes the insulation is already up to the party-wall grade, or not far below it, but the householder wants still better insulation. The solution with most chance of success is to increase substantially the weight of the party wall. A further skin of brickwork, half- or one-brick thick, tight up against the existing wall, and mortared to it, may help to reduce both flanking and direct transmission. But this solution calls for new foundations, and it is not possible to say how extensive the heavier wall needs to be to give worth-while results; it might be necessary to thicken up the whole party wall to obtain only limited benefits.

Further reading

National Building Studies, Research Paper No 33, *Field measurements of sound insulation between dwellings*, P H Parkin, H J Purkis and W E Scholes; HMSO: 1960.

BRE Digests

143 *Sound insulation: basic principles*

Sound insulation of traditional dwellings—2

Concrete floors

Various types of floor construction are listed in Table 1 with the insulation grading that each is likely to give in practice for airborne and impact sound insulation. In cases where the gradings under these two headings differ from each other, the lower grading is to be taken as the overall grade; thus, only floors that give Grade I both for airborne and impact sound insulation will give Grade I overall.

Floor finishes

The effect of the floor finish on insulation against impact noise is shown in Table 1; for example, in items (4)–(6) a 'hard' floor finish gives worse than Grade II but a 'soft' finish upgrades the same construction to Grade I. Ordinarily, no floor finish adds significantly to the airborne sound insulation, but the impact insulation of most concrete floors can be raised to Grade I simply by adding a finish or covering that is soft enough. A fitted carpet on an underlay of hair felt or sponge rubber will nearly always give Grade I impact insulation. It is obviously undesirable, however, to rely on a floor covering that is under the control of the occupant of the flat above the floor to provide the insulation required by the occupant of the flat below.

There is no standard grading of floor finishes for impact noise but resilience and thickness both have some effect. The terms hard, medium and soft are used in Table 1 and elsewhere only to indicate the properties of the floorings in relation to impact noise; the following examples should help to interpret this classification:

Hard: *concrete, terrazzo, clay and stone; pitchmastic and mastic asphalt; magnesium oxychloride; thermoplastic tiles; 2·5 mm linoleum; wood block, parquet and mosaic.*

Medium: *thin carpet without underlay; cork flooring not less than 6 mm thick; wood board and strip; chipboard.*

Soft: *thin carpet with underlay, thick carpet with or without underlay; rubber or plastics on spongerubber or felt backing of combined thickness not less than 4·5 mm; cork tile not less than 8 mm thick.*

This list is not comprehensive and in particular many new floorings have not been included, either because no examples have been measured by the Station in field tests or because their long-term behaviour is not known. On constructions which are shown in Table 1 as achieving Grade I only marginally, it would be inadvisable to use floorings of which the performance is in doubt.

Floating floors

Among the constructions shown in Table 1, there are two forms of floating floor, items (2) and (3), for use on any concrete structural base. Each is capable of giving Grade I overall sound insulation, with any floor finish.

The principle underlying the design of a floating floor is its isolation from any other part of the structure. To ensure this, it is recommended that the resilient layer on which it rests should be turned up at all edges which abut walls, partitions or other parts of the structure. Partitions should be built off the structural floor so that the floating screed or raft is self-contained within each room. There must be no continuity between the floating element and the structural base either by fixings or by bridging of the resilient layer.

Concrete screed In this form of construction the screed rests on a resilient layer which is laid over the structural floor. Details of the design of floating screeds—thickness, mix proportions, bay sizes, etc.—are given in Digest 104, *Floor screeds.* A thickness of at least 65 mm is required and if the area of screed is greater than about 15 m² there will be a considerable risk of curling or cracking unless special precautions are taken. In the past, it has been suggested that the screed should be divided into bays, but because of the difficulty of dealing with joints where adjacent bays have moved out of alignment it is now suggested that a floating screed should be lightly reinforced rather than divided into bays. Reinforcement will not, however, eliminate curling.

The concrete screed in its wet state must not be allowed to penetrate the resilient layer, of which the joints are the most vulnerable in this respect. A layer of waterproof paper or plastics sheeting should therefore be used to prevent this. Failure to do so will allow solid bridges to form which would reduce very considerably the effectiveness of the construction for sound insulation.

It is usual to lay wire netting (eg 20–50 mm mesh chicken wire) over the quilt and sheeting to protect them from mechanical damage during the operation of placing the concrete.

Synthetic anhydrite screed A satisfactory material for floating screeds is synthetic anhydrite; shrinkage and curling are practically eliminated even with very large areas. The material is more expensive than an equal volume of concrete but the permissible reduction in thickness (to about 30 mm) helps to offset the higher cost. It is normally placed on a layer of waxed paper which, if adequately lapped, avoids the risk of solid bridging mentioned above. The material has the further advantage of faster drying than concrete screeds.

Wood raft floating floors on concrete (Table1 item 3) A wood raft floating floor consists simply of wood flooring, fixed to battens to form a raft which rests on a resilient quilt laid over the structural floor slab. The battens must not be fixed to or in direct contact with the slab. For structural reasons, floorboards should preferably be tongued and

grooved and not less than 20 mm thick; 18 mm plywood or chipboard of flooring grade is equally suitable. The battens should be at least 40 mm deep (preferably 50 mm or more) and not less than 50 mm wide; they are usually spaced at about 400 mm centres.

The use of chipboard sheets tongued and grooved together and placed on the resilient layer without battens is sometimes suggested; little is known about the insulation properties of this method and there may be difficulty in joining the sheets together to form an isolated raft that is structurally stable. Chipboard, factory bonded to an expanded polystyrene layer and with provision for forming tongued and grooved joints, is also available.

Softwood is generally used for floating floors in dwellings; special precautions are necessary with hardwood because this is usually supplied kiln-dried to a low moisture content. If the moisture content at the time of laying is lower than it will be when the building is in use, the wood will swell as it takes up moisture and the floating raft, being unrestrained, may buckle. Therefore, when using hardwood for this purpose it should be at a higher moisture content than that at which it is usually supplied and species of low movement value should be specified.

Resilient layers The resilient layer is a very important part of floating floor construction for which only reliable material should be used.

Glass wool and mineral wool are in common use; quilts of the long-fibre type have been found the most satisfactory. As a basic guide to a suitable quilt, the long-fibre glass-wool type PF 225 with a nominal thickness of 13 mm has proved satisfactory. The prefix means that it is paper-faced (one side); it has an uncompressed density of 36 kg/m³.

The thickness and density quoted are minimum figures for this type of quilt; increased thickness or density tend to improve the sound insulation performance because the resilience increases. The type 600 resin-bonded glass-wool quilt or slab, 25 mm thick, of density about 100 kg/m³, which is sometimes used under heated floor screeds for thermal insulation, is eminently satisfactory from the sound insulation viewpoint. Long-fibre mineral-wool quilts are likely to prove just as satisfactory for sound insulation in floating floors as their glass-wool counterparts. For design purposes, it may be assumed that the 13 mm '225' quilt compresses in service to 3 mm under wood battens or to 6 mm under a concrete screed and that the 25 mm '600' quilt, normally used under concrete screeds, does not compress to less than 22 mm.

The above quilts are equally suitable for use with concrete-screed or wood-raft floating floors, on

Table 1 — Sound insulation grading of concrete floors between flats

Construction	FLOOR FINISH [1]	GRADE AIRBORNE	GRADE IMPACT
(1) Concrete floor The basic floor construction of items (1) to (5) is assumed to weigh not less than 220kg/m^2 (including any integral screed and plaster finish) and may be dense or lightweight reinforced concrete, hollow concrete beams or concrete beams with hollow clay block infilling	Hard	II	4dB worse than Grade II
	Medium	II	II
	Soft	II	I [2]
(2) Concrete floor with floating screed Any floor finish — Screed — Wire mesh — Paper — Resilient layer — Not less than 220kg/m^2	Any	I	I
(3) Concrete floor with floating wood raft Wood flooring — Battens — Resilient layer — Not less than 220kg/m^2	Any	I	I
(4) Concrete floor with suspended ceiling Floor finish — Not less than 220kg/m^2 — Battens wired to slab — Absorbent quilt — Heavy ceiling e.g. plaster on expanded metal lathing	Hard	I [2]	2dB worse than Grade II
	Medium	I [2]	II
	Soft	I [2]	I
(5) Concrete floor with lightweight screed Floor finish — Dense topping (see text) — 50mm lightweight screed — Not less than 220kg/m^2	Hard	I [2]	4dB worse than Grade II
	Soft	I [2]	I [2]
(6) Heavy concrete floor Floor finish — Not less than 365kg/m^2 including screed and plaster	Hard	I	4dB worse than Grade II
	Soft	I	I

Notes: (1) refer to paragraph ' Floor finishes on page 1   (2) the grades shown may be obtained only marginally.

concrete slabs or on wood joists. When laying paper-faced quilts on concrete slab floors, the paper face should be upwards (essential under floating screeds unless additional waterproof paper is used), whilst on wood-joist floors the paper face should be downwards. Although glass wools and mineral wools are the most tried materials for resilient layers under floating floors, other materials can be used if they have adequate resilience in a stable or permanent form. Resilience is best checked by measurement in the field against the Grade I performance standard. Permanence is more difficult to check and is likely to be verified only by experience.

Expanded polystyrene board has been used on a fairly wide scale as an alternative to glass-wool and mineral-wool quilts. Complete reassurance that its creep properties are insignificant is still lacking but after a number of years experience there is no evidence of any serious deterioration in performance. Therefore, although the Station has not itself investigated those properties of the material that affect its resilience, it is suggested as a resilient layer for floating floors provided the following specification is adhered to:

(a) density to be in the range 15–25 kg/m³

(b) nominal thickness to be not less than 13 mm

(c) the board should be pre-compressed to half its initial thickness with a rapid recovery to at least 90 per cent of the initial thickness.

Expanded polyurethane shows some promise as a resilient layer but has not been much used as yet and not fully investigated.

Some other materials, eg cane fibreboards and hair felts, have been found unsatisfactory because of compaction under continuous loading.

Pipes and conduits in floating floors It is often necessary for services such as electric conduits, gas and water-pipes, etc, to traverse a concrete floor. Whenever possible these pipes should be accommodated within the thickness of the floor slab and integral screed, if any, but sometimes they have to be laid on top of the slab and contained within the depth of a floating floor. This need not cause trouble with a floating screed provided that the pipes do not extend more than about 25 mm above the base, that they are securely fixed so as not to move whilst the floating floor is being laid and are haunched up with mortar on each side to give continuous support to the resilient quilt. When two pipes cross, one of them should be sunk into the base slab. The resilient quilt should be carried right over the pipes. If a wood-raft floating floor is being used and the pipes have to be laid above the slab, the pipes can of course be readily accommodated parallel to and between the raft battens, but in the other direction the battens will have to be notched over them; the battens must be thick enough to allow for this.

Lightweight concrete screeds
Lightweight concrete is sometimes used for floor screeds without a resilient quilt (see Table 1 (5)). In many instances this construction has given Grade I airborne sound insulation, but with no improvement in impact insulation. A soft floor finish (see p. 109) is therefore necessary in order to bring the overall rating up to Grade I.

Although not all the details of the construction have yet been investigated, the essential requirements appear to be as follows:

The density of the concrete screed should be not more than 1100 kg/m³.

The thickness of the screed should be at least 50 mm, exclusive of any dense topping (see next paragraph).

An impervious or airtight layer should be provided above the lightweight screed. The dense concrete topping often required on a lightweight screed to ensure a satisfactory base for the floor finish will serve this purpose.

It is not yet known whether some types of lightweight concrete screed are better than others for sound insulation purposes, assuming that the density requirements are met.

Recommendations for the design and laying of lightweight aggregate concrete screeds are given in Digest 104.

Suspended ceilings
Suspended ceilings are chiefly of benefit against airborne sound and are comparable with lightweight concrete screeds in that they can be used to raise the sound insulation of a normal concrete floor to Grade I, provided that a soft floor finish is also used to give the necessary improvement in impact insulation. One form of construction is shown in Table 1, item 4. Not all the suspended ceilings that have been measured have given a satisfactory improvement of sound insulation and the requirements for a successful system of construction are not all known precisely; the following features appear to be significant:

The ceiling should be not less than about 24 kg/m²; plaster on expanded metal lathing or on plasterboard can provide this weight.

The ceiling should be essentially airtight so as to eliminate direct sound penetration via air-paths, such as would occur with open-textured materials or with open joints.

The points of suspension from the floor structure should be as few and as flexible as possible.

The air space above the ceiling may range in depth from 25 to 300 mm or more—the deeper the better —and should preferably contain sound-absorbent material.

Table 2 Sound insulation grading of wood-joist floors between flats

Construction				Grade	
Floor	Ceiling	Pugging	Walls	Airborne	Impact
Plain joist	Plasterboard and single-coat plaster	None	Thin	8 dB worse than Grade II	8 dB worse than Grade II
			Thick	4 dB worse than Grade II	5 dB worse than Grade II
		15 kg/m²	Thin	4 dB worse than Grade II	6 dB worse than Grade II
			Thick	Possibly Grade II*	Possibly Grade II*
	Heavy lath and plaster	None	Thin	Probably 4 dB worse than Grade II*	Probably 6 dB worse than Grade II*
			Thick	Grade II	Grade II
		80 kg/m²	Thin	Grade II	Grade II
			Thick	Grade II or possibly Grade I*	Grade II
Floating	Plasterboard and single-coat plaster	None	Thin	4 dB worse than Grade II	3 dB worse than Grade II
			Thick	Possibly Grade II*	Possibly Grade II*
		15 kg/m²	Thin	2 dB worse than Grade II	2 dB worse than Grade II
			Thick	Grade II or possibly Grade I*	Grade II or possibly Grade I*
	Heavy lath and plaster	None	Thin	2 dB worse than Grade II	Grade II
			Thick	Grade II or I†	Grade I
		15 kg/m² (as Fig 2)	Thin	Possibly Grade II*	Grade II*
			Thick	Grade II or I†	Grade I
		80 kg/m² (as Fig 1)	Thin	Probably Grade I	Probably Grade I
			Thick	Grade I	Grade I

*Assumed from other measurements   † May give Grade I with very thick walls

Although useful for sound absorption in the acoustic treatment of rooms, ceilings of soft insulating fibreboard are not recommended for sound insulation between rooms because of their light weight and porous nature.

Wood joist floors
The influence of wall thickness Indirect or flanking transmission always has some effect on overall sound transmission and the performance of wood-joist floors is influenced by the amount of flanking sound transmitted via the walls. If the sound energy passing up or down the walls is greater than that passing through the floor, then the walls and not the floor will control the sound insulation between the rooms. Further treatment of the floor will be of little value unless the sound transmitted via the walls can be reduced correspondingly. This means reducing the vibration of the walls by one of the following means:

(a) making the walls thicker
(b) making the floors heavy enough and stiff enough laterally to restrain vibration of the walls.

Concrete floors are heavy and stiff enough to restrain vibration of the walls but most wood-joist floors are not and the maximum net sound insulation is controlled by the thickness of the walls, even though the floor may have potentially higher sound insulation. Therefore insulation values for wood-joist floors can only be given in conjunction with the wall system, as in Table 2.

To be classed as a thick wall system, three or more of the walls below the floor must be at least one brick thick or of similar weight; the walls above the floor need not be so thick. The overall thickness of a cavity wall does not count in this respect, but only the thickness of that leaf which constitutes the vertical flanking path and radiates sound into the room below. Metal anchorages connecting floor joists to external walls are sometimes employed in order to give lateral support to the walls, but the additional stiffness imparted to the walls by this means is insufficient to give any improvement of sound insulation.

Wood-joist floors with thin walls Unless in conjunction with a thick wall system, most wood-joist floors, even if designed for sound insulation, will fall short of Grade II by at least 2 dB, because of transmission via the walls. To reach a higher standard than this, the floor must be heavy enough and stiff

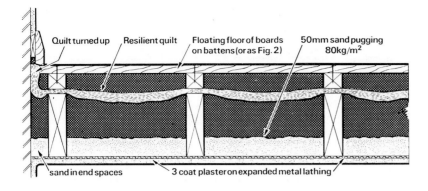

Quilt turned up   Resilient quilt   Floating floor of boards   50mm sand pugging
                                      on battens (or as Fig. 2)   80kg/m²

sand in end spaces   3 coat plaster on expanded metal lathing

Fig 1 Insulated wood-joist floor with heavy pugging

enough to restrain the walls laterally. The only satisfactory method known at present is to provide a ceiling of expanded metal lath and three-coat plaster, loaded directly with a pugging* of 50 mm of dry sand or other loose material weighing not less than 80 kg/m² together with a properly constructed floating floor. The construction is shown in Fig 1.

The pugging will be most effective if it is supported by the ceiling and not on separate supports (pugging boards). The metal lathing must be securely fixed so as to support the combined weight of the pugging and plaster. The sand must not be omitted from the narrow spaces between the end joists and the walls; it should be as dry as possible when it is placed in the floor and should not contain deliquescent salts.

There is not much evidence at present about the use of alternatives to metal lath and plaster ceilings for supporting the sand pugging. However, it seems likely that strong board ceilings (eg thick plasterboard or asbestos board) will give similar results provided that they are firmly bonded to the walls, the joints between boards are solid and airtight and the combined weight of ceiling and pugging is the same, ie not less than 120 kg/m².

Wood-joist floors with thick walls When the walls are thick, it is possible to use a lighter form of floor construction than the one just described without falling below Grade II insulation—though the heavier construction is still to be preferred because with thick walls it can be relied on to give Grade I insulation. In the lighter form of construction (see Fig 2) the floating floor remains an essential feature but the pugging is reduced in weight to 15 kg/m²; the ceiling may be of plasterboard with a single-coat plaster finish, but a heavier ceiling is preferred. The pugging should be supported direct on the ceiling

and not independently. Wire netting separately stapled to the joists is sometimes inserted above the ceiling to retain the pugging in position in order to ensure that the floor attains a full half-hour fire resistance; the netting must not be allowed to prevent the pugging from bearing fully on the ceiling. Alternatively, the same fire resistance can be achieved by increasing the plaster finish on the plasterboard ceiling to 13 mm thickness. The pugging material normally recommended is high-density slag wool (about 200 kg/m³), laid to a thickness of about 75 mm. Loose or pelleted mineral-wool puggings of 110–150 kg/m³ density need to be 100–130 mm thick. Other pugging materials can be employed provided they are of loose wool or granular type and the thickness used is sufficient to give the stipulated weight of 15 kg/m²; very light materials such as glass wool or exfoliated vermiculite are not suitable. Old wood-lath and plaster ceilings, which are usually quite thick, can provide very good sound insulation; they are often associated with thick walls, in which case the floors are likely to be Grade II without further treatment. The addition of a floating floor may sometimes result in Grade I insulation; therefore in conversion work, wood-lath and plaster ceilings should be retained whenever possible.

Floating floors for wood-joist floors The floating floor consists of wood-board or strip, plywood, or chipboard flooring, nailed to battens to form a raft, which rests on resilient quilt draped over the joists. The raft must not be nailed down to the joists at any point and it must be isolated from the surrounding walls, either by turning up the quilt at the edges (which is the better practice) or by leaving a gap round the edges—this can be covered by a skirting.

* 'Deafening' Scotland

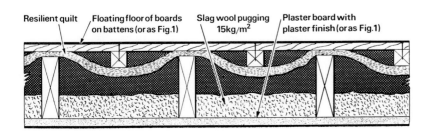

Resilient quilt   Floating floor of boards   Slag wool pugging   Plaster board with
                  on battens (or as Fig.1)   15kg/m²   plaster finish (or as Fig.1)

Fig 2 Insulated wood-joist floor with light pugging

Skirtings should be fixed only to the walls and not to the floors. The flooring should be not less than 20 mm thick, preferably tongued and grooved. The battens should be 50 mm wide and preferably 40 or 50 mm deep to form a stable raft. They should be parallel with the joists because battens that cross the joists provide too small a bearing area and overload the resilient quilt; the whole area of the top edges of the joists should share the loads transmitted from the floating raft. If it is essential for the battens to lie across the joists, then specially designed resilient pads must be used instead of the normal quilt or in addition to it.

There are two common methods of constructing the raft; one is to place the battens on the quilt along the top of each joist and to nail the boards to the battens in the normal way, as shown in Fig 1. The other method is to prefabricate the raft in separate panels the length of the room and up to 1 m wide, with the battens across the panels positioned so that they will lie between the joists when the panels are placed in position; the battens of adjoining panels should be staggered on alternate sides of the centre line between the joists and should project beyond the flooring sufficiently as to enable the panels to be screwed together to form a complete raft of the type shown in Fig 2.

Because the flooring is not nailed down to the joists, particular care must be taken to level up joists that are to carry a floating floor; in particular, end joists must not be lower than the others as this produces a tendency for the floating raft to tip and for furniture next to the walls to rock. This could encourage householders to spoil the insulation by nailing down the floating floor.

Partitions on floating floors It has become a practice in some methods of construction, particularly for two-storey dwellings, to build the internal partition walls up to first-floor level and then to construct the wood-joist floor continuously over the whole area of the dwelling, building the upper partitions on top of the flooring. This practice has sometimes been adopted even on a floating floor but is not recommended. Partitions should be supported either on the partitions below or on the floor joists, the floating floor being constructed as a separate independent raft within the confines of each room.

Suspended ceilings An independent ceiling is not very effective as a means of improving the sound insulation of a wood-joist floor. When used in addition to a floating floor and pugging, a suspended ceiling gives little further improvement; to be of much value when used alone, say for improving the insulation of an existing floor, it needs to be so heavy that it might not be practicable to construct it.

Although the airborne insulation at high frequencies could be improved by a comparatively lightweight suspended ceiling, it is at the lower frequencies that wood-joist floors are mainly deficient and more weight is generally required to remedy this deficiency. Suspended ceilings are not of much benefit for impact insulation. If a floating floor is being built it is usually a simple matter to pug the ceiling, and nothing worth while is then gained by adding a suspended ceiling.

Sound insulation of lightweight dwellings

This digest replaces Digest 96 which is now withdrawn.

This digest sets out the Station's present knowledge of sound insulation in dwellings gained from field measurements of industrialised systems of lightweight construction. The trend towards lightweight construction has made it necessary to develop ways of achieving adequate sound insulation other than simple recourse to weight, as in traditional houses. The alternative principles employed, notably structural isolation, are outlined. Practical forms of construction embodying these principles are discussed, against the background of established grades or standards of performance and the requirements of the building regulations. It is suggested that Digest 143 'Sound insulation: basic principles' should be read in conjunction with this digest.

Digests 102 and 103 are concerned with sound insulation between dwellings of basically traditional construction, in which the weight of the structural elements (separating and flanking elements) is by far the most important factor; performance standards or grades relating to walls and floors are suggested, and various traditional constructions are reviewed and graded.

Strictly speaking, no firm recommendation is made in Digests 102 and 103 respecting the choice of a standard, although for walls between houses the fact that only one criterion (the house party-wall grade) is nominated seems to imply that a lower standard would not be considered satisfactory. The British Standard Code of Practice* recommends the house party-wall grade in houses, but selects Grade 1 for walls and floors in flats.

Building regulations
The Building Regulations 1972 require only that a party wall or floor shall, in conjunction with its associated structure, provide 'adequate' resistance to the transmission of airborne sound and that a party floor between dwellings shall also (in conjunction with its associated structure) provide adequate resistance to the transmission of impact sound.

* CP 3: Chaper III: 1972, 'Sound insulation and noise reduction'.

For walls, the requirement is deemed-to-satisfy if measurements made and normalised in accordance with Sections 2A and 3A of BS 2750 : 1956* show the transmission of airborne sound to be reduced by the amounts shown in Table 1. A tolerance not exceeding 23 dB total adverse deviation from the values shown, measured at all sixteen frequencies, is permitted.

Floors must meet the standards of Table 2, again with a permissible total adverse deviation of 23 dB each for airborne and impact sound.

The Building Standards (Scotland) (Consolidation) Regulations 1971, operative only in Scotland, set out in Part VIII precise minimum sound insulation performance standards for new dwellings, equivalent to the BRS house party-wall grade between houses and Grade I between flats.

The two sets of regulations differ in that, for England and Wales all party walls are expected to satisfy the house party-wall grade, while for Scotland this standard is only required between houses, the slightly lower standard of Grade I being acceptable in flats for party walls as well as for party floors.

* BS 2750 : 1956, 'Recommendations for field and laboratory measurement of airborne and impact sound transmission in buildings'.

Prepared at Building Research Station, Garston, Watford WD2 7JR
Technical enquiries arising from this Digest should be directed to Building Research Advisory Service at the above address.

Sound insulation grades

The three grading curves, party-wall grade, Grade I (Airborne sound) and Grade I (Impact sound), are shown in Tables 1 and 2. Grade II curves have no relevance to the present regulations, but they can be derived from the Grade I curves, being 5 dB worse at all frequencies for airborne sound and 6 dB worse at all frequencies for impact sound. In measured field results some spread is inevitable for any given construction, and a tolerance not exceeding 23 dB total adverse deviation from the grade curve measured at all sixteen ⅓-octave frequency bands between 100 and 3150 Hz is, therefore, permitted in determining compliance with a particular grade.

The term 'party-wall grade' is used in this digest to denote the highest of the BRS grades, which was based on the average performance of normal one-brick solid party walls in traditional houses.

The BRS grades are related to field measurements, not to laboratory test results, which may be very different from field results of the same nominal constructions. The allocation of a grading to a type of construction should, whenever possible, be made on the average results of at least four examples, gradings of single examples being regarded as tentative.

Measurements of sound insulation for grading purposes can only be made between two rooms or enclosed spaces of a reverberant nature. Normally, the rooms will be on opposite sides of the party wall or floor, and it is convenient to quote the results as being the sound insulation of the wall or floor, although what is measured is not the insulation of the separating element alone, but the net reduction between the pair of rooms or dwellings via whatever transmission paths may happen to be present. Sometimes flanking transmission exceeds direct transmission and controls the measured reduction. This is the principal reason for large discrepancies between laboratory and field results for particular separating wall constructions. Measurements are affected by local conditions, such as furnishings, and for grading purposes all results are 'corrected' to conform to standardised furnished conditions, i.e. 0·5 sec. reverberation time in the receiving room.

Lightweight house construction

The most influential factor for sound insulation in conventional residential buildings is the weight of the structure, not only of the party wall or floor but of the flanking walls or floors as well, but it is well known that double-leaf walls can give insulation in excess of their weight contribution. Cavity party walls of plastered lightweight concrete blocks can show a saving of about 40 per cent of the weight of brick walling but this is not enough to satisfy modern demands. Moreover, the current move towards maximum factory prefabrication with rapid

Table 1 Party-wall grade

(Hz)	Minimum airborne sound reduction through wall (dB)
100	40
125	41
160	43
200	44
250	45
315	47
400	48
500	49
630	51
800	52
1000	53
1250	55
1600	56
2000	56
2500	56
3150	56

(Maximum total adverse deviation 23 dB)

Table 2 Grade 1

(Hz)	AIRBORNE SOUND minimum reduction through floor (dB)	IMPACT SOUND maximum octave-band sound pressure level under floor (dB)
100	36	63
125	38	64
160	39	65
200	41	66
250	43	66
315	44	66
400	46	66
500	48	66
630	49	65
800	51	64
1000	53	63
1250	54	61
1600	56	59
2000	56	57
2500	56	55
3150	56	53

(Maximum total adverse deviation 23 dB)

site erection calls for dry-construction techniques as well as for the lowest possible weight. Dry construction adds to the acoustic problem, because it is vital for sound insulation that all air-paths should be eliminated. Wet plastering, which could seal all joints and cracks, is currently out of favour and new developments in industrialised house construction have had to bear in mind the special need for efficient sealing. Sound insulation factors other than weight have had to be explored to the full.

The essential requirements for good acoustic performance of dry lightweight construction are:

(a) an independent structure for each house;
(b) a double-leaf party wall construction;
(c) a certain minimum weight in each leaf;
(d) wide cavity separation between the two leaves.

The houses should be regarded basically as detached houses alongside each other. There is no objection to disguising them as a terraced row by

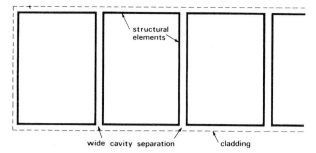

Fig. 1 Schematic plan of 'detached' terraced housing.

continuous claddings of suitable type with minimum acoustic continuity, but the houses should be self-contained structurally, as shown schematically in Fig. 1. They may be skeletal framed structures lined with prefabricated panels inside and out, or they may be constructed of loadbearing panels. The commonest system at present is the timber-framed structure, factory-made in storey-height sections and site assembled, linings being added either in the factory or on site. Typical details of a party wall of this construction are given in Fig. 2; the graph in Fig. 3 shows average performance and variability based on recent measurements of 28 party walls of this type.

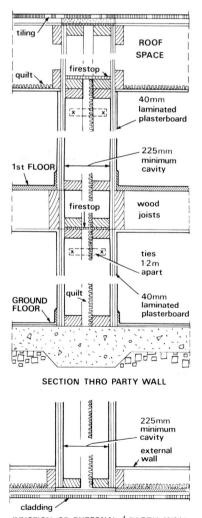

Fig. 2 Typical construction of party wall in isolated house system.

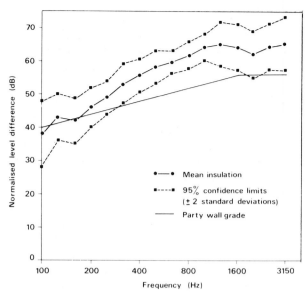

Fig. 3 Typical acoustic performance of isolated house systems.

A precise minimum weight for the leaves of the party wall cannot be given, but about 25 kg/m² is suggested as the minimum for each leaf, i.e. about 50 kg/m² for the double wall. This weight is exclusive of any framework, whether in timber, steel or concrete; it is the continuous weight of the linings or panels. The most common lining the Station has encountered, especially with timber-framed systems, is 40 mm thick laminated plasterboard. This may consist of two layers of 20 mm plasterboard, three layers of 13 mm, or four layers of 10 mm. The most common form is three layers of 13 mm plasterboard, two layers being glued together in the factory and pre-fixed to the timber-framed sections, the third layer being fixed *in situ* with the sheets laid in the opposite direction to the pre-fixed sheets. Recent tests on walls using one layer of 13 mm and one layer of 19 mm plasterboard seem to indicate that this also is satisfactory. This method should help to minimise the risk of joints between panels opening up in course of time and forming air paths through the party wall.

Other panel materials of equal weight are likely to behave acoustically in much the same way, with minor variations in performance at different frequencies depending on the resonant properties; for practical purposes, the differences between the materials in normal use are not thought to be significant. Considerations such as thermal and moisture movements should be kept in mind, however, because of the serious acoustic effects that could arise from the joints opening up; obviously, the material must be robust enough to withstand mechanical damage.

The width of the separating cavity cannot be fixed very precisely, but 225 mm seems to be about right as a minimum dimension. Making the cavity wider is beneficial. Cavity width is measured between the backs of the lining panels; supporting framework up to, say, 15 per cent of the panel area may project into the cavity, providing it does not touch the opposite frame. It is quite possible that the weight of the leaves and the width of the cavity are linked, since both are relevant to the critical problems of sound insulation performance at low frequencies, i.e. the bottom octave (100–200 Hz) of the range we are concerned with. If so, it may be possible to reduce the cavity a little provided the weight of the leaves is increased to compensate; however, it may be necessary to increase the weight by at least 50 per cent each time the cavity is reduced by 25 mm.

Cavities or air-spaces are more effective for sound insulation if they contain a reasonable amount of sound absorbent material, especially if the cavity is a wide one designed to make a significant contribution to the total performance. As most panels will present a fairly hard reflecting surface to the cavity, the insertion of a sound absorbent quilt or blanket is recommended to provide the necessary absorption. Almost any standard glass-wool or mineral-wool quilt 13–25 mm thick will serve, except that any paper covering should be limited to one side of the quilt only. The quilt is usually suspended in the cavity by stapling it to the frame on one side. Its position in the cavity is not important. In platform-type timber-framed construction the provision of two quilts, one on each side of the cavity and firmly attached to the timber frame, contributes to fire-resistance and may dispense with the need for ties bridging the cavity.

Structural isolation

The important matter of the structural independence of adjoining houses needs to be considered in more detail. For effective sound insulation, complete isolation above ground level should be the aim, but this often conflicts with other functional needs, such as stability and fire protection, which may call for bridging of the cavity. Weather protection at the ends of the party wall may also bridge the cavity. Fortunately some forms of bridging, applied with restraint, give the required standard of sound insulation. For example, normal roof tiling can safely be carried across the top of the party wall provided the cavity extends up to the underside of the tiling. Fire protection regulations require the top of the wall to be sealed against the passage of fire from one side to the other; to comply with this, roof tiling needs to be bedded on non-combustible material (e.g. mortar) which is usually supported over the cavity by asbestos-board. This construction is

acceptable acoustically. Similarly, an asbestos-board bridge inserted across the cavity at first-floor level as a fire-stop can be accepted; best results are likely to be obtained if it is fixed to the structural frame on one side only or not fixed at all but simply left resting on the horizontal members of the timber frames.

In two-storey housing, the fire requirement across the party wall is one hour's fire resistance. Within a two-storey house of single occupancy, the required fire-protection standard for the first-floor construction is half an hour. This difference raises a special problem in timber-framed houses of 'platform' construction, because collapse of the floor at the end of half an hour could cause premature collapse of the supporting leaf of the party wall, exposing to the fire the timber frame of the other leaf. Structural stability is thus a contributory factor in fire resistance. A method of safeguarding stability employed in some systems is to tie the timber frames of the party wall together by a row of metal straps 1·2 m. part just below each ceiling level. The straps are usually 40 mm by 3 mm flat steel bar nailed or screwed to the sides of the timber studs. These ties seem to have no serious effect on sound insulation, though this may not continue to be the case if the number of ties is increased beyond the dozen or so used in the houses tested.

Floors in flats

The type of house construction discussed so far can, of course, be used for two- or three-storey flatted dwellings, but in that case the floors also have to reach a certain standard of sound insulation, and requirements for impact sound reduction as well as for airborne sound reduction must be satisfied. If the building is timber-framed or a very light steel-framed structure, concrete floors may not be feasible and a joisted floor may be the only appropriate type.

It is known that a Grade I standard of sound insulation is unlikely to be achieved with wood joist floors in traditional structures (except with very heavy supporting walls) because of flanking transmission. It is not clear whether flanking transmission is also a problem in timber-framed structures. Some tests done by BRE have shown that it is possible to achieve Grade I performance with timber joist floors in timber-framed structures but there is not yet sufficient evidence on which to base detailed guidance.

Insulation against external noise—1

The performance of the external envelope, walls and roof, in reducing the level of noise transmitted from outside to inside a building, depends mainly on the mass of the envelope, its continuity and the extra insulation afforded by double-leaf construction. In addition, there are certain planning measures that will minimise the exposure to outside noise.

The proportion of wall area that is occupied by window glazing has an appreciable effect on the overall performance of the window/wall area unless the windows are designed to a standard that matches the wall construction. Open windows greatly reduce the overall performance.

The next digest, No. 129, discusses the ventilation needs of buildings in noisy locations; it shows how to apply the principles already dealt with and how to obtain a reasonable estimate of the noise level inside a building.

The measurement and prediction of external noise levels, especially road traffic noise, is dealt with in Digests 185 and 186.

The broad problem

The general method of protecting against ambient noise is by means of enclosure. Therefore, outdoor noise protection focuses attention on the external envelope or shell of the building and its effective or net sound reduction. In assessing or designing for sound insulation, three principal factors need to be considered:

(a) the mass (weight per unit area) of the envelope, which basically determines the sound insulation;

(b) the continuity, uniformity or completeness of the enclosure, since gaps or areas of reduced weight weaken the potential insulation provided by the main mass;

(c) the extra insulation, afforded by double-leaf construction, which is especially useful for any lightweight areas.

Enlarging on these three points, sound insulation is closely related to mass, but on a logarithmic ratio scale, as shown in Fig 1. Only the average performance for the range of frequencies 100–3150 Hz is shown, though sound insulation varies with frequency or pitch of the sound, increasing very roughly by 5 dB for each octave rise.

After establishing the basic performance due to mass, the modifying effect of gaps or other areas of inferior sound insulation can be readily calculated from

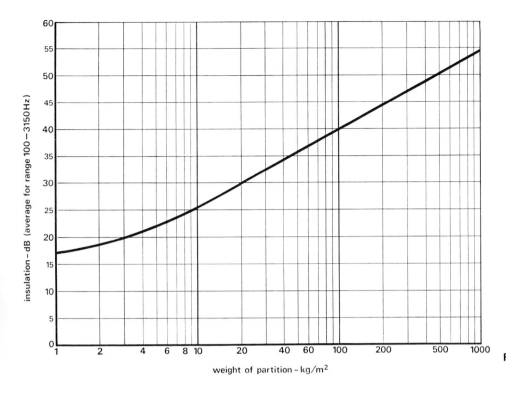

Fig 1 Relationship of sound insulation to mass

Fig 2, which is based on the relationship of area to sound reduction factor. The sound reduction factor for holes can be assumed for general design purposes to be 0 dB, though the loss of insulation may be even greater than this assumption predicts. The ratio of the areas of gap to solid clearly plays a major role in sound insulation. As a simple example, a gap of the area ratio 1 : 1000 limits the wall in which it occurs to a maximum insulation of 30 dB, though the net insulation may be still less if the rest of the wall is poor. In the same way, a window of 25 dB insulation in a wall of 50 dB (difference 25 dB) with an area ratio of 1 : 4 will cause a loss of 18 dB from the higher value, resulting in a net performance of 32 dB.

Double-leaf construction may add anything up to 20 dB or more to the insulation expected from the total mass of the two leaves. The extent of the improvement depends on the width of the separating cavity and on the degree of isolation between the two leaves, with sound absorption within the cavity as a subsidiary contributing factor. The wider the cavity and the more complete the structural isolation, the better the sound insulation. However, ordinary brick or block external cavity walls, with 50–75 mm cavities bridged at frequent intervals by wall-ties or other connections give no more than is accounted for by the mass-law. Reasonably effective separation of leaves is necessary for any worth-while improvement, and it is only when complete structural isolation is resorted to, coupled with wide, absorbent-lined cavities, that a very high performance can be expected. Further comment on this special field is made in the next digest (No 129).

Design measures

In addition to these structural factors, planning has a part to play in noise protection. For example, the area of the structure actually separating the protected room or space from the noise-infected environment will have an effect, since a large common transmitting area will obviously let in more noise energy than a small common area. Deep planning, so that rooms of a given size have less external walling, helps to protect by reducing the area of exposure.

The total sound absorption within the room will have a modifying effect, because it determines the level of reverberant sound to which noise of a given energy content within the confines of the room will build up by reflection from the room surfaces.

When the noise source is restricted to one side of a building, rooms most sensitive to noise should, if possible, be sited on the remote or quieter side. The building then serves as its own noise barrier. If the building is reasonably extensive in relation to the noise source, the level of noise on the screened side may well be 10–15 dB(A) below the level on the exposed side. Internal courts give similar protection from noise on all sides, but not from noise above, such as low-flying aircraft. With aircraft in flight as the source, roofs become vulnerable as well as walls, and there is little that localised planning, as distinct from siting, can do to aid noise protection apart from putting the noise-sensitive rooms on the lower floors so that they are shielded by more noise-tolerant rooms above.

The next stage is to consider the resistance of the different elements of the shell to noise penetration.

Elements of the shell

External walls

A typical traditional external wall may consist of brickwork or blockwork, sometimes comprising the whole wall area but much more often containing a proportion of window glazing. The proportion varies, occasionally approaching 100 per cent (ie fully glazed, perhaps curtain walling) but more commonly ranging from 25–75 per cent. The sound reduction of the brick or block portions of external wall lies in the range 45–50 dB, whereas closed single windows normally have reductions of only 20–25 dB. In these circumstances the windows dictate the insulation, but the net reduction will vary by 5 dB over the stated 1:3 range of 25–75 per cent window area. Thus to a limited extent the area of window affects the average insulation. If double windows are used, their insulation is much closer to that of the brick or block wall itself, and the net reduction will not vary so much with the proportion of glazed area, but the variation over the 1:3 range will still be 3–4 dB. Obviously, when the window and wall insulations match, the area relationship does not affect the sound insulation.

When windows are open, only the area of opening is significant; if this is 10 per cent of the total area, the basic noise reduction will be about 10 dB whatever type of window or wall construction occupies the remaining 90 per cent. Thus, although with heavy walls the type of window is normally the most important factor in the external wall construction so far as noise protection is concerned, the area of window is also significant, and the openable area even more so.

Windows

In respect of sound insulation performance, the range of choice of windows can be set out in a series of insulation steps, as in Table 1.

When designing windows with sound insulation in mind, the factors set out below should be taken into account (but see also Digest 140):

(a)   Sound insulation increases with thickness of the glass (*see* Fig 1) in both single and double windows, subject to other limiting factors, particularly air-gaps.

(b)   Air paths severely restrict sound insulation performance. For example, if windows have openable lights without weather-stripping there is no

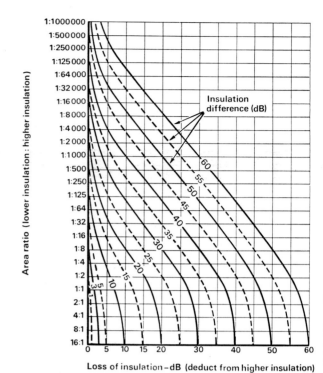

Fig 2 The effect of non-uniform insulation

acoustical advantage in using glass thicker than 4 mm; with weather stripping, 6 mm glass may be worth while, but glass above 6 mm thick is only justified in fixed windows.

(c)   If one leaf of a double window is fixed and sealed, gaps in the other leaf are not critical; but if both leaves have openable lights, weather-stripping should, if possible, be provided on both sides.

(d)   In double windows, sound insulation increases with the width of the cavity between the glazings, particularly at low frequencies. The width should not be less than 150 mm if possible, though cavities of 100 mm may sometimes be worth while; very small cavities, as in double glazing units for thermal insulation, are of no real value for sound insulation.

(e)   The sound insulation performance of multi-leaf windows seems to depend mainly on the total airspace available rather than the number of leaves; experiments have indicated that converting double windows to triple windows by inserting an extra pane to subdivide the air-space does not increase the average insulation, though it may change its character.

(f)   Other features that contribute to sound insulation to some extent are sound-absorbent linings, eg acoustic tiles, on the reveals between double windows, and methods of edge-mounting the glass to promote resonance damping, eg mounting in neoprene gaskets.

(g)   The best results may be obtained when all these factors are applied, but the limitations of each case should be recognised and there is no advantage in designing windows to a higher performance standard than the associated conditions permit.

Table 1   Sound insulation of windows

Description	Sound Reduction (av 100–3150 Hz)
Any type of window when open	about 10 dB
Ordinary single openable window closed but not weather-stripped, any glass	up to 20 dB
Single fixed or openable weather-stripped window, with 6 mm glass	up to 25 dB
Fixed single window with 12 mm glass	up to 30 dB
Fixed single window with 24 mm glass	up to 35 dB
Double window, openable but weather-stripped, 150–200 mm airspace, any glass	up to 40 dB
Double window in separate frames, one frame fixed, 300–400 mm airspace, 6–10 mm glass, sound-absorbent reveals	up to 45 dB

123

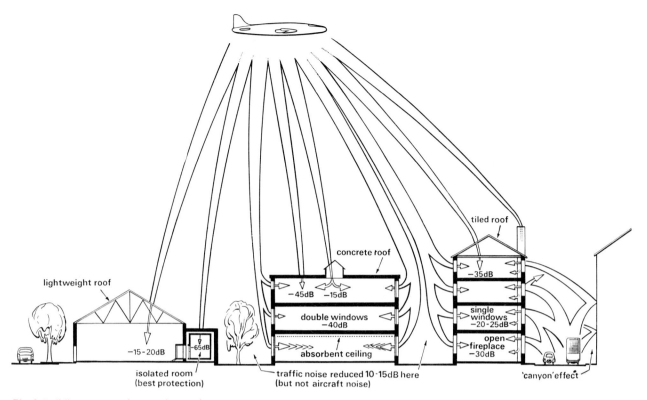

Fig 3 Buildings exposed to outdoor noise

Roofs

Roofs form part of the enclosure of every building, though not of every room. Moreover, their exposure to the environmental noise may differ from that of walls. In some circumstances roofs have a significant part to play in noise protection, *see* Fig 3, and their design must take this into account. In respect of sound insulation, roofs fall mainly into four types:

(i) *The pitched roof of single sheeting* is usually lightweight, seldom air-tight, and rarely gives more than 15–20 dB sound reduction.

(ii) *The flat concrete roof* commonly has a basic weight of 200 kg/m² or more, giving a sound reduction of about 45 dB which is enough to deal with most ordinary noise reduction problems.

(iii) *The flat joisted roof* with lightweight ceiling and/or decking, varies according to whether there is a ceiling below the joists in addition to the decking above, and the precise weight of each membrane; without a ceiling the insulation may be no more than 20–25 dB; with a ceiling, the likely range is 30–35 dB.

(iv) *The pitched slated or tiled roof* with an under-ceiling usually gives about 35 dB reduction but sometimes as much as 40 dB.

Roof glazing

Any of these roof types may contain glazing, though roof-lighting is not often met with in type (iv) dormer windows to attic rooms being a different case. With type (i), day-lighting usually takes the form of patent roof-glazing following the general slope of the roof. The glazed areas, having about the same mass as the sheeting, do not affect the sound insulation significantly, except that the gaps so often associated with this type of glazing tend to make the net insulation nearer to 15 dB than 20dB. If transparent corrugated plastics sheeting is used for roof-lighting the low weight of this product will also tend to keep the insulation down to about 15 dB. Glazing in type (ii) or type (iii) roofs normally takes the form of upstand lantern-lights or dome-lights; as these are generally single-glazed and often contain deliberate permanent ventilation, they reduce the potential roof insulation in nearly all cases—especially the basically superior insulation of the type (ii) roof. If 10 per cent of the ceiling area is composed of single-glazed roof-lights, the insulation of both type (ii) and type (iii) roofs is limited to about 25 dB.

Insulation against external noise—2

Part 1, Digest 128, discussed the performance of the external envelope in reducing noise transmission and how some planning measures can minimise the exposure to outdoor noise. The effect of open windows on overall performance was discussed.

Part 2, this digest, describes some of the special ventilation measures that may be needed in noisy locations. It summarises the technical requirements of the London (Heathrow) Airports Noise Insulation Grants Scheme and discusses the application of the principles already outlined. It shows how a reasonable approximation to the transmitted noise level inside a building can be obtained by subtracting the average net sound insulation of the envelope from the external noise level.

Ventilation

The previous digest, No. 128, showed that a major part of the insulation-loss caused by windows and roof-lights is due to the gaps or air-paths present, which permit direct leakage of sound. Even small gaps, such as the cracks around openable window-lights, can reduce sound insulation significantly, whilst open windows, which may be required for normal ventilation in summer or for the control of solar heat gain, virtually preclude anything more than minimal sound reduction. If double windows are installed for sound insulation, it is important to remember that when they are open to meet ventilation needs they do not function acoustically. It is, therefore, vital to consider noise exclusion and ventilation needs together, because the two functions conflict. The primary function of the window is light transmission, but traditionally it has been adapted for ventilation as well. If noise control is to be added then (with a possible exception mentioned later) separate

provision should be made for ventilation. Solar heat gain (*see* pp 58, 141) depends on a number of factors such as aspect, weight of structure, size of windows and openings for ventilation. Heavy structure and small window size help to reduce both solar heating and noise transmission, but free natural ventilation to control solar heating seriously undermines noise protection.

A complete solution to the ventilation problem without inhibiting noise protection is full air-conditioning. This method of ventilation has its own noise problems (mainly internal generation and transmission) but with careful location of the air intake and, if necessary, the addition of a sound attenuator, external noise can be excluded without much difficulty. Mechanical ventilation without cooling is a compromise which may function adequately for most of the time ; the same precautions to exclude noise as for air-conditioning need to be taken.

For existing buildings which have been subjected to an increase of noise (eg near new motorways or expanding airports) forced ventilation of the whole building from a centralised system may not be practicable. An alternative solution is to provide an individual room ventilator, acoustically designed for the purpose, in each room requiring protection. This will provide some ventilation without also letting in the noise, thus allowing the window to be kept shut most of the time. If the noise situation demands it, the window can be modified to give increased sound reduction, usually by conversion to a double window. The Government Grants Scheme (administered by the British Airports Authority) for noise protection of existing dwellings in the vicinity of London's Heathrow Airport is based on this method.

The technical requirements of this scheme, and conditions of eligibility for grant payments, are set out in Statutory Instruments.* Briefly, to qualify technically for a grant, each room included in a claim has to be fitted with a double window with an air-space having a minimum width related to the glass thickness used, as follows:

24 oz (3 mm) glass
> not less than 8 in (200 mm) cavity

32 oz (4 mm) glass
> not less than 6 in (150 mm) cavity

$\frac{1}{4}$ in (6 mm) glass
> not less than 4 in (100 mm) cavity

In addition, an approved ventilator has to be installed which meets specified performances for ventilation and acoustics; the air delivery capacity must be not less than 65 ft³/minute (1·84 m³/min), sound reduction must be at least 49 dB at 500 Hz, with matching performance at other frequencies based on the ISO grade curve, and the noise level with the fan in operation at the specified setting must not exceed 35 dB(A) in the room. Certain other measures are specified, designed mainly to safeguard performance against flanking noise infiltration such as noise entering through roofs, unblocked air-bricks or chimney flues; these measures are optional, because the need for them varies with the circumstances, but if carried out their cost can be included in the claim, subject to the overall limit laid down.

Application to building types

The Heathrow Airport grants scheme for noise protection is concerned only with dwellings. Other buildings, such as schools, hotels and offices, can be protected by similar methods, though with larger rooms or higher density of occupation the individual room ventilators may need to have greater air-flow capacities. Alternatively, it may be more economical to provide mechanical ventilation on the basis of

*SI 424:1966 London (Heathrow) Airport Noise Insulation Grants Scheme; HMSO 1966
SI 1842:1968 London (Heathrow) Airport Noise Insulation Grants (Amendment) Scheme; HMSO 1968

groups of rooms (or, of course, the whole building) rather than room by room; this involves the additional provision of ducting but simplifies the problem of silencing. The windows usually need to be made double, with as wide an air-space as is practicable, as for housing.

The performance of various types of roof has already been discussed; concrete roofs usually present no problems apart from roof-lights, but timber roofs are more limited in the noise protection they can achieve. Most office blocks have adequate roof protection by reason of their concrete roof construction, but schools may be vulnerable through their roofs. The proportion of roof area is often relatively high in school design, and the roofs are frequently of light-weight construction; roof-lights are prevalent as a means of balancing daylight in classrooms and obtaining cross ventilation. In new schools near aircraft flight paths, the roofs should if possible be of heavy construction; features normally having poor insulation, such as roof-lighting, should be kept to a minimum and carefully designed with noise in mind.

Open-plan offices

The special case of the open-plan office merits separate consideration. The term 'open-plan' is used here to mean not the ordinary large accounting office or typing pool but the really large mixed-personnel office (sometimes referred to as landscaped offices or *bürolandschaft*) which would normally have been designed as a group of separate offices but has been opened up in the interests of flexibility. The loss of partitions is compensated by an increase of acoustic surface treatments, etc, so as to give maximum sound attenuation with distance. Such offices (if properly designed) are necessarily planned in considerable depth, with low ceilings, and air-conditioning is essential. Perimeter windows can therefore be fixed and, if thick glass is used, single windows will usually suffice, though double windows may be necessary for very noisy exposures. Moderately good sound insulation is thereby assured even at the perimeter of the office space; any intruding noise rapidly diminishes towards the middle of the office space, besides being masked by fairly continuous office background noise.

Masking

This matter of masking one noise by another merits some further remarks, because loose thinking on this point may lead to wrong conclusions. Noise may be annoying either because it is unpleasant in character or because it is too loud and interferes with the proper hearing of wanted messages. Traffic noise is usually regarded as impersonal in its nature, but in towns it may be so loud that, without good sound-insulating measures, the intruding noise results in undesirable masking of speech, etc. At some reduced level, the incoming traffic noise in, say, an office may not be sufficient to interfere with conversation or telephone use but may nevertheless create an unexceptionable background noise loud enough to mask other extraneous noises such as conversation from the next office. The traffic noise then becomes a welcome masking noise, an acceptable 'meaningless' sound obscuring the unwanted message of a 'meaningful' sound. It is clearly a matter of balance of the noise sources in a given situation, involving indoor and next door noise in a satisfactory ratio. The deliberate introduction or raising of background noise for masking purposes is not being advocated as a policy, but it is a method sometimes employed to meet complaints of disturbing noise which cannot be reduced at source or *en route.* A final point to bear in mind is that when the three noise sources we have been considering—wanted, unwanted and masking or background—are out of balance and a noise problem exists, then increasing the sound absorption in the room by acoustic linings is no solution, because, in general, the *balance* of the noises is not thereby changed but only their levels.

Ventilated windows

Windows cannot protect against noise and provide adequate ventilation at the same time, because letting in air also lets in noise. There are occasions when one or both of these requirements are in abeyance and a compromise solution may then serve. Double windows are sometimes suggested having small ventilator lights or openings on each side but diagonally opposed and as far apart as possible; but this is a somewhat poor compromise, since the small amount of ventilation gained is at the expense of a big drop in sound insulation, even if the window reveals are lined with sound-absorbent. The situation can be improved, ie ventilation increased and insulation loss reduced, by installing a small window-fan in one window, thus (for a given air-flow) restricting the area through which sound can leak. The reduced area is certainly helpful acoustically, but it is doubtful whether the second window is justified for the few decibels gained by the dog-leg air-path, and of course the fan could itself be a source of irritating noise. It seems more sensible to separate the window and the fan, when the latter can be served by a silencer unit which will suppress both the fan noise and the incoming outdoor noise.

However, there is a particular set of conditions when the requirements of ventilation and noise exclusion need not be at variance, simply because neither is required the whole of the time. When noise is intermittent it is possible to keep the window open at other times for ventilation, but to close it automatically whenever the noise level demands it. Changes in the external environmental noise can be monitored by a microphone linked to an acoustic switch which controls a suitable power drive connected to operating arms on the window, so that when the noise rises above a pre-set level the window closes, reopening as the noise falls again below that level. Although there is no window of this type on the market at the time of publication, a prototype window has operated successfully for several years in a school near London Airport, proving the system to be feasible.

Curtain walling

Buildings that are clad externally with curtain-walling systems may have no more mass in the unglazed panels that in the windows, so that increasing the insulation of the windows alone will not improve significantly the sound insulation of the façade. The wall panels as well as the windows must be insulated, and in principle the best method is to make the whole wall or cladding double. A double system should present no great difficulty in new designs, but modifying existing buildings may pose problems of awkward opening lights, means of support, and so on. If space permits, a possible solution is to keep the new cladding at least 600 mm from the existing face of the building, thus allowing space between the two claddings for a supporting framework as well as access for cleaning and maintenance. Very high sound insulation could be obtained in this way, in spite of the low weight of the structure.

Structural isolation

In the majority of buildings insulation against external noise in excess of about 45 dB is not called for, and this reduction can just be achieved by double windows of suitable design coupled with traditional forms of external wall construction, eg 275 mm cavity brick. In a few cases higher insulation is required, either because of very loud noise sources (as encountered in close proximity to airports) or because of very sensitive internal requirements, eg concert halls, acoustic research laboratories, recording studios, etc. The constructional technique demanded is not difficult, though it may be expensive. It involves a complete double structure of adequate weight, a wide and preferably absorbent separating air-space, and if necessary the isolated mounting of one of the double elements—usually the inner room. Isolation by spring mountings has been used for several of the acoustic laboratories at BRS, whilst another has a double envelope though without resilient mounting. The auditorium at the Royal Festival Hall also has a double envelope without spring mountings, but with additional shielding by deep surrounding foyers and other rooms. The two auditoria of the Queen Elizabeth Hall rely mainly on a very massive structure, combined with some surrounding rooms for greater depth of enclosure and an isolated plant room structure. Buildings of this nature are of course special design problems calling for advice from specialist consultants; in ordinary buildings isolation techniques are rarely necessary for protection against external noise.

Calculation of indoor noise

A reasonable approximation to the transmitted noise level inside a building can be obtained by subtracting the average net sound insulation (in dB) of the window-wall from the external noise level just outside. The levels are expressed in dB(A), ie corresponding to the use of the 'A' scale on a sound level meter, which gives a measure of loudness for many typical urban noises. This scale takes some account of the subjective assessment of loudness by suitably attenuating the lower frequencies and slightly accentuating the higher frequencies. Restating this approximation briefly:

L inside, dB(A), = L outside, dB(A), − av insulation (dB). This will hold generally for domestic rooms, offices and school classrooms. A worked example is given in Fig 1. For a more accurate estimate, other factors have to be taken into account. The problem of protecting buildings against outdoor noise is a very complex one, logically involving detailed knowledge of the external noise exposure, the insulation performance of the building structure and permissible noise levels in each room within the building. The example in Fig 1 is included primarily to illustrate the claim that if dB insulation values are subtracted from typical outdoor noise levels in dB(A), reasonably accurate figures for the internal levels in dB(A) are obtained. Moreover, apart from Fig 1, the insulation performances of windows, walls, roofs, etc, are given here only as single-figure averages in dB. This is because for most design purposes the approximate insulation performance is sufficient, since noise exposure levels and/or permissible indoor noise levels will not often be available in any detailed form but simply as single-figure totals in dB(A).

Methods of measuring and predicting the levels of road traffic noise, mainly fast-flowing traffic as on motorways, are described in Digest 185.

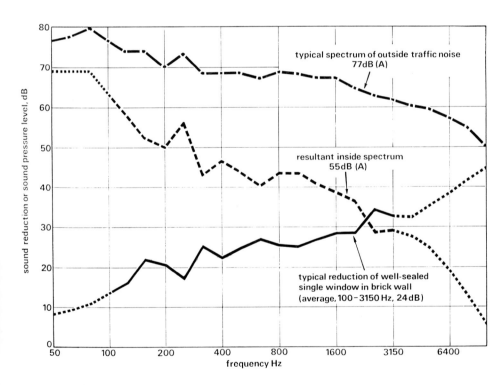

Fig 1 Noise reduction due to window/wall

Prediction of traffic noise: part 1

This digest describes prediction methods for a wider range of conditions than Digest 153 which was concerned with motorway noise and which is now withdrawn.

This digest is published in two parts. Part 1 deals with the prediction procedures; Part 2 (published February 1976) illustrates some of the procedures by a worked example and gives information on barrier design, definitive formulae on which the charts are based and information on traffic speeds to be used in the context of the regulations.

The L_{10} noise index has come into general use for the assessment of traffic noise. The L_{10} (18-hour average) index has been adopted by Government for planning and to determine entitlement of dwellings to sound insulation treatment. This digest sets out the recommended method for predicting L_{10} (1 hour) and L_{10} (18 hour) noise levels due to road traffic at points up to 300 m from the road. It also describes how to allow for the effects of obstructions and roadside barriers.

Space does not permit the charts to be reproduced here at a size that can be used for accurate prediction; much-reduced illustrations are included merely as an aid to comprehension of the techniques. The charts are reproduced full-size in 'Predicting road traffic noise' [1].

Introduction

This digest is based on the Department of the Environment's publication 'Calculation of road traffic noise'[2] which should be consulted for a more detailed exposition than is given here. Prediction is for typically adverse noise conditions ignoring non-permanent site features which might influence received noise levels.

The designation L_{10} denotes the level in dB(A) which is exceeded for one-tenth of any specified hour. The L_{10} (18-hour) level is used in regulations[3] and in certain planning considerations; it is the arithmetic mean of the 18 separate one-hourly values of L_{10} covering the period 06.00 to 24.00 hours, on a normal working-day.

Predictions are made in a series of steps, each involving the use of a formula which in most cases is also presented graphically. In many situations the use of the charts will give acceptable accuracy, even though the errors in chart reproduction and reading may result in errors in the final prediction of up to 1 dB(A). When high accuracy is required the calculation should be based on the definitive formulae

given in Part 2 of the digest. Each step in the calculation should be accurate to 0.1 dB(A). The overall result should be rounded off to the nearest whole number, 0.5 dB(A) being rounded up.

Outline of procedure

To predict the noise at the reception point, first predict the noise at 10 m away from the nearside road edge using the traffic data: flow, speed and percentage of heavy vehicles, allowing for the road gradient and surface where appropriate.

Then apply a series of corrections to this 10 m noise level to allow for propagation—distance of the reception point from the road, nature of the intervening ground and screening—and for complexities such as non uniformity of the road, partial screening, angle of view from the reception point, multiple sources and reflection effects, all appropriate to the particular situation.

The limits of the prediction method are indicated by the scales on the charts. Extrapolation is likely to cause errors.

Prepared at Building Research Station, Garston, Watford WD2 7JR
Technical enquiries arising from this Digest should be directed to Building Research Advisory Service at the above address.

The procedure

In this section, the prediction procedure is described simply in a series of steps. Where fuller guidance on a particular input might be useful, it is included in the footnote below the procedure.

1 Prediction of 10 m noise level

From the traffic flow[a] determine a noise level 10 m from the road corresponding to no heavy vehicles, a mean speed of 75 km/h, zero gradient and a conventional road surface, see Fig 1 [b]

Apply a correction to take into account speed[c] and percentage of heavy vehicles[d], see Fig 2
Apply a correction for road gradient[e] of
  0.3 G when actual mean speed is used, or
  0.2 G when using design speed of road,
  where G is the percentage gradient.

Apply the third correction, for road surface, only if road is concrete with 5 mm deep random grooving. The correction is (4—0.03 p) where p is the percentage of heavy vehicles.

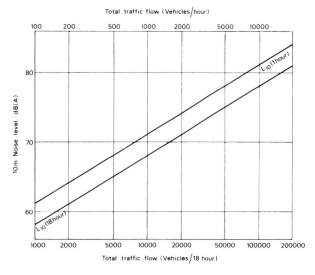

Fig 1 10 m noise level, L_{10} (1 hour) or L_{10} (18 hour) (V = 75 km/h; p = 0; G = 0)

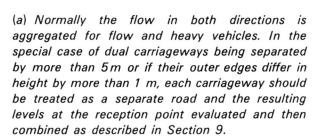

(a) *Normally the flow in both directions is aggregated for flow and heavy vehicles. In the special case of dual carriageways being separated by more than 5 m or if their outer edges differ in height by more than 1 m, each carriageway should be treated as a separate road and the resulting levels at the reception point evaluated and then combined as described in Section 9.*
(b) *If L_{10} (1 hour) is required, enter hourly flow q veh/h into upper flow scale on Fig 1 and use the upper chart line to determine the noise level. For L_{10} (18 hour) enter flow Q, veh/18 h day into the lower flow scale on Fig 1 and use the lower chart line.*
(c) *The speed is the mean speed of all the vehicles in the flow.*
(d) *Heavy vehicles are defined as having an unladen weight exceeding 1525 kg (30 cwt approx).*
(e) *Where it is necessary to treat a dual carriageway as separate roads (under Note a) the gradient correction is applied only for the upward flow.*

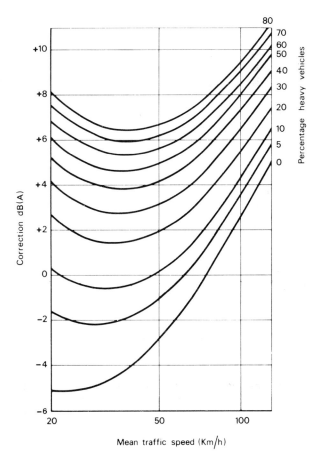

Fig 2 Correction for mean traffic speed, V, and percentage heavy vehicles, p

2 Open ground propagation—flat situations

Simple flat ground situations are not common. Only rarely will there be an unobstructed view of a straight, level and uniform road, where the intervening ground is uniform and flat and at the same level as the road. Nevertheless, procedures must be specified for idealised situations such as this as these are required in the treatment of more complex practical situations.

Figures 3 and 4 give the corrections to be applied to the 10 m noise level for propagation over hard intervening ground and grassland. The corrections take into account distance from the road edge [f] and height above ground of the reception point. The hard ground correction from Fig 3 is used for the distance correction when screening is involved, see Section 4. The grassland correction, from Fig 4, applies to all other non-hard ground surfaces [g].

3 Open ground propagation—sloping or undulating ground

To determine the distance correction, the contours should be rotated about the source position S [f], so that the base line of the contours passes through a point situated h′ vertically below the reception point, where h′ = 2 x (the average height of the propagation path above the intervening ground, source to reception point) minus 0·5m. If the terrain is such that the reception point is screened from the source, the treatment given in Section 5 applies instead.

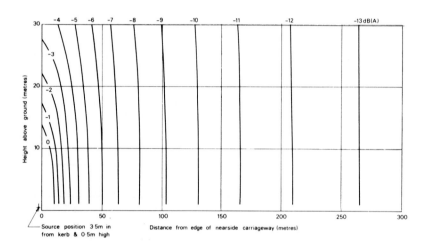

Fig 3 Propagation over hard ground: correction in dB(A) as a function of horizontal distance from edge of nearside carriageway (d) and height above ground (h)

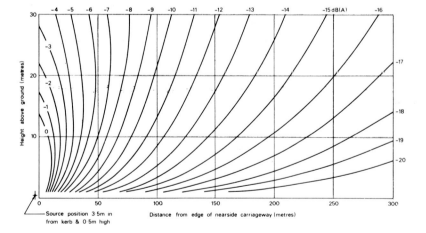

Fig 4 Propagation over grassland: correction in dB(A) as a function of horizontal distance from edge of nearside carriageway (d) and height above ground (h)

(f) Throughout the prediction method, the line of the source of traffic noise is taken to be 0·5 m high and 3·5 m in from the nearside edge of the road. If it is necessary to treat the different carriageways of a dual carriageway road as separate roads (note a) in the evaluation of the remote carriageway, the zero distance point on Figs 3 or 4 should be located at

7 m in from the far edge of that carriageway. For the near carriageway in such cases the zero distance point is located at the near edge of that carriageway, as for normal roads.

(g) In cases where the intervening ground is a mixture of hard ground and grassland, the correction for the type of ground which predominates applies.

4 Screening by very long simple barriers

Again this is an idealised situation corresponding to the open flat ground situation described in Section 2, but with an imperforate uniform barrier of constant height and distance from the road edge, long enough to obscure practically 180° of the view of the road.

The correction for screening to be applied to the noise level at the reception point, calculated as in Section 2 using the hard ground propagation correction of Fig 3, is determined from Fig 5. The correction is read from the curve at the appropriate value of the path difference $(a + b - c)$[h], see inset sketch in Fig 5.

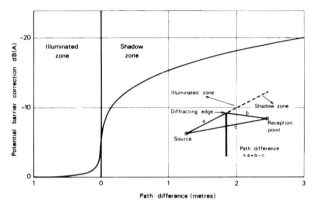

Fig 5 Potential barrier correction (very long barrier) as a function of path difference

An alternative to Fig 5 for the determination of the screening correction and one which does not need the calculation of path difference, is provided by the nomogram, Fig 6.

The nomogram is in two parts. The upper part deals with situations where the reception point is in the geometrical shadow of the barrier and the lower part corresponds to the 'illuminated zone' of Fig 5[h]. To evaluate the screening correction using Fig 6, determine a, b and h in metres from a scale drawing, see inset sketches in both parts of the nomogram[f]. Enter the horizontal scale on the left of the nomogram at (a + b) metres, the slant distance between the source and receiver. Draw a vertical line from this point on the horizontal scale to a point corresponding to the effective barrier height, h, using the height scale provided by the sloping lines.

(h) *The 'illuminated zone' part of Fig 5 is used to evaluate the small screening correction for reception points which can just see the source over the top of the barrier.*

From this point draw a horizontal line into the right-hand side of the nomogram. Enter the horizontal scale on the right-hand side of the nomogram at

$$\left(\frac{a}{a + b} \times 100 \right)$$

and draw a vertical line from this point. The intersection point of this vertical with the horizontal line from the left-hand side of the nomogram is used to read off the potential barrier correction from the curves in the right-hand side of the nomogram, using interpolation if necessary.

5 Screening by low barriers, crash barriers, buildings and earthmounds

In some cases the correction for screening by low barriers may be less than the difference between the hard ground and soft ground distance corrections. In cases where the propagation is over soft ground the noise levels at the reception point should be evaluated for grassland, as in Section 2, ignoring the barrier, and also as in Section 4, using the hard ground distance correction and the barrier correction. The result which yields the lower noise level should be used.

As a single exception for the need for barriers to be imperforate, a conventional low crash barrier of double corrugated beam construction, with the mounting height of the centre of the beam no more than 610 mm above road level, should be treated as a solid imperforate barrier whose height above the ground is equal to the width of the beam. The treatment is then as in Section 4 and the remarks on low barriers in this section apply. All other types of safety fence should be ignored.

When buildings or earth mounds screen the reception point from the road, the height and position of the equivalent simple barrier may be determined by the method indicated in Fig 7.

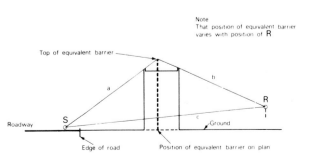

Fig 7 Equivalent barrier position

132

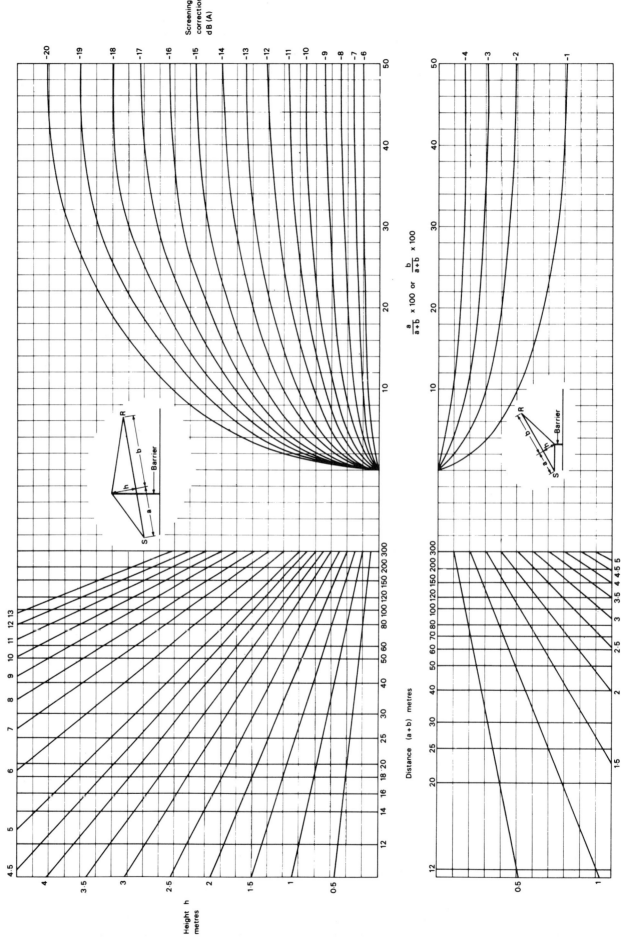

Fig 6 Nomogram for approximate potential barrier correction

6 Screening by roads on embankments and elevated roads

The correction for screening in these cases is evaluated in the same way as that for simple barriers using Fig 5 or 6. The top of the equivalent barrier will be the shoulder of the embankment or the outer edge of the elevated deck or the top of the parapet wall if this is solid.

7 Screening by cuttings

The screening correction for cuttings is determined from the path difference, as in Section 4, using the method indicated in Fig 7 to determine the equivalent simple barrier. In the special case of retained cuttings with near-vertical reflecting retaining walls, the screening effect is degraded by multiple reflections between the cutting walls. A further correction is given in Section 13.

The foregoing has dealt with the prediction of the 10 m noise level and the determination of the corrections which are added to the 10 m noise level, to allow for distance and screening in the rare situations where both the road and the propagation conditions are uniform throughout the whole 180° view of the road from the reception point. Reflection effects have been ignored. In most practical situations the configuration will be more complex in that the road may not be uniform and straight, the propagation conditions from one part of the road may be different from the propagation conditions from other parts of the road and reflections may influence the received noise level. The procedures for dealing with those more complex situations are given in Section 8.

8 Complex situations

In many cases the angle of view of the road from the reception point includes a range of different configurations, for example, bends in the road, changes in the elevation of the road, partial screening by short barriers or other changes in propagation conditions. In such cases the field of view from the reception point must be divided into a number of segments so that within each segment the conditions are uniform and simple. With a straight road for example, part of the angle of view may be screened by a building, another segment may have soft ground between the road and the reception point and for a third segment the road may be in a cutting. The received noise contribution is calculated for each segment as described above by first assuming that the simple configuration obtaining for the segment applies to the whole 180° angle of view and then by adding a correction for the actual angle of that segment using Fig 8. In this, distance is measured perpendicularly from the extension of the line of the road within the segment.

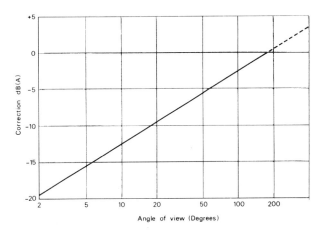

Fig 8 Correction for angle of view

Similarly, for screening calculations the path difference is evaluated using a section perpendicular to the extension of the line of the barrier. Curved roads are dealt with by approximating them to a series of shorter sections of straight road each within its segment (but see below for guidance on the choice of the number of segments necessary).

Having evaluated the contribution of each segment, these are combined successively in pairs to give the total received noise level by using Fig 9.

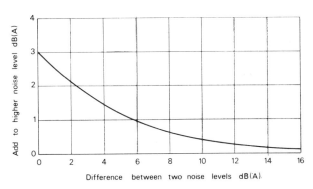

Fig 9 Combination of noise levels

In some cases, where the road bends away from or round the reception point, the angle subtended by the road will not be 180°. The procedure given above still applies but the sum of the individual segment angles will be different from 180°. The application of this segment principle allows for predictions in all situations within the limits of traffic parameters indicated by the scales in the charts but in some particularly complex situations the need for the configuration within each segment to be uniform and simple could lead to a large number of segments. The number of segments required is a matter of judgment. For practical purposes, small changes in propagation conditions within a segment are acceptable. As a guide, the extremes of the propagation conditions within a segment should not correspond to uncertainties of more than ±1 dB(A) in the noise levels received from that segment. The mean contribution from the segment should be used.

The procedures given so far cover the evaluation of the corrections to be added to the 10 m noise level, in situations where the configuration can be divided into segments, each segment covering a simple and uniform source and propagation condition. Remaining topics are the procedures for dealing with more than one road, road junctions, screening by more than one barrier, screening by rows of buildings with gaps between, noise levels down side roads and the effects of reflection.

9 Multiple sources and road junctions

The method used for combining the noise levels from the segments of a complex configuration applies to the combination of the noise levels from two or more roads and Fig 9 is used. The contributions are calculated using the appropriate traffic data and propagation conditions. For road junctions any speed change in the vicinity of the junction should be ignored.

10 Multiple barriers

In situations where more than one barrier screens the reception point from the road, the screening corrections for each of the barriers should be determined separately. Only the correction resulting in the lowest noise level is used.

11 Screening by a row of buildings

For a reception point partially screened by a row of houses or other buildings parallel to the road and with gaps between the buildings, strict application of the segment principle will often lead to a large number of segments and tedious calculations. In such cases a simplified procedure may be used. This is based on using only two segments, one to cover the contributions through the gaps between the buildings and the other to cover the contribution from the parts of the road screened by the buildings.

The angle of the single segment covering the gaps between the buildings is 180° x Z and the angle of the other segment is 180° x (1 − Z), where $Z = \dfrac{R}{R + b}$, R is the mean gap between the buildings and b is the mean length of the buildings.

For each segment apply corrections appropriate to the propagation conditions.

12 Side roads

Noise levels down side roads from main roads are calculated using the segment principle. In this special case, the contributions of the segments within which the main road is screened from the reception point by housing in the side road are ignored. In most cases this means that the only segment to be considered is that defined by the aperture of the junction of the side road with the main road and the distance of the reception point down the side road. Propagation down the side road should be calculated using the hard ground distance correction.

13 Reflections

All the procedures given above have ignored the influences of reflections and in many situations further corrections need to be applied as follows:

(i) *Façade effect*
For reception positions 1 m from a building façade a correction of +2·5 dB(A) is required.

(ii) *Façade opposite*
Where there is a substantial reflecting surface on the far side of the traffic flow a further correction of +1 dB(A) applies. In this context the substantial reflecting surface could be a wall, a noise barrier or buildings lining the road on the far side of the traffic flow. The +1 dB(A) correction applies if the percentage opening due to gaps in the reflecting surface on the far side of the traffic flow does not exceed 50 per cent. If gaps predominate then this correction is zero.

(iii) *Side roads*
At all positions between façades down a side road, whether 1 m from a façade or not, a correction of +2·5 dB(A) is required. This correction is not in addition to that under (i) above.

Correction for the opposite façade applies to a side road only when there is a substantial reflecting surface along the main road opposite the end of the side road and within the angle of view of the main road from the reception point.

(iv) *Retained cutting*
For retained cuttings, the degradation of the screening correction due to reflections depends upon both the depth of the cutting and the angle to the vertical of the retaining walls. If the walls are vertical, an additional correction to the prediction (having already taken screening into account) of D dB(A) is required, where D is the depth of the cutting in metres. If the walls of the cutting slope back, the correction is (F x D) dB(A), where F is given by Fig 10.

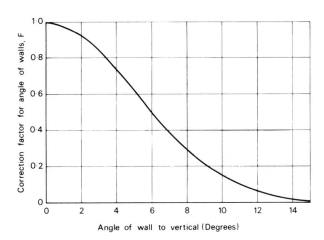

Fig 10 Correction factor, F, for reflection effect of hard-walled cuttings (overall correction for degradation of screening by retained cutting = F × D dB(A))

This correction includes that necessary for opposite façade, (*iii*) above, and therefore the separate +1 dB(A) for the far retaining wall does not apply in the case of retained cuttings.

The above procedures will calculate traffic noise in the vast majority of situations. They apply to typical traffic flows. However, for unusual traffic characteristics, eg severe congestion, for traffic flows outside the ranges covered by the scales of Fig 1 and for distances outside the ranges in Figs 3 and 4, use of the procedures may lead to errors and a measurement may be necessary. Specifications of the measuring equipment to be used, when necessary, in determining entitlement to sound insulation treatment are given in the regulations[3].

References

1 Predicting road traffic noise; 1976; HMSO.
2 Calculation of road traffic noise; 1975; HMSO; £1·70
3 Noise insulation regulations; 1975; HMSO.

Prediction of traffic noise: part 2

Definitive formulae

Calculations using even the large working-size charts referred to in Part 1 of this digest may result in an overall error of around 1 dB(A) due to the errors in chart reproduction and chart reading. For many situations this kind of error will be acceptable but when a higher degree of accuracy is required, the calculation should be based on the definitive formulae given below. Such a case would occur in the context of the Noise Insulation Regulations 1973 if the evaluation gave a result in the critical region of 66 to 69 dB(A) or in assessing a 1 dB(A) increase.

Where appropriate the chart numbers in Part 1 of this digest associated with the respective formulae are given. All the corrections are to be added to the 10 m noise level.

10 m noise level (Fig 1)

(i) L_{10} (18 h) = $28 \cdot 1 + 10$ log Q. dB(A)

(ii) L_{10} (1 h) = $41 \cdot 2 + 10$ log q. dB(A)

where Q = total vehicle flow, $06 \cdot 00$ to 24.00 hours
q = total vehicle flow within the hour.

Correction for mean traffic speed and traffic composition (Fig 2)

$$\text{Correction} = 33 \log \left(v + 40 + \frac{500}{v} \right)$$
$$+ 10 \log \left(1 + \frac{5p}{v} \right) - 68 \cdot 8 \text{ dB(A)}$$

where v = mean speed of all the traffic, km/h, during relevant period

p = percentage of heavy vehicles in traffic flow during relevant period.

Gradient

Correction = $0 \cdot 3$ G when actual mean speed is used
Correction = $0 \cdot 2$ G when design speed of road is used, see Traffic speed,

where G = percentage gradient.

Correction for surfaces with 5 mm or more deep random grooving

Correction = $4 - 0 \cdot 03p$

where p = percentage of heavy vehicles.

Distance (hard ground) (Fig 3)

$$\text{Correction} = -10 \log \frac{d'}{13 \cdot 5} \text{ dB(A)}$$

where d' = slant distance, in metres, between reception point and source line ($0 \cdot 5$ m high and $3 \cdot 5$ m in from nearside edge of road) measured perpendicularly to source line.

ie $d' = \left[(d + 3 \cdot 5)^2 + (h - 0 \cdot 5)^2 \right]^{1/2}$

where d = horizontal distance, in metres, between reception point and nearside edge of road

h = height, in metres, of reception point above road surface

This correction is valid for $d' > 4$ m.

Distance (grassland) (Fig 4)

(i) for $h > \dfrac{d + 3 \cdot 5}{3}$ Correction =

$$-10 \log \frac{d'}{13 \cdot 5} \text{ dB(A)}$$

(ii) for $1 \le h \le \dfrac{d + 3 \cdot 5}{3}$ Correction =

$$-10 \log \frac{d'}{13 \cdot 5} + 5 \cdot 2 \log \left[\frac{3h}{d + 3 \cdot 5} \right] \text{ dB(A)}$$

Screening by very long barrier (Fig 5)

$$\text{Correction} = A_0 + A_1 x + A_2 x^2 + \ldots \ldots \ldots + A_n x^n \text{ dB(A)}$$

where $x = \log (a + b - c)$, $(a + b - c)$ being the path difference in metres (see Fig 5), and

$A_0, A_1 \ldots \ldots A_n$ are constants given in the following table.

	Shadow zone	Illuminated zone
A_0	$-15 \cdot 4$	0
A_1	$- 8 \cdot 26$	$+0 \cdot 109$
A_2	$- 2 \cdot 787$	$-0 \cdot 815$
A_3	$- 0 \cdot 831$	$+0 \cdot 479$
A_4	$- 0 \cdot 198$	$+0 \cdot 3284$
A_5	$+ 0 \cdot 1539$	$+0 \cdot 04385$
A_6	$+ 0 \cdot 12248$	
A_7	$+ 0 \cdot 02175$	
Range of validity	$-3 \le x \le + 1 \cdot 2$	$-4 \le x \le 0$

Prepared at Building Research Station, Garston, Watford WD2 7JR
Technical enquiries arising from this Digest should be directed to Building Research Advisory Service at the above address.

Outside these ranges of validity the correction is defined as follows

Shadow zone	Illuminated zone
For x < −3 A = −5·0	For x < −4 A = −5·0
For x > 1·2 A is not defined	For x > 0 A = 0

The illuminated zone in the table refers to the region in which the source can just be seen from the reception point, over the top of the barrier. In this case the use of constants above A_5 is not necessary. In the shadow zone it is not necessary to use constants beyond A_7.

Angle of view (Fig 8)

$$Correction = 10 \log\left(\frac{\theta}{180}\right) dB(A)$$

where θ = the angle of view of the segment, in degrees.

Combination of noise levels (Fig 9)

Given two noise levels, L_1 and L_2 dB(A), L_1 being higher than or equal to L_2, the total combined noise level is determined by adding a correction to L_1.

$$Correction = 10 \log\left[1 + 10^{\frac{(L_2-L_1)}{10}}\right] dB(A)$$

Degradation of screening by retained cuttings (Fig 10)

Correction = F × D   dB(A)

where D = depth of cutting, in metres
F = exp $(-0.019 \phi^2)$
ϕ = the angle of the cutting walls to the vertical, in degrees.

Barrier design

The principle of barrier design is to provide as large an acoustic shadow as possible. The further the reception point is within the shadow the greater will be the noise reduction. For maximum noise reduction, barriers should be as high and as long as possible and as close to the source or the receiver as possible. Barriers to reduce traffic noise will generally be constructed close to the road. Screening corrections, appropriate to particular configurations, may be determined using the methods given in Part 1 of this digest. Throughout, it is assumed that the sound transmitted through the barrier is negligible compared with that passing over or round it: the barrier should, therefore, resist the transmission of sound. It should be largely imperforate but not necessarily of massive construction. The potential screening A (from Figs 5 and 6) is used to determine the minimum desirable mass per unit area, M, of a single leaf barrier or of one of the leaves of a multi-leaf barrier.

In general, $M = 3 \times 10^{-\left(\frac{A+10}{14}\right)}$ kg/m²

Note that the potential screening correction, A, is negative, so for
A = −17 dB(A), $M = 3 \times 10^{1·2} = 9·5$ kg/m²

As a rough practical guide the following values of M are suggested:

For A between 0 and −10 dB(A)   M = 5 kg/m²
For A between −10 and −15 dB(A)   M = 10 kg/m²
For A between −15 and −20 dB(A)   M = 20 kg/m²

Thus, for a barrier to achieve its full potential screening of, say, −10 dB(A) it would require a superficial mass of about 5 kg/m²; 12 mm plywood, at 6 kg/m² would, therefore, be more than adequate.

Other factors to be considered in barrier design include appearance, safety, durability and wind loading.

Traffic speed

General information on future mean speeds for different types of road is given in 'Calculation of road traffic noise'[2]. These data are to be used in determining eligibility for sound insulation treatment under the regulations[3] except that where based on knowledge of particular local conditions, the highway authority's estimate of traffic speed at a location differs significantly from that prescribed, then the highway authority's estimate must be used.
The generalised estimates of traffic speed from reference[2] are given below.

Roads not subject to a speed limit of less than 60 mph

Special roads (rural) excluding slip roads 108 km/h
Special roads (urban) excluding slip roads 97 km/h
All-purpose dual carriageways excluding
slip roads   97 km/h
Single carriageways, more than 9 m wide 88 km/h
Single carriageways, 9 m wide or less   81 km/h
(Slip roads are to be estimated individually)

Roads subject to a speed limit of 50 mph
Dual carriageways   80 km/h
Single carriageways   70 km/h

Roads subject to a speed limit of less than 50 mph but more than 30 mph
Dual carriageways   60 km/h
Single carriageways   50 km/h

Roads subject to a speed limit of 30 mph or less
All carriageways   50 km/h

It is only in urban areas where mean speeds of less than 50 km/h will be found over 18 hours and in most cases the range within this period is sufficiently variable to allow the use of the 50 km/h mean speed.

Worked example

The application of the prediction procedure, given in Part 1 of this digest, is illustrated for the fairly simple situation shown below. It is assumed that the L_{10} (18 hour) level at the reception point R is required. R is 1 m from the facade at a height of 6 m.

The traffic flow, Q, is taken to be 30,000 v/18 hour day, of which the percentage of heavy vehicles, p, is 35 per cent. The mean speed, v, is 90 km/h. It is further assumed that the whole site is flat and mainly grassland and that there is no road gradient. The barrier is 3 m high and 5 m from the edge of the single carriageway road. There is a bend in the unscreened part of the road.

This situation requires three segments: X, Y and Z.

Segment X, angle 70°, covers the part of the road screened by the barrier.

Segment Y, angle 60°, covers the road between the end of the screened section and the bend

and Segment Z, angle 40°, covers the road beyond the bend.

Step 1: 10 m noise level

From Fig 1: for Q of 30,000 v/18 hour day, L_{10} (18 hour)	= 72·8 dB(A)
From Fig 2: for v = 90 km/h and p = 35%; correction	= + 6·3 dB(A)
There is no road gradient, therefore 10 m noise level	= 79·1 dB(A)

Step 2: Contribution of segments

(i) Segment X

Horizontal distance between reception point and extension of the edge of the screened part of the road

	= 44m;
reception point height	= 6 m;
barrier height	= 3 m;
angle of segment	= 70°

Because screening applies in this segment, the hard ground distance correction must be applied to the 10 m noise level together with the screening correction to obtain the screened noise level; a further correction must be applied for angle of view because the screening does not operate over the full 180°.

From Fig 3 (hard ground, because there is screening): distance and height correction	= −5·5 dB(A)

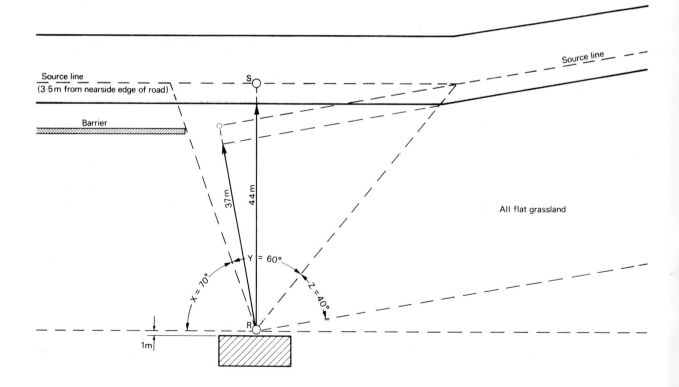

All flat grassland

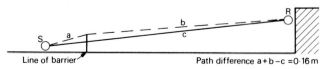

Section RS

Path difference a + b − c = 0·16 m

From Fig 5: calculate path
difference from a section
perpendicular to extended line of
barrier, 0·16 m, and obtain screening
correction (see inset on figure)　　= −10·4 dB(A)

From Fig 8: enter 70° (angle of
segment X) to determine angle of
view correction　　　　　　　= −4·0 dB(A)

Total correction to be applied to
10 m noise level for segment X　= −19·9 dB(A)
Contribution of segment X
　　　　= 79·1 −19·9　　= 59·2 dB(A)

(ii) Segment Y
Horizontal distance from reception point to edge of
road　　　　　　　= 44 m
Reception point height = 6 m
Angle of segment　　= 60°

From Fig 4 (grassland): distance
and height correction　　　　= −7·5 dB(A)

From Fig 8: enter 60° (angle of
segment Y) to determine angle of
view correction　　　　　　= −4·8 dB(A)

Total correction to be applied to
10 m noise level for segment Y　= −12·3 dB(A)

Contribution of segment Y
　　　　= 79·1 −12·3　　= 66·8 dB(A)

(iii) Segment Z
Horizontal distance from reception point to the ex-
tension of the edge of the road within this segment
　　　　　　　= 37 m
Reception point height = 6 m
Angle of segment = 40°

From Fig 4 (grassland): distance
and height correction　　　　= −6·4 dB(A)

From Fig 8: enter 40° (angle of
segment Z) to determine angle of
view correction　　　　　　= −6·5 dB(A)

Total correction to be applied to
10 m noise level for segment Z　= −12·9 dB(A)

Contribution of segment Z
　　　　= 79·1 −12·9　　= 66·2 dB(A)

Step 3: Combination of contributions from segments
　　Segment X contributes 59·2 dB(A)
　　Segment Y contributes 66·8 dB(A)
　　Segment Z contributes 66·2 dB(A)

Combining the contributions from segments X and Y
first (the order of combination is unimportant):
Difference between X and Y
　　= 66·8 −59·2　　= 7·6 dB(A)
From Fig 9: correction to be added
to higher level　　　　= +0·7 dB(A)
Combined level = 66·8 + 0·7　= 67·5 dB(A)

Combine this level with Z:
Difference between combined level
and Z　　= 67·5−66·2　　= 1·3 dB(A)

From Fig 9: correction to be added to higher level
　　= +2·4 dB(A)
Total for all segments
　　= 67·5 + 2·4　　= 69·9 dB(A)

This is the free field noise level with no allowance
for reflections.
Step 4: Reflections
The reception point is 1 m from a facade and there
are no buildings lining the road on the far side. A
correction of +2·5 dB(A) applies for the facade
effect.

Final result: L_{10} (18 hour) at reception point = 69·9 + 2·5 = 72·4 dB(A).

In this example, the working-size charts published in
'Predicting road traffic noise' were used. Figure 5
was used to determine the screening correction,
which involves a subsidiary calculation of the path
difference (a + b − c) and which gives a correction
of −10·4 dB(A). Alternatively, the use of the
nomogram, Fig 6, and a scale drawing results in a
screening correction of −10·3 dB(A). Such a close
agreement may not always be obtained; generally
an accurate determination of the path difference
and the use of Fig 5 will give a more accurate result
than the use of Fig. 6. The highest accuracy will
result from the use of the definitive formulae with an
accurate determination of path difference. In this
worked example, the final result using the formulae
throughout is 72·2 dB(A) compared with 72·4 dB(A)
from the charts (irrespective of whether Fig 5 or Fig
6 is used).

References
1 Predicting road traffic noise; HMSO
2 Calculation of road traffic noise; 1975; HMSO; £1.70
3 Noise insulation regulations; 1975; HMSO

Traffic noise and overheating in offices

A comfortable environment in offices relies partly on acceptable levels of noise, temperature and ventilation. Traffic noise may force measures to be taken that will reduce natural ventilation and tend to raise internal temperatures to an unacceptably high level, particularly in summer. Design standards for noise, temperature and mechanical ventilation rates must be established before the means to achieve a comfortable environment can be devised. This Digest discusses these standards and shows their use in design.

Traffic noise

Standards

The noise level in any traffic situation fluctuates considerably. An index, L_{10}, has therefore been developed to take account of this fluctuation. L_{10} is the arithmetic average hourly value of the level of noise in dB(A) at 1 metre from the façade of a building, exceeded for 10% of the time. Its adoption by Government in relation to motorway noise and dwellings is described in Digest 153.

A level of 70 dB(A) on the L_{10} (18-hour) index has been recommended as the upper limit to which residential development should be subjected. This corresponds roughly to 60 dB(A) internally with windows open. Recent investigation of experience of traffic noise in small offices shows that complaints increase markedly if the traffic noise, internally, is above 60 dB(A). This suggests that the standard for offices should not be worse than for dwellings. A further point is that in many dwellings there is some choice of living on the quieter side of the house, even if for only part of the time, but this choice is not usually open to office workers.

The Wilson Committee suggested an upper limit of 55 dB(A) for buildings in which communication by speech is important. These buildings were considered to 'include many offices and committee rooms'. Occasional slight difficulty occurs in telephone conversations with the level of 55 dB(A), but the limit can be reasonably taken as applying to maximum continuous sound where telephones are to be used. A level of 48 dB(A) was recommended for industrial business offices and 43 dB(A) for private or semi-private offices.

Thus at one end of the scale a standard of, say 45 dB(A), might be chosen. Any further improvement would have only a minimal subjective effect; it could be expensive to implement and its benefit could not at present be quantified. At the other end of the scale, noise levels above 60 dB(A) are likely to give rise to considerable complaint. An intermediate value of 55 dB(A) during normal office hours might therefore be taken as a reasonable standard for two-person offices.

Prediction and measurement

L_{10} (18-hour) can be predicted with reasonable accuracy for free-flowing traffic if the traffic density, composition, speed, etc, are known (the necessary information is given in Digest 185) and this normally obviates the need for measurement. 2 dB(A) should be added to the predictions because the traffic is heavier during office hours.

In the centres of towns and cities the traffic does not flow freely. At present, a noise level cannot be predicted and site measurement is necessary.

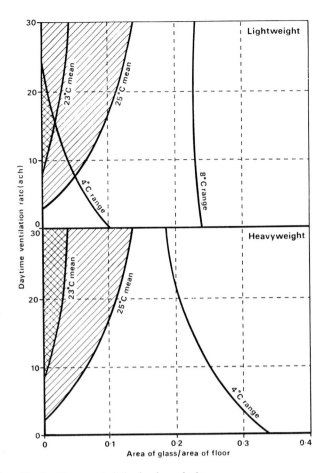

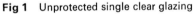

Fig 1 Unprotected single clear glazing

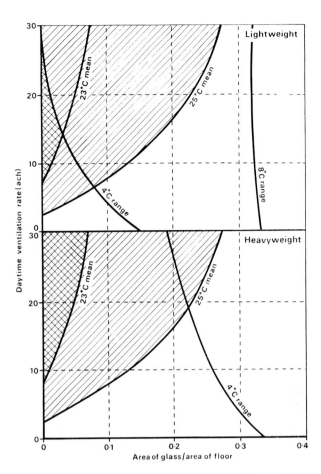

Fig 2 Single clear glazing with internal venetian blind

Temperature

Summer-time standards

Temperatures suitable for a level of activity typical of office work and for two modes of clothing, (a) shirt and trousers, and (b) lightweight suit, are approximately 25°C and 23°C respectively, if the air movement is slight. The method of calculation is set out in BRE current paper CP 14/71. Some temperature variation is acceptable during the day. An acceptable standard might be a daily mean temperature of 25°C with a range of 8°C. However, even if lightly clad, nearly half the occupants would feel too hot in the peak temperature of 29°C. A much better standard would be a daily mean of 23°C, with a range of 4°C. This would reduce thermal discomfort to a low level where men are expected to wear jackets; few would be uncomfortable at the peak temperature of 25°C. Intermediate mean temperatures and ranges may be chosen, depending upon the clothing and activity levels that seem most appropriate. Whether the chosen standard is likely to be achieved can be discovered by interpolation on Figs 1–4, described below.

Calculation

Overheating is most likely to occur in hot sunny weather but in the variable climate of the UK it is difficult to obtain useful figures by measurement; calculation is therefore recommended.

The admittance procedure set out in the IHVE *Guide* (1970) can be used to calculate the variations of internal temperature in buildings that are not air-conditioned. The critical factors are: window size, shading devices, the thermal response of the building, ventilation rate with outside air and the internal heat gains from occupants and lights. When the windows are closed to reduce traffic noise, sun blinds or curtains are the only control that the occupant has over his environment.

Calculations have been made to determine the quantities of fresh air that are needed to achieve acceptable thermal conditions in buildings that are occupied predominantly by day. The calculated values of mean temperature and temperature swing are those likely to be experienced in occupied hours in two-person offices with one glazed façade.

Within the latitudes of the UK, orientation is not of prime importance and the results, although calculated for a south-west aspect, are representative of values between NW, through S, to NE in the summer months. For orientations NW–N–NE, overheating is not likely to occur and there are no limits on window area from this point of view.

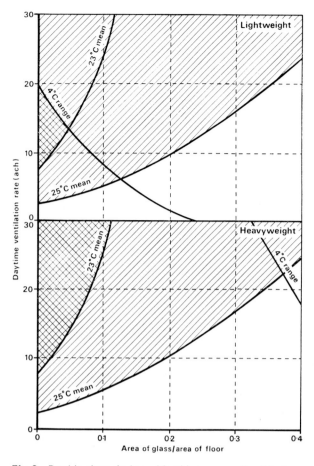

Fig 3 Double clear glazing with mid-pane venetian blind

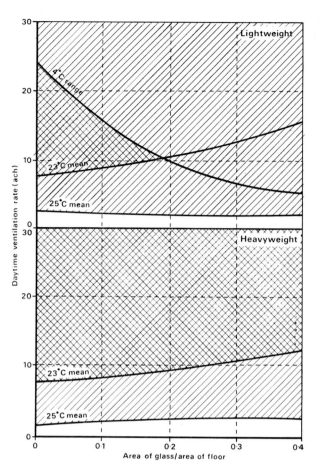

Fig 4 Single clear glazing with external blind

With design use in mind, plots of temperature against ventilation rate and window size were prepared: Figs 1–4 are examples. These assume that lights are not in use during the period considered; they relate to a room height of 2·7 m but are applicable with little loss of accuracy to other heights in the range 2·4–3·0 m. Ventilation rates are those occurring during the daytime (08·00–18·00 hours BST). A night-time rate of 0·5 ach is assumed. The double-hatched areas are bounded by lines representing 23°C mean temperature and 4°C range and indicate the range of conditions within which this temperature standard may be obtained when ventilating with outside air. The single-hatched areas show the wider

Table 1 Typical descriptions of two weights of room

	Ceiling	Internal walls	External wall	Floor
LIGHTWEIGHT	Suspended	Dry partitions	Light-weight	Carpeted or cork-tiled
HEAVYWEIGHT	Exposed slab	100 mm brick	Cavity brick	Concrete, linoleum-covered

range of conditions that will produce the minimum acceptable standard with limits of 25°C mean and 8°C range. A mean temperature of 24°C may be interpolated midway between the 23° and 25° curves.

The thermal design data relate to 'heavy' and 'light' rooms, see Table 1. Most modern offices tend towards the latter, having dry partitions and suspended ceilings, whilst floors are made effectively lightweight by carpeting. The weight and U-value of the external wall is not particularly important in this context.

The diagrams give limitations on window size for the more common clear glass situations and have been selected to illustrate a span of window types including extreme conditions. For other window types, eg tinted and reflective glazing, curtains and roller blinds, it is practicable to interpolate between the limits given using relative values of solar gain factor from Book A of IHVE *Guide* 1970 for the purpose.

The mean temperatures derived from the diagrams are likely to be exceeded on average for about 2 working days per year. In practice, a higher 'failure rate' may be acceptable. For example, if the indicated mean temperatures are reduced by 2°C, the diagrams then relate to conditions which are likely to be exceeded for about 6 working days per year.

Fresh air requirements

The requirements for the provision of oxygen and removal of carbon dioxide are much lower than the requirement for odour removal which may therefore be assumed to dominate. The volume of fresh air required for this depends on the density of occupation of the room. For an average two-person office, the minimum requirement would be 0·5 air changes/hour (ach). But if smoking is to be permitted, the IHVE *Guide* recommends a minimum ventilation rate of 0·9 ach whilst the GLC Byelaws for offices require 1·5 ach, possibly in anticipation of a higher density of occupation. For an office with openable windows not weatherstripped, only the lowest rate of 0·5 ach could be achieved by infiltration. Generally, therefore, some form of mechanical ventilation is needed to satisfy the fresh air requirement throughout the year as well as for summer-time temperature control.

Where mechanical ventilation is required, the location of the fan is primarily a matter of cost and convenience. It may be in the window or in the wall, or installed as a central unit with ducting to a number of offices; whatever its position, some means of noise attenuation may be necessary. An adequate outlet for the air from each office is also needed. Installing a normal fan in an otherwise sealed office is useless, but a two-way fan in which the outlet is in the fan itself may be useful.

General guidelines

The information given may be used to develop solutions to the problem of traffic noise in existing offices, taking account of ventilation, thermal and lighting requirements. It can be used also in the preliminary design of new buildings as it enables a quick appreciation to be made of possible design solutions.

Decisions must be taken on the standards for room temperature and the internal noise levels to be aimed for. A standard of 60 dB(A) may be the best that can be achieved with existing offices but for new offices 55 dB(A) or even 45 dB(A) may be preferred, bearing in mind that standards are tending to improve and that buildings erected in the 1970s will still be in use at the end of the century. The level of traffic noise on the L_{10} index must also be determined.

The difference between L_{10} and the required internal noise standard will influence the choice of window type. Any of the following will achieve an internal noise standard of 55 dB(A):

where L_{10} = 75 dB(A) a closed, weatherstripped window

L_{10} = 80 dB(A) a fixed, single window with 12 mm glass

L_{10} = 90 dB(A) a closed double window

Table 1 of Digest 128 gives further information from which the above have been derived. The window insulation has been taken as 5 dB less than the maximum given in the Table so as to allow for variations in performance according to the manner of use.

The area of glass/area of floor should be calculated and reference to the appropriate section of Figs 1–4 will then show the rate of ventilation with outside air required to avoid overheating. The effects of modifying the window type, position of blinds, and window areas can easily be determined.

The ventilation requirement to avoid overheating is normally greater than that for fresh air, but the latter must, of course, be borne in mind when specifying the fan. Air conditioning provides an alternative solution but the illustrations give some guidance on how to achieve the better temperature standard without recourse to air-conditioning.

It is assumed that fans and shading devices will be used effectively. Advice on this should be given to the occupants of treated offices.

Further reading:

BRE Digests
41 and 42 *Estimating daylight in buildings*
128 and 129 *Insulation against external noise*

Current papers
M A Humphreys and J F Nicol—*Theoretical and practical aspects of thermal comfort* BRE CP 14/71
A F E Wise—*Designing offices against traffic noise* BRE CP 6/73
N O Milbank—*A new approach to predicting the thermal environment in buildings at the early stage of design* BRE CP 2/74

Demolition and construction noise

Excessive noise can damage people's health or hearing and can cause disturbance to living or working environments.
This digest describes briefly some of the legislation and codes of practice aimed at controlling the levels of noise emitted from sites, and discusses methods of estimating and reducing plant noise.

Legislation

Legislation for the protection of people and the environment against noise has recently been extended by two Acts of Parliament.

The Health and Safety at Work etc Act, 1974,[1] provides for the protection of workers against risks to their safety and health, including the risk of hearing damage. On construction and demolition sites this mostly concerns plant operators or drivers, but men working close to plant must also be considered.

The Control of Pollution Act, 1974,[2] gives local authorities powers to protect the community against noise. Under Part III of this Act, a local authority may specify its own requirements to limit noise on construction and demolition sites by serving a notice that may in particular:

a. specify the plant or machinery which is, or is not, to be used;
b. specify the hours during which the work may be carried out;
c. specify the level of noise which may be emitted from the premises in question or at any specified point on those premises or which may be so emitted during specified hours; and
d. provide for any change of circumstances.

In specifying their requirements the local authority shall have regard:

a. to the relevant provision of any code of practice issued under this Part of this Act;
b. to the need for ensuring that the best practical means are employed to minimise noise;
c. before specifying any particular methods or plant or machinery, to the desirability in the interests of any recipient of the notice in question of specifying other methods or plant or machinery which would be substantially as effective in minimising noise and more acceptable to them;
d. to the need to protect any persons in the locality in which the premises in question are situated from the effects of noise.

The Act also provides for a developer or contractor to ascertain from the local authority the noise restrictions that will apply to a contract at the same time as, or subsequent to, a request for approval under building regulations under Part II of the Public Health Act 1936, or in Scotland a warrant under Section 6 of the Building (Scotland) Act 1959.

Codes of Practice

Code of practice for reducing the exposure of employed persons to noise[3] gives recommendations for the maximum safe daily dosages of noise for the unprotected human ear. The accepted limit for an eight-hour exposure to a steady sound is 90 dB(A). If the exposure is for a period other than eight hours, or if the sound level is fluctuating (as is usually the case with building works), the equivalent continuous sound level, Leq, should not exceed 90 dB(A). Practical rules for the calculation of Leq are given in the Code. If the daily dosage exceeds the safe limits, the contractor should provide the operatives with ear protectors. Advice on the selection of protectors is also given in the Code.

British Standard 5228, Code of Practice for noise control on construction and demolition sites[4] outlines the application of The Control of Pollution Act, 1974, and considers the role of the local authority as well as that of the developer, the designer and the contractor. In setting a noise limit, a local authority may specify a permissible noise level at a point 1 m from the exterior surface of the nearest noise-sensitive building. The limit may be in terms of a maximum noise level or of an equivalent con-

Prepared at Building Research Station, Garston, Watford WD2 7JR
Technical enquiries arising from this Digest should be directed to Building Research Advisory Service at the above address.

tinuous sound level, Leq, for a given period. Although the Code suggests limiting values of Leq for a 12-hour period (eg 0700 to 1900 hours), these values have no legal significance.

Planning

The developer and the designer have a part to play at the early stages of planning a project. Early consultation should be made with the local authority to ascertain the noise restrictions that will apply. The design of the whole project should then be considered so as to avoid, as far as possible, noisy processes and machines. It may also be necessary to make estimates of the total noise emission from the site to check if proposed construction methods will be acceptable.

Details of noise restrictions should be included in the tender documents, so that the tenderer can take this into account in estimating plant requirements and at the same time consider whether any special noise control measures will be needed.

A more detailed check on noise emission can be made when the contract is awarded. For example, a production programme can indicate the locations and periods of operation of plant as well as the numbers and types of machine needed, so that noise levels can be predicted and compared with the limits set. Such a prediction will be possible only if the actual noise output of each machine is known; this information is also needed to determine whether operatives should wear ear protectors. If prediction shows that infringement is likely, means for controlling the spread of noise or the use of quieter machines or methods should be considered.

The Code[4] gives detailed guidance on the prediction and monitoring of noise as well as on methods for controlling the spread of noise.

In general, an extensive knowledge of acoustics is not needed for the prediction of noise levels on construction and demolition sites; a familiarity with some of the general principles and the terminology is, however, necessary.

Noise limits and measurement

Sound is transmitted as pressure disturbances in the air which vibrate the ear drum to cause the sensation of hearing. A sound is quantified by its sound pressure level: a ratio of the pressure of the sound to a reference sound pressure, usually that of the quietest audible sound. This ratio is expressed in decibels (dB) and is measured with a sound level meter. To represent subjective reactions to noise, the dB measurements have to be modified or 'weighted' to take account of the response of the human ear which varies with the frequency of sound. This 'weighting' is achieved by setting the

sound level meter to the 'A' scale; the sound level is then indicated in dB(A). A sound level is only meaningful when the location of the point of measurement is specified; for example, the noise level from a building site measured at a point 1 m outside the nearest occupied buildings, or the noise of a machine measured at a point 1.5 m above ground at a distance of 10 m.

The noise from a machine can also be expressed as a sound power level: a measure of the total sound energy emitted per unit time. For practical purposes the sound power level is calculated:

Sound power level, dB(A) =
    sound level, dB(A) + 8 + 20 log R
where R is the distance in metres from the centre of the machine at which the sound level is measured.

Sound power is a measure of the noise output of a machine (or source) and is therefore not dependent on distance from the source, except that the distance should not be too small because of near field effects.

An equivalent continuous sound level (Leq) is a notional steady level which would, over a given period of time, deliver the same sound energy as an actual fluctuating sound. Instruments are available that indicate Leq directly. If Leq is calculated or measured for a period of 12 hours it is referred to as Leq (12 hr) and the value is given in dB(A).

Limitation of site noise by Leq (12 hr) permits a higher Leq for shorter periods provided the background noise during the remainder of the period is at least 10 dB(A) below the 12-hour limit. In that case an increase of 3 dB(A) can be allowed for each halving of the period. Thus, for a limit of Leq (12 hr) 75 dB(A), an Leq of 78 dB(A) would be permissible for a working period of 6 hours, or 81 dB(A) for a period of 3 hours. If it is known that a particular process will produce high noise levels for short periods, it may be more appropriate to restrict the maximum noise level rather than Leq.

Both codes[3, 4] give further information on noise units and recommendations concerning measuring instruments.

Plant noise

The main problem of noise in construction and demolition arises from the operation of plant and powered hand-tools. Some machines are already produced in a sound-reduced form and methods for obtaining greater reductions and for reducing the noise of other plant are under investigation. It remains to be seen how much noise reduction is physically and economically feasible.

In making predictions of site noise it is essential to have reliable information on the sound output of the plant that will be used. The noise produced by a machine depends on many factors such as the speed of operation and the work performed. Some manufacturers publish noise levels for their machines, but the conditions under which the noise levels were obtained are not necessarily the same for all types of plant. Standard procedures for noise tests representative of plant in operation are currently being developed by the International Organization for Standardization and the European Economic Community. The only document so far published is ISO 2151 — *Measurement of airborne noise emitted by compressor/prime-mover — units intended for outdoor use.** Some countries have already set upper limits for the noise outputs of certain types of plant: EEC Directives on this subject are also in preparation. Until standard test procedures have been adopted and the noise outputs of machines are quoted on this basis, use can be made of published figures such as those in Table 5 of BS 5228. The values in this table are maximum sound power levels taken from limited measurements on site and from various published data. In practice, values could be found which fall outside the ranges quoted. Alternatively, plant owners can make their own measurements.

In noise tests that have already taken place on plant such as excavators, loading shovels and dumpers it has often been the practice to measure noise levels at a distance of about 10 m from the centre of the machine (in standard procedures for testing it is likely that this distance will depend on the dimensions of the plant) and at a height of 1·5 m above ground. Measurements are necessary on four sides of a machine and, perhaps, at points above the machine as well, because differences can occur, due for example, to the position of the engine or the exhaust.

There can be considerable variation in the noise output of a machine during working. Measurements 10 m from a mobile crane, for example, gave levels of: 87 dB(A) hoisting; 83 dB(A) slewing; and 73 dB(A) engine idling.

Estimating noise emissions from sites

If the noise outputs of the machine to be used are known and production has been planned so that the locations of the machines and their periods of operation can be forecast, the resultant maximum noise level can be estimated at a point outside the site where a maximum noise limit may be specified.

In theory, and under specific conditions, the level of sound from a point source of noise is reduced by 6 dB(A) when the distance from the source is doubled. Although construction machines cannot be

regarded as point sources of noise and conditions on sites vary considerably from those on which the theory is based, site studies[5] have shown that this figure can be used to give sufficiently accurate estimates of distance attenuation in practice.

If the noise output of a machine is given as a noise level measured at a radius x from the centre of the machine, the reduction in this noise level at a greater radius y is equal to $20 \log_{10} \frac{y}{x}$.

If, however, the noise output of a machine is given in terms of sound power level, the noise level at any radius up to 700 m from the centre of the machine can be quickly obtained from the graph, Fig 1. For example, if the sound power level of a machine is 120 dB(A), the noise level at a radius of 30 m is $120 - 37 = 83$ dB(A).

Often several machines will be operated simultaneously and the total noise at a point where a maximum noise limit applies has to be estimated. The first step is to determine the noise level of each machine separately at that point. The following table then shows how to predict the total effect of two sound levels:

Difference between the two sound levels dB(A)	Amount to be added to the higher sound level to obtain the total sound level dB(A)
0 to 1	3
2 to 3	2
4 to 9	1
10 and above	0

If four machines individually produce noise levels of 70, 75, 76 and 77 dB(A) the total noise is obtained by combining the levels in pairs as follows:

70 and 75 dB(A), combined level is $75 + 1 = 76$ dB(A)
76 and 76 dB(A), combined level is $76 + 3 = 79$ dB(A)
79 and 77 dB(A), combined level is $79 + 2 = 81$ dB(A)

Thus the total noise from the four machines is 81 dB(A).

When estimating the noise level at the façade of a building, the Code[4] recommends that an addition of 3 dB(A) should be made to the levels calculated to allow for the effect of reflection. Where multiple reflections occur, due for example to a recess in a building, sound levels may be increased even further.

The estimation of site noise emission in terms of an equivalent continuous sound level (Leq) is more complex because a more precise forecast of the overall working time of each machine is necessary as well as the fluctuations in level that occur during the working cycle of the machine. The Code[4] deals with methods of calculation in estimating Leq and gives examples of their use. As yet, however, there is no published information on practical experience in the use of Leq for the control of noise from construction and demolition sites.

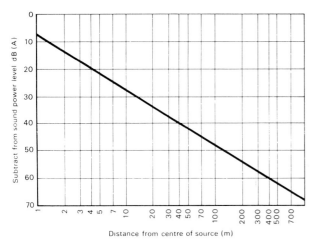

Fig 1 Amount to be subtracted from sound power level to determine the sound level at some distance from the source

Note Sound levels determined by reference to this graph are unreliable at distances less than 3 major dimensions of the source or if the source radiates sound in a marked directional manner

Reducing noise emission
When estimates of site noise emission show that infringement of a noise limit is likely, or when an infringement has occurred, one or several of the following means for reducing noise should be considered:

Siting of plant
By siting plant or operations as far as possible from a noise-sensitive area, or so that mounds of earth, stacks of materials or buildings on site reduce the transmission of sound.
By positioning a machine so that the quietest side faces the noise-sensitive area.

Operation of plant
By operation at a low speed.
By shutting down when not in use.
By keeping engine or machinery covers closed.
By reducing the number of machines in simultaneous operation.
By regular maintenance, notably to avoid rattling noises due to loose fixings or frictional noises due to lack of lubrication and to keep exhaust systems in good order.

Alternative plant or method
By using noise-reduced plant or by modifying the construction method so that noisy plant is unnecessary.

Screening
By erecting a noise barrier between the source of noise and the noise sensitive area (suitable for mobile plant).
By providing an acoustic enclosure for a machine, paying due regard to cooling and ventilation (suitable for a static machine).
By using an acoustic shed in which, for example, a man can work with a concrete-breaker. The operator is then exposed to a greater risk of hearing damage and should wear ear-protectors.

References
1 The Health and Safety at Work etc. Act, chapter 37, HMSO
2 The Control of Pollution Act 1974, chapter 40, HMSO
3 Department of Employment Code of practice for reducing the exposure of employed persons to noise, HMSO
4 BS 5228, Code of Practice for noise control on construction and demolition sites, British Standards Institution
5 Construction site noise, BRE Current Paper CP 57/75

Further reading
Thermic boring, BRE Current Paper CP 58/75
The Nibbler: a new concept in concrete breaking, BRE Current Paper CP 83/74

4 Daylight

Estimating daylight in buildings—1

The method of estimating daylight in buildings with the aid of BRS Daylight Protractors involves two discrete stages: the first deals with the direct sky component and the externally reflected component and is explained in this Digest; the second deals with the internally reflected component only (see Digest 42).

The amount of daylight received in buildings is most conveniently expressed in terms of the percentage ratio of indoor to outdoor illumination, the ratio being called the 'daylight factor'. For any given situation, the value of this factor depends on the sky conditions, the size, shape and position of the windows, the effect of any obstructions outside the windows and the reflectivity of the external and internal surfaces. The daylight factor can be predicted at the design stage, either by actually measuring the daylight received in a model of the building, or, more frequently, by calculation, using drawings and other necessary data. Some of the more commonly used aids to calculation which simplify what would otherwise be a tedious operation are described here.

Source of daylight

The quality and intensity of daylight varies with latitude, season, time of day and local weather conditions. In contrast with the tropics, direct sunlight in temperate zones cannot be relied on for lighting the interiors of buildings and therefore reliance has to be placed on the light received from the sky. In so doing, this Digest is concerned solely with daylighting in this country and in those others where the sky conditions are strictly comparable.

The seasonal and daily variations in daylight can be seen from Figs. 1 (*a*) and (*b*), which show the illumination received on a horizontal surface out of doors from the sky as a whole, averaged for each month.

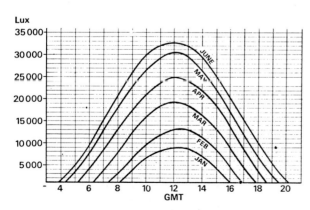
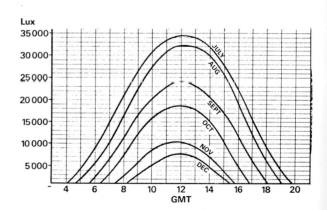

Fig. 1 (*a*) and (*b*). Illumination in lux, received from the sky, on a horizontal surface out of doors, plotted against time.

Table 1 Sky components (CIE overcast sky) for vertical glazed rectangular windows

Ratio H/D = Height of window above working plane: distance from window

Ratio W/D = Effective width of window to one side of normal: distance from window

W/D \ H/D	0	0·1	0·2	0·3	0·4	0·5	0·6	0·7	0·8	0·9	1·0	1·1	1·2	1·3	1·4	1·5	1·6	1·7	1·8	1·9	2·0	2·2	2·4	2·6	2·8	3·0	3·5	4·0	5·0	∞
0·1		0	0	0·1	0·1	0·2	0·2	0·3	0·4	0·5	0·6	0·6	0·7	0·8	0·8	0·9	0·9	0·9	1·0	1·0	1·0	1·1	1·1	1·1	1·1	1·2	1·2	1·2	1·2	1·3
0·2	0	0	0·1	0·1	0·2	0·4	0·5	0·7	0·8	1·0	1·1	1·3	1·4	1·5	1·6	1·7	1·8	1·9	1·9	2·0	2·0	2·1	2·2	2·2	2·3	2·3	2·4	2·4	2·4	2·5
0·3	0	0·1	0·1	0·2	0·3	0·5	0·7	1·0	1·2	1·5	1·7	1·9	2·1	2·3	2·4	2·6	2·7	2·8	2·9	3·0	3·1	3·2	3·3	3·4	3·4	3·5	3·6	3·6	3·7	3·7
0·4	0	0·1	0·1	0·3	0·4	0·7	1·0	1·3	1·6	1·9	2·2	2·5	2·7	2·9	3·2	3·3	3·5	3·6	3·8	3·9	4·0	4·1	4·3	4·4	4·5	4·5	4·6	4·7	4·8	4·9
0·5	0	0·1	0·1	0·3	0·5	0·8	1·2	1·5	1·9	2·2	2·6	3·0	3·3	3·6	3·8	4·0	4·2	4·4	4·6	4·7	4·8	5·0	5·2	5·3	5·4	5·5	5·7	5·8	5·9	5·9
0·6	0	0·1	0·1	0·3	0·6	1·0	1·3	1·7	2·2	2·6	3·0	3·4	3·8	4·1	4·4	4·6	4·9	5·1	5·3	5·4	5·6	5·8	6·0	6·2	6·3	6·4	6·6	6·7	6·8	6·9
0·7	0	0·2	0·2	0·4	0·7	1·0	1·5	1·9	2·4	2·8	3·3	3·8	4·2	4·5	4·8	5·1	5·4	5·6	5·8	6·0	6·2	6·4	6·6	6·8	7·0	7·1	7·3	7·4	7·6	7·7
0·8	0·1	0·2	0·2	0·4	0·7	1·1	1·6	2·1	2·6	3·1	3·6	4·1	4·5	4·9	5·2	5·6	5·8	6·1	6·3	6·5	6·7	7·0	7·3	7·5	7·6	7·8	8·0	8·2	8·3	8·4
1·0		0·1	0·2	0·4	0·8	1·2	1·7	2·2	2·7	3·3	3·8	4·3	4·8	5·2	5·6	5·9	6·2	6·5	6·7	6·9	7·1	7·4	7·7	7·9	8·1	8·2	8·5	8·7	8·8	9·0
1·2		0·1	0·2	0·4	0·8	1·3	1·8	2·3	2·9	3·4	4·0	4·6	5·0	5·5	5·9	6·2	6·5	6·8	7·1	7·3	7·5	7·9	8·1	8·4	8·6	8·7	9·0	9·2	9·4	9·6
1·4		0·1	0·2	0·5	0·9	1·4	1·9	2·5	3·1	3·7	4·3	4·9	5·4	5·9	6·4	6·8	7·2	7·5	7·8	8·1	8·3	8·7	9·1	9·3	9·6	9·8	10·1	10·3	10·5	10·7
1·6		0·1	0·2	0·5	0·9	1·4	1·9	2·5	3·2	3·8	4·5	5·1	5·7	6·2	6·7	7·1	7·5	7·8	8·2	8·5	8·7	9·1	9·5	9·8	10·0	10·2	10·6	10·9	11·1	11·6
1·8		0·1	0·2	0·5	0·9	1·4	2·0	2·6	3·3	3·9	4·6	5·3	5·9	6·4	7·0	7·4	7·8	8·2	8·5	8·8	9·1	9·6	10·0	10·2	10·5	10·7	11·1	11·4	11·7	12·2
1·9		0·1	0·2	0·5	1·0	1·4	2·0	2·6	3·3	4·0	4·7	5·4	6·0	6·6	7·2	7·6	8·1	8·5	8·8	9·2	9·5	10·0	10·4	10·8	11·1	11·3	11·8	12·0	12·3	12·6
2·0		0·1	0·2	0·5	1·0	1·5	2·0	2·6	3·3	4·0	4·7	5·4	6·1	6·7	7·3	7·8	8·2	8·6	9·0	9·4	9·7	10·2	10·7	11·1	11·4	11·7	12·2	12·4	12·7	13·0
2·5		0·1	0·2	0·5	1·0	1·5	2·1	2·6	3·3	4·0	4·8	5·5	6·2	6·8	7·4	7·9	8·4	8·8	9·2	9·6	9·9	10·5	11·0	11·4	11·7	12·0	12·6	12·9	13·3	13·7
3·0		0·1	0·2	0·5	1·0	1·5	2·1	2·7	3·4	4·1	4·8	5·6	6·2	6·9	7·5	8·0	8·5	8·9	9·3	9·7	10·0	10·7	11·2	11·7	12·0	12·4	12·9	13·3	13·7	14·2
4·0		0·1	0·2	0·5	1·0	1·5	2·1	2·7	3·4	4·1	4·9	5·6	6·3	6·9	7·5	8·0	8·6	9·0	9·4	9·8	10·1	10·8	11·3	11·8	12·2	12·5	13·2	13·5	14·0	14·6
6·0		0·1	0·2	0·5	1·0	1·5	2·1	2·8	3·4	4·2	5·0	5·7	6·3	6·9	7·6	8·1	8·6	9·1	9·5	9·9	10·2	10·9	11·4	11·9	12·3	12·6	13·2	13·6	14·1	14·9
∞		0·1	0·2	0·5	1·0	1·5	2·1	2·8	3·4	4·2	5·0	5·7	6·3	7·0	7·6	8·1	8·6	9·1	9·5	9·9	10·3	10·9	11·5	11·9	12·3	12·7	13·3	13·7	14·2	15·0
Angle of obstruction	0°	6°	11°	17°	22°	27°	31°	35°	39°	42°	45°	48°	50°	52°	54°	56°	58°	60°	61°	62°	63°	66°	67°	69°	70°	72°	74°	76°	79°	90°

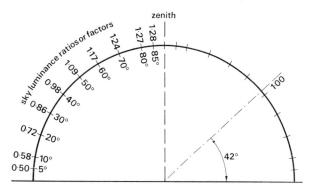

Fig. 2. Luminance distribution of densely overcast sky compared with the average luminance taken as unity.

These values are based on measurements of total sky illumination made by the National Physical Laboratory for a wide range of sky conditions but excluding direct sunlight.

The amount of daylight available is least when the sky is heavily overcast, and it is this type of sky which is accepted in this country as the basis for daylighting calculations and the framing of minimum standards. The overcast sky is amenable to mathematical analysis because its luminance (i.e. brightness) varies regularly from the horizon to the zenith and does not vary appreciably in azimuth. Fig. 2 shows that the range of luminance from horizon to zenith of a densely overcast sky is about 1 : 3. Because of the uniformity of luminance horizontally no account need be taken of the orientation of a particular window, and simple methods can be used to allow for variations of sky luminance, according to the altitude of the patch of sky visible from the room.

The illumination provided by the sky varies throughout the day, but for a heavily overcast sky it is most likely to be in the region of 5000 lux. This value is generally assumed as the basis for daylighting calculations in this country, but it is in fact exceeded for about 85 per cent of the normal working time throughout the year.

The daylight factor

The amount of light received inside a building is usually only a small fraction of that received out of doors from the whole sky, and it varies with the outdoor illumination. It is therefore impracticable to express interior daylighting in terms of the illumination actually attainable indoors at any one time, for within a few minutes, perhaps, that figure will change with changes in the luminance of the sky. For practical purposes, therefore, use is made of the daylight factor, a ratio which can be expressed as:

$$\frac{\text{Instantaneous illumination indoors}}{\substack{\text{Simultaneously occurring} \\ \text{illumination outdoors}}} \times 100\%$$

(direct sunlight is excluded from both indoor and outdoor illuminations).

Fig. 3 shows that daylight reaching a reference point in a room can be made up of three components:

(1) *The sky component:* the light received directly from the sky.

(2) *The externally reflected component:* light received after reflection from the ground, buildings or other external surfaces.

(3) *The internally reflected component:* the light received after being reflected from surfaces inside the room.

In calculating the daylight factor the three components of daylight are calculated separately and simply added together. Corrections for glazing materials other than clear glass, dirt on glass and reductions caused by the window framing can be applied to the total daylight factor as described in Digest 42.

Much of the work in calculating (1) and (2) can be avoided by the use of the simple aids described below. Calculation of (3) is usually rather more complex but can be made easier by the methods described in Digest 42.

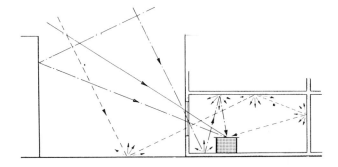

Fig. 3. Ways in which daylight reaches indoor position.

Sky component

Three of the aids in common use in this country for calculating the sky component are:

The BRS daylight tables

The grid method known as the 'Waldram Diagram'

The BRS Daylight Factor Protractors.

BRS simplified daylight tables

These tables, published in full in *National Building Studies Special Report 26*, have been prepared to enable all the components of the daylight factor to be determined when scale drawings are not readily available, and are thus most appropriate for use in the early stages of a design. Table 1, reproduced here, gives sky components from an overcast sky for rooms with vertical rectangular windows glazed with clear clean glass. Allowance can be made for simple external obstructions, the use of other types of glass, the presence of dust on the glass, etc., as explained in the next Digest.

The following information is needed to use the table:

H, the effective height of the window head above the working plane after allowing for any obstructions;

W_1, W_2, the effective widths of the window on each side of a line drawn from the reference point normal to the plane of the window, taken separately;

D, the distance from the reference point to the plane of the window.

The ratios H/D, W_1/D and W_2/D can be worked out and the sky components can then be read directly from the table. In general, the sky component at any other reference point can be obtained by addition or subtraction. Comprehensive details for the use of the table are given in the Special Report referred to.

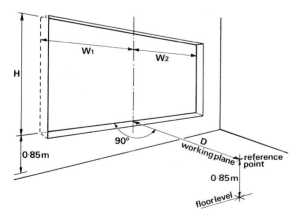

Fig. 4. Application of BRS simplified daylight tables

154

Waldram diagram

Grid methods of calculating sky components can be applied with accuracy to a wide variety of circumstances but are inclined to be tedious. They are particularly useful when obstructions and the window are complex in outline. The most widely used grids are based on the Waldram diagram, which, in its latest form (Fig. 5), is based on the luminance distribution of an overcast sky and allows for glass losses.

The diagram comprises a grid representing 50 per cent sky component (i.e. half the hemisphere of sky) and is so constructed that equal areas of grid represent equal sky components. The area of sky visible through the window from the reference point is plotted on the grid in lines of angular co-ordinates from information obtained from a scaled plan and section of the room being examined. The area of the patch of sky plotted on the diagram is then proportional to the sky component at the reference point.

The diagram can be of any convenient height and length but in practice it is most convenient to make it some simple multiple of 50 units in area. As the complete diagram represents half the hemisphere of unobstructed sky, one unit represents 1 per cent sky component. The outline of window and external obstruction is distorted in shape when transferred to the diagram but the area enclosed relative to the area of the whole grid gives an accurate direct measure of the sky component. The superimposed curved lines on the grid—known as droop lines—correspond to the horizontal edges of obstructions parallel to and at right angles to the plane of the window.

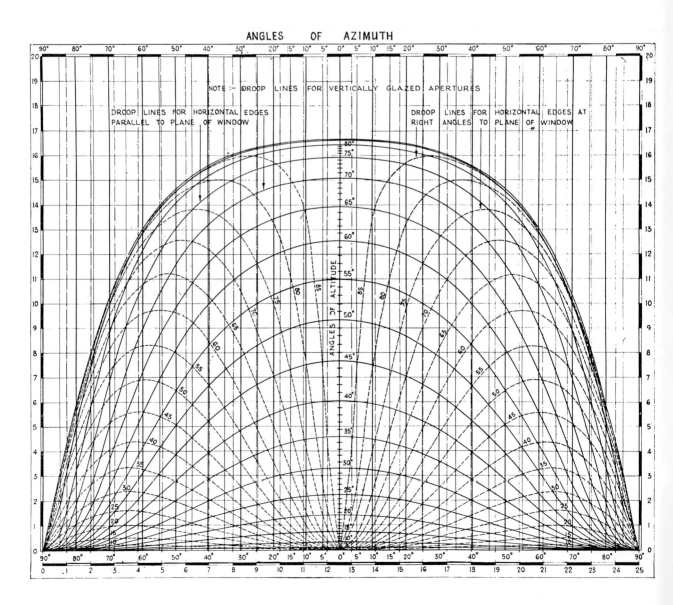

Fig. 5. Waldram diagram for CIE. Overcast sky—corrected for glass losses.

BRS daylight factor protractors

Of the methods for determining the sky components at the design stage, the Daylight Factor Protractors designed by the Building Research Station are the most widely used by designers in this country. The protractors are quicker to use than grid methods although they may not be so accurate when the obstructions and window are of complicated outline. For the majority of cases it is usually possible to assume an average simple outline for external obstructions without serious loss of accuracy.

The protractors were designed originally for use with drawings to determine the sky components from a sky of uniform luminance; a correction in the luminance of the CIE overcast sky was then made by reference to Fig. 2 or published tables. The revised protractors now available (second series) include a

series which incorporates an allowance for the CIE overcast sky and thus can be applied to obtain sky components from such a sky without further correction. A full description of these protractors is given in *BRS Daylight Protractors* by J. Longmore, published by HMSO, 1968.

The new protractors are circular, two semi-circular scales, one for sky component and an auxiliary scale to correct for windows of finite length, being combined in the same instrument. An ordinary angle protractor is included to help in determining the required angle of elevation of the patch of sky to be estimated. The new protractors have been prepared on the same principle as the original series, but more accurate techniques have been used in their preparation and more recent data on glass transmission have been incorporated.

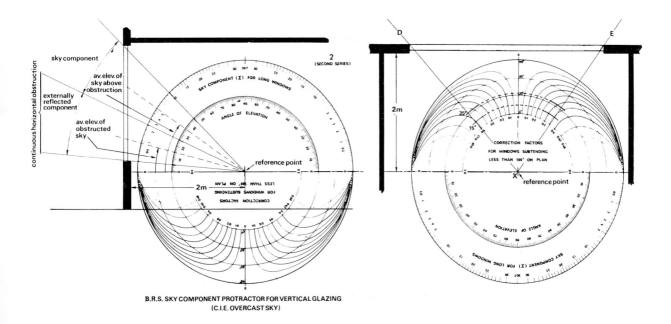

B.R.S. SKY COMPONENT PROTRACTOR FOR VERTICAL GLAZING
(C.I.E. OVERCAST SKY)

Fig. 6. Building Research Station protractors.

The full set of protractors comprises:

Protractor	Type of sky	Slope of glazing
1	Uniform	Vertical
2	CIE	Vertical
3	Uniform	Horizontal
4	CIE	Horizontal
5	Uniform	30°
6	CIE	30°
7	Uniform	60°
8	CIE	60°
9	Uniform	Unglazed
10	CIE	Unglazed

Example—Application of protractor 2

Assume that the room being examined has one rectangular glazed window and that there is a long horizontal obstruction, parallel with the plane of the window (see Fig. 6).

On a section of the room drawn to *any scale*, draw sight-lines from the reference point to the horizontal boundary edges of the patch of sky visible through the window. In this case, the lines are drawn to the head of the window and to the top edge of the external horizontal obstruction. Place the primary scale on the section, with its centre at the reference point and its base along the working plane, and note the two values where the sight-lines intersect the protractor scale. The *difference* between the two readings, 9·8 per cent and 2·5 per cent in this example, gives the sky component for a window of infinite length, i.e. 7·3 per cent.

This sky component value must be corrected for the finite length of the window by applying the auxiliary scale on a plan of the room drawn to any scale. First, however, find the *average* angle of altitude of the sight-lines on the section—about 35° in the example. Next, take the plan and draw two further sight-lines (*XD* and *XE*) from the reference point to the vertical boundary edges of the patch of sky—in this case, the sides of the window opening. Place auxiliary scale on the plan, with its centre at the reference point and base parallel to the plane of the window. Observe where the sight-lines intersect the semicircular scale on the auxiliary protractor corresponding to the average angle of altitude of 35°. The *sum* of these two readings, 0·38 and 0·33 in this example, gives the correction factor for the length of window shown, i.e. 0·71.

> From readings on the primary scale, sky component for a window of infinite length = 9·8−2·5 = 7·3%.
>
> Average angle of elevation of unobstructed sky = 35° from readings on the auxiliary scale, correction factor = 0·38 + 0·33 = 0·71.
>
> Sky component for windows shown in Fig. 6 = 7·3×0·71 = 5·2%*.

> *Note*—In practice there is little to be gained by working to more than two significant figures.

Externally reflected component

The externally reflected component can be calculated by considering the external obstructions visible from the reference point as a patch of 'sky' whose luminance is some fraction of that of the sky obscured. In other words, the 'equivalent sky component' is first calculated by one of the methods already described and is then converted to the externally reflected component by allowing for the reduced luminance of the obstructing surfaces compared with the luminance of the sky. The luminance of obstructions varies widely in practice but, unless the actual value is known, it is assumed to be uniform, with a luminance of one-tenth of the *average luminance* of the sky.

When using the protractors for a uniform sky it is sufficient to calculate equivalent sky component and divide by 10. When using the protractors applicable to the CIE overcast sky, however, the equivalent sky component is divided by 5 instead of 10. This is because the luminance of a densely overcast sky near the horizon is approximately half the average luminance and this correction has already been incorporated in the protractors for the overcast sky. This correction also applies when using the BRS Sky Component Table and the Waldram Diagram modified for overcast sky, for obstructions below 20°. Thus, in the example in Fig. 6 using Protractor No. 2 applicable to vertical glazing and the CIE sky:

> Readings on primary scale corresponding to sight-lines drawn to the sill and the top of the obstruction, are 0·10% and 2·50%.
>
> Equivalent sky component = 2·50−0·10 = 2·40%.
>
> Average angle of elevation of obstruction = 15°.
>
> Readings on auxiliary scale are 0·41 and 0·36.
>
> Correction factor = 0·41 + 0·36 = 0·77.
>
> Equivalent sky component for obstructed part of sky visible from reference point = 2·4×0·77 = 1·85%.
>
> Assuming luminance of obstruction to be one-fifth that of the overcast sky, externally reflected component = $\frac{1·85}{5}$ = 0·37%.

Estimating Daylight in Buildings—2

The internally reflected component

Having calculated the amount of light reaching a reference point, first from the sky directly, and secondly, after reflection from surfaces outside the window by one of the methods described in the previous Digest, the next stage is to calculate the light reaching the reference point after reflection and inter-reflection from surfaces inside the room. This depends on the reflection factors of the walls, ceiling and floor, and on the amount of light received on them from the sky and the obstructions and ground outside. This process of reflection and inter-reflection is complex, but three methods of calculation have been devised which are sufficiently accurate and yet simple enough for general application:

(1) The inter-reflection formula.

(2) Tables based on the formula.

(3) Nomograms based on a modified form of the formula.

Table 1

Effect of external obstruction on internally reflected component

Angle of obstruction measured from centre of window (degrees above horizontal)	Coefficient C (see equation 1)
No obstruction	39
10°	35
20°	31
30°	25
40°	20
50°	14
60°	10
70°	7
80°	5

In practice, the amount of inter-reflected light varies according to the distance of the reference point from the window. For most purposes, however, it is sufficient to assume an *average* internally reflected component over the greater part of the room with a lower *minimum* value at points far from the window. To comply with statutory regulations the minimum internally reflected component should be deter-mined. In all the methods, with the exception of Nomogram III, values for both average and minimum internally reflected components may be calculated.

The BRS inter-reflection formula

The formula is applicable where relatively high accuracy is required in estimating the internally reflected component of daylight factor for side-lit rooms. It is given in the form:

Average internally reflected component

$$= \frac{0 \cdot 85 W}{A\,(1-R)} \times (\mathrm{CR_{fw}} + 5\mathrm{R_{cw}})\,\% \ldots\ldots(1)$$

where

W = Area of window

A = Total area of ceiling, floor and walls including area of window

R = Average reflectance of ceiling, floor and all walls, including window, expressed as a fraction

R_{fw} = Average reflectance of the floor and those parts of the walls below the plane of the mid-height of the window (excluding the window wall)

R_{cw} = Average reflectance of the ceiling and those parts of the walls above the plane of the mid-height of the window (excluding the window wall)

C = A coefficient having values dependent on the obstruction outside the window.

Table 1 gives values of C for various obstructions. The values of C in the table depend on the following assumptions:

(*a*) The sky luminance distribution is that of the CIE overcast sky.

(*b*) The ground and any obstructions have a luminance one-tenth of the average sky luminance.

(*c*) The obstructions are continuous and are horizontal and parallel to the window wall. (If the obstructions do not fulfil these conditions, an estimate should be made of the equivalent horizontal obstruction.)

(*d*) The glass losses are those appropriate to clean clear window glass.

Where it is required to find the *minimum* internally reflected component of the daylight factor, the conversion factors in Table 2 can be applied to the average internally reflected component (R is expressed here as a percentage instead of a fraction).

Table 2

Conversion of average to minimum internally reflected component

Average reflectance R(%)	Conversion factor
30	0·54
40	0·67
50	0·78
60	0·85

The BRS tables

Table 3 was formulated to allow rapid assessment of the *minimum* internally reflected component of the daylight factor where certain limitations about the size of room and scheme of decoration can be accepted. The table was designed primarily for rooms about 40 m² in floor area and 3 m ceiling height, with a window on one side extending from a sill height of 0·9 m to the ceiling; by means of simple conversion factors the internally reflected component can be calculated for rooms of 10 and 90 m² floor area, with ceiling heights ranging from 2·5 to 4 m.

Changes in the height of sill and of window head will not as a rule affect the estimate of the internally reflected component, neither will the shape of the room, so long as the ratio of length to width does not exceed 2 : 1.

In Table 3, the ceiling reflection factor is assumed to be 70 per cent, but other values can be allowed for by means of conversion factors. A continuous horizontal external obstruction of 20° measured from the centre of the window is assumed. Where there is more than one window in a room, the internally reflected components are calculated separately for each window and then added together, remembering that the minimum internally reflected component for each window occurs at points in the room farthest from the window. For very large rooms, and/or different external obstructions, the inter-reflection formula or the nomograms should be used.

For rooms of 10 m² and 90 m² the following conversion factors can be applied to the results obtained by Table 3; conversion factors for intermediate sizes can be obtained by interpolation.

Floor area	Wall reflection factor, %			
	20	40	60	80
10 m²	0·6	0·7	0·8	0·9
90 m²	1·4	1·2	1·0	0·9

For rooms with ceiling reflection factors other than the 70 per cent assumed in Table 3, the following conversion factors can be applied:

Ceiling reflection factor %	Conversion factor
40	0·7
50	0·8
60	0·9
70	1·0
80	1·1

The values given in Table 3 are *minimum* values at points farthest from the window. Where the *average* value of the internally reflected component is required the following conversion factors are applied:

Wall reflection factor %	Conversion factor
20	1·8
40	1·4
60	1·3
80	1·2

The BRS nomograms

Three nomograms have been devised for determining the internally reflected component of the daylight factor:

Nomogram I for the *average* internally reflected component in side-lit rooms.

Nomogram II for the *minimum* internally reflected component in side-lit rooms.

Nomogram III for the *average* internally reflected component in top-lit rooms.

They offer a more comprehensive method of computation than the tables although their precision is not so great as the original formula from which they

Table 3

Minimum internally reflected component of daylight factor (%)*

*Assuming ceiling reflection factor=70%, angle of external obstruction=20°.

Window area as percentage of floor area	Floor reflection factor											
	10%				20%				40%			
	Average wall reflection factor (excluding window)											
	20%	40%	60%	80%	20%	40%	60%	80%	20%	40%	60%	80%
2	—	—	0·1	0·2	—	0·1	0·1	0·2	—	0·1	0·2	0·2
5	0·1	0·1	0·2	0·4	0·1	0·2	0·3	0·5	0·1	0·2	0·4	0·6
7	0·1	0·2	0·3	0·5	0·1	0·2	0·4	0·6	0·2	0·3	0·6	0·8
10	0·1	0·2	0·4	0·7	0·2	0·3	0·6	0·9	0·3	0·5	0·8	1·2
15	0·2	0·4	0·6	1·0	0·2	0·5	0·8	1·3	0·4	0·7	1·1	1·7
20	0·2	0·5	0·8	1·4	0·3	0·6	1·1	1·7	0·5	0·9	1·5	2·3
25	0·3	0·6	1·0	1·7	0·4	0·8	1·3	2·0	0·6	1·1	1·8	2·8
30	0·3	0·7	1·2	2·0	0·5	0·9	1·5	2·4	0·8	1·3	2·1	3·3
35	0·4	0·8	1·4	2·3	0·5	1·0	1·8	2·8	0·9	1·5	2·4	3·8
40	0·5	0·9	1·6	2·6	0·6	1·2	2·0	3·1	1·0	1·7	2·7	4·2
45	0·5	1·0	1·8	2·9	0·7	1·3	2·2	3·4	1·2	1·9	3·0	4·6
50	0·6	1·1	1·9	3·1	0·8	1·4	2·3	3·7	1·3	2·1	3·2	4·9

Table 4

Use of Nomogram III for horizontal rooflights

Variation of function K with angles of obstruction from centres of rooflight (degrees above horizontal)

Obstruction on one side of rooflight	Obstruction on second side of rooflight								
	0°	10°	20°	30°	40°	50°	60°	70°	80°
0°	88	87	87	85	82	78	72	65	57
10°	87	87	87	85	82	77	71	64	56
20°	87	87	86	85	82	77	71	64	56
30°	85	85	85	83	80	75	69	62	54
40°	82	82	82	80	77	72	66	59	51
50°	78	77	77	75	72	68	62	54	47
60°	72	71	71	69	66	62	56	48	41
70°	65	64	64	62	59	54	48	41	33
80°	57	56	56	54	51	47	41	33	25

are derived. For the majority of cases, however, the nomograms provide probably the most useful method of calculation, especially as one of them enables calculations for rooflights to be made.

Nomograms I and II are designed on the following assumptions: (a) that the reflection factors of the ceiling and floor are 70 and 15 per cent respectively, and (b) that the external obstructions are horizontal and continuous in relation to the window and have a luminance one-tenth the average sky luminance. If the obstructions are not horizontal and continuous the equivalent horizontal obstruction should be assumed.

To use the nomograms it is necessary to know the ratio of window area to total surface area (scale A); the average reflection factor of all the interior surfaces (scale B); and where applicable, the angle of the external horizontal obstruction or its equivalent measured from the centre of the window. The average reflection factor can be estimated in a number of cases from the table drawn on Nomogram I (strictly applicable only where the ceiling and floor reflection factors are 70 and 15 per cent respectively and the ratio of window area to total surface area is 0·05).

The following example illustrates the use of Nomogram I.

Assume a room 9 m × 6 m × 3·6 m high.

Reflection factor of internal surfaces:

$$\text{Walls} = 40\%$$
$$\text{Ceiling} = 70\%$$
$$\text{Floor} = 15\%$$

Net glass area in window = 15 m²

Angle of external obstruction (as measured from centre of window) = 40°

Total area of internal surfaces (including window) = 2 (9×3·6 + 6×3·6 + 9×6) = 216 m²

$$\text{Ratio,} \frac{\text{window area}}{\text{total area}} = \frac{15}{216} = 0\cdot069$$

Total wall area = 2 (9×3·6 + 6×3·6) = 108 m²

$$\text{Ratio,} \frac{\text{wall area}}{\text{total area}} = \frac{108}{216} = 0\cdot5.$$

The average reflection factor of the room, from the table in Nomogram I, is seen to be mid-way between 36 and 44 = 40 per cent. Lay a straight-edge across Nomogram I from the graduation of 0·069 on scale A (ratio of window area: total surface area) to the graduation of 40 per cent on scale B (average reflection factor). Note where the line intersects scale C, in this case 1·2 per cent. If there is no external obstruction, this figure represents the average internally reflected component. As there is an external obstruction of 40°, place the straight-edge across the graduation of 1·2 per cent on scale C to the graduation of 40° on scale D. The straight-edge intersects scale E at 0·67 per cent which is the *average* internally reflected component for the room with a 40° external obstruction.

Nomogram III for rooflights is similar to the other nomograms but introduces the factor K, given in Tables 4–7, to allow for various angles of obstruction. The following example explains the procedure to be followed using Nomogram III.

Consider the shed roof building shown in Fig. 1, 12 m × 9 m × 4·2 m high (average).
Reflection factor of internal surfaces:

$$\text{Walls} = 30\%$$
$$\text{Roof} = 70\%$$
$$\text{Floor} = 15\%$$

Table 5

Use of Nomogram III for rooflights sloping at 30° to the horizontal (see Fig. 1)

Variation of function K with angles of obstruction from centre of rooflight (degrees above horizontal)

Obstruction facing slope of glazing	Obstruction opposed to slope of glazing					
	0°–30°	40°	50°	60°	70°	80°
0°	82	82	81	79	74	69
10°	81	80	80	77	73	67
20°	78	78	77	74	70	64
30°	74	73	72	70	66	60
40°	68	68	67	64	60	54
50°	61	61	60	58	53	48
60°	54	54	53	50	46	40
70°	46	45	45	42	38	32
80°	37	37	36	33	29	23

Net area of roof glazing = 19·8 m²
Slope of glazing = 30° (approx.)
Angle of obstruction facing glazing = 20°
Angle of obstruction opposing glazing = 35°
Total area of internal surfaces, including roof light =
  2 (12×3 + 9×4·2) + (12×9) + 2 (12×5·2)
  = 147·6 + 108 + 124·8 = 380·4 m²

Ratio, $\dfrac{\text{roof glazed area}}{\text{total area}} = \dfrac{19\cdot8}{380\cdot4} = 0\cdot052$

Total wall area = 147·6 m²

Ratio, $\dfrac{\text{wall area}}{\text{total area}} = \dfrac{147}{380\cdot4} = 0\cdot39$

Average reflection factor = 37·0%
K value, from Table 5 = 78.

Lay a straight-edge across Nomogram III between the graduation 0·052 on scale *A* (ratio of roof glazing : total surface area) to the graduation of 37·0 per cent on scale *B* (average reflection factor of room); mark the point where the straight-edge intersects scale *C*—in this case 6·00—then run the straight-edge from this point to the value 78 on scale *D* (value K, Table 5). The straight-edge intersects scale *E* at 2·3 per cent, which is the average internally reflected component of the daylight factor.

Additional corrections

Having estimated the three components of daylight factor, the separate values are simply added together. Corrections may need to be applied to allow for deterioration of the reflectance of the decorations, dust or dirt on the glazing, types of glazing material other than clear glass, and obstructions caused by the window framing. With the exception of deterioration of the decorations it is usually more convenient to apply these corrections to the sum of the three components, i.e. the daylight factor. In all cases the product of the correction factors is taken.

Table 6

Use of Nomogram III for rooflights sloping at 60° to the horizontal

Variation of function K with angles of obstruction from centre of sloping rooflight (degrees above horizontal)

Obstruction facing slope of glazing	Obstruction opposed to slope of glazing		
	0°–60°	70°	80°
0°	70	70	68
10°	65	65	64
20°	59	59	58
30°	52	52	51
40°	44	44	43
50°	36	36	34
60°	28	28	26
70°	21	21	20
80°	16	16	15

Deterioration of the decorations

Except for domestic buildings, an allowance for deterioration of the decorations may need to be made. An average correction factor of 0·75 is usually assumed and is applied to the value for the internally

reflected component only. Alternatively, where the location and type of work done in the building are known the following maintenance correction factors may be used:

Type of location	Type of work	Depreciation correction factor
Clean	Clean	0·9
Dirty	Clean	0·8
Clean	Dirty	0·7
Dirty	Dirty	0·6

Allowance for dirt on glass

The allowance for dirt on glass is applied to the total daylight factor. In practice it will vary depending on the degree of air pollution, the slope of glazing, and how often the windows are cleaned. In rural areas and for domestic buildings little or no allowance is usually necessary, but the rate of dirt deposition for factories and offices in the centres of cities may be rapid and there may be infrequent cleaning. The following correction factors are suggested according to the location and slope of the glazing:

Type of location	Vertical glazing	Sloping glazing	Horizontal glazing
Clean	0·9	0·8	0·7
Dirty (e.g. industrial)	0·7	0·6	0·5
Very dirty	0·6	0·5	0·4

Allowance for window framing and window bars

This can be made by multiplying the total daylight factor by the ratio:

$$\frac{\text{actual glass area of typical window}}{\text{area of window aperture}}$$

Where the net area of glass is difficult to estimate it is often sufficient to allow a correction factor of 0·75.

Allowances for alternative types of glazing materials

The methods described in this Digest and the previous one, with the exception of BRS Daylight

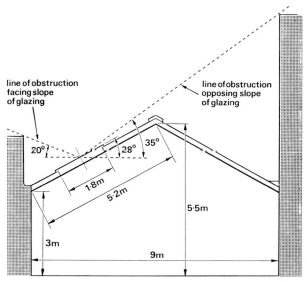

Fig. 1. Example with sloping rooflight, for application of Nomogram III

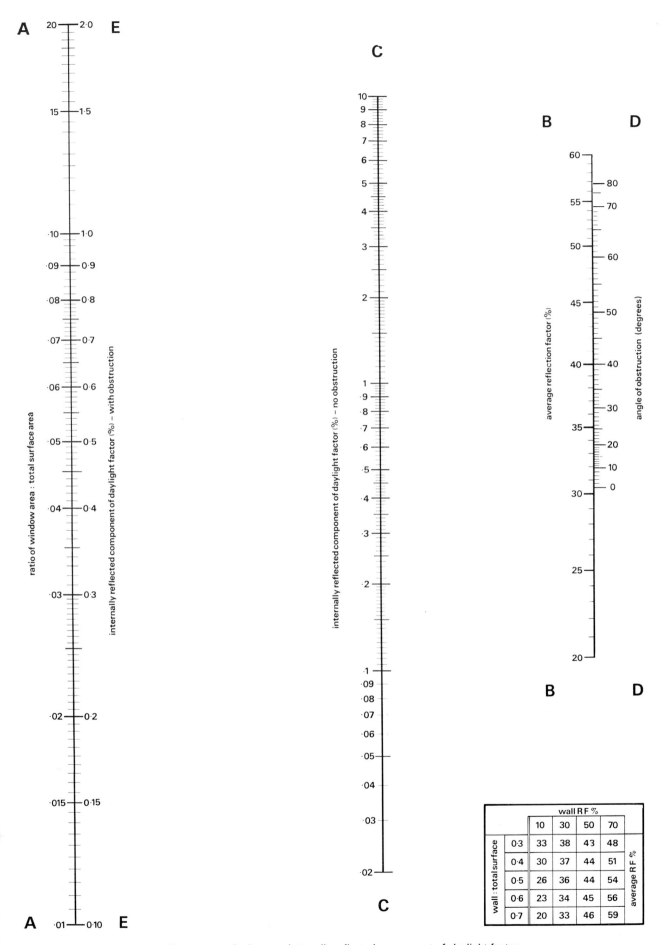

Nomogram I. Average internally reflected component of daylight factor

161

162

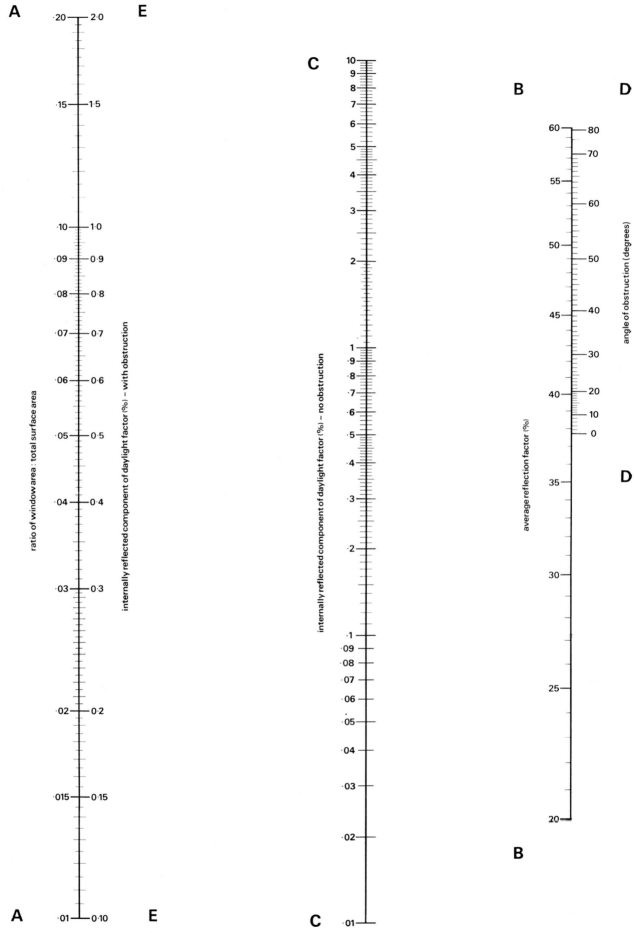

Nomogram II. Minimum internally reflected component of daylight factor

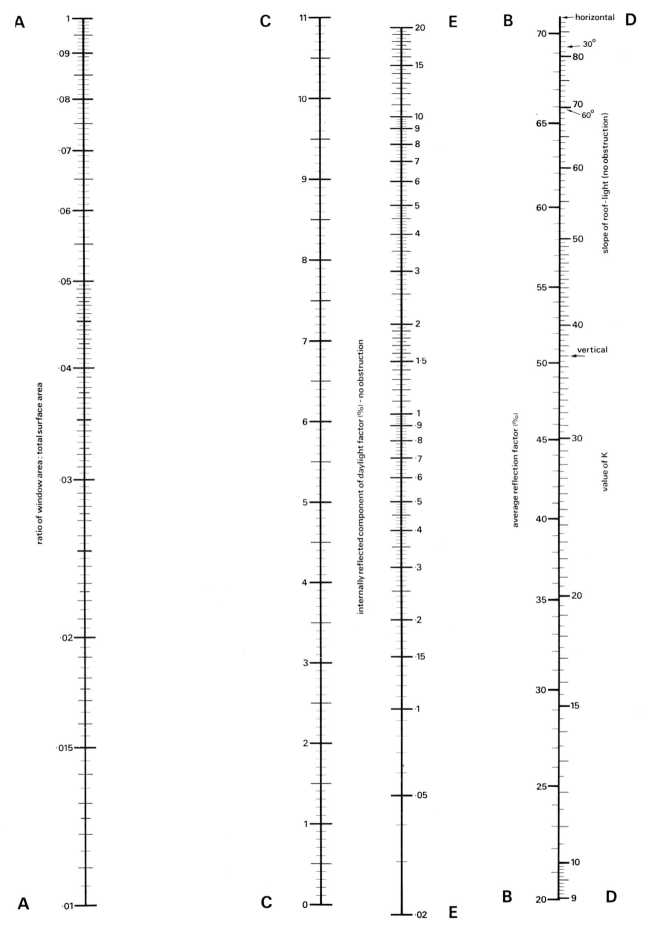

Nomogram III. Average internally reflected component of daylight factor for top-lit rooms

164

Table 7

Use of Nomogram III for vertically glazed rooflights

Variation of function K with angle of obstruction (overcast sky conditions)

Angle of obstruction from centre of window (degrees above horizontal)	K
0° (No obstruction)	37
10°	34
20°	30
30°	26
40°	21
50°	16
60°	12
70°	10
80°	9

Protractors Nos. 9–10, have incorporated an allowance for the transmission losses when ordinary clean clear sheet glass or plate glass is used. Further corrections to the daylight factor may be required to allow for other types of glazing material.

The generally accepted average diffuse light transmission values for some of the alternative glazing materials are given in Table 8 together with the correction factor to be applied in each case to the total daylight factors calculated as above.

When *double* glazing in ordinary clean clear glass is used, the transmission will vary with the angle of the incident light, but for most practical purposes a correction factor of 0·9 applied to the total daylight factor will suffice.

Alternative method

A slide rule known as a Daylight Factor Calculator enables daylight factors to be read off directly for a wide range of wall, ceiling and floor reflection factors and ceiling heights. The slide rule and an example of its operation are illustrated in Principles of Modern Building, Vol. 1.

Table 8

Correction factors for various glazing materials

Material	Diffuse transmission factor	Recommended correction factor to apply to daylight factor
Transparent glasses		
Flat drawn sheet		1·0
6 mm Polished plate		1·0
6 mm Polished wired		0·95
Patterned and diffusing glasses		
3 mm Rolled		0·95
6 mm Rough cast		0·95
6 mm Wired cast		0·90
Cathedral	The transmission factors have been omitted in accordance with the amendment which was published in the 1970 edition of Digest 41	1·00
Hammered		1·00
Arctic		0·95
Reeded		0·95
Small Morocco		0·90
Special glasses		
Heat-absorbing tinted plate		0·90
Heat-absorbing tinted cast		0·60–0·75
Laminated insulating glass		0·60–0·70
Plastic sheets		
Corrugated resin-bonded glass-fibre reinforced roofing sheets		
Moderately diffusing		0·90
Heavily diffusing		0·75–0·90
Very heavily diffusing		0·65–0·80
Diffusing opal 3 mm acrylic plastic sheets		0·65–0·90

5 Electricity

Electricity distribution on sites

Digests 87 and 88 (1967) described the use of 'cubicles' of various patterns and functions to distribute electricity on building sites. Supply, distribution, transformer and outlet units of this type are now specified in a British Standard (BS 4363:1968) and their use is fully described in CP 1017:1969. The two earlier digests are therefore withdrawn and the present one is restricted to a brief description of the standard units and their use on building and construction sites.

Safety requirements and supply voltage

Construction work makes considerable use of electrical supplies. The risk of serious accident by electrical causes is greatest in damp conditions, such as are common on building sites; in addition, equipment and cabling are subject to frequent repositioning, and this increases the danger.

For site lighting other than floodlighting, and for portable and hand-held tools and lamps, the accident risk is reduced if a lower operating voltage is employed. A 110 V supply, with the centre-point of the secondary winding of a transformer earthed, so that the nominal voltage of the sub-circuit to earth (the usual shock current path) is only 55 V, has been selected as the most satisfactory compromise between safety requirements and working efficiency. It is still necessary even at this voltage, to take precautions, which should include a high standard of routine maintenance and periodic testing of all apparatus used.

With a properly designed system, voltage drop is not serious and the slightly heavier cables needed do not prove to be a disadvantage. It is now possible to obtain the whole range of portable power tools for 110 V working.

Contract planning

Electrical supplies for construction sites should be planned with as much care as is normally given to permanent installations. Electricity is required for motive power, heating and lighting for construction work and for space and water heating, cooking and lighting for staff and office accommodation. Demands on electricity are extensive, as the examples in Table 1 show; smaller portable tools such as saws, hammers, drills, sanders and grinding wheels are also widely used. The use of electricity for aggregate heating and drying-out processes to facilitate winter working and speedier construction is increasing; site lighting is an appreciable part of the total load in winter months.

Depending on the total load requirement for the site, and by arrangement with the Area Board, a 415 V supply for the site can be provided either direct or

Table 1. Electrically operated plant (415V three phase)

Plant	kW range
Tower crane hoist motor	2·25–60
Tower crane slew motor	1·00–11·00
Tower crane travel motor	0·75–7·50
Tower crane crab or derricking motor	0·75–7·50
Weigh-batching unit	6·00–15·00
Concrete mixer	1·00–15·00
Goods hoist	1·50–33·50
Passenger/goods hoist	5·50–33·50
Compressor	5·50–75·00
Conveyor	2·25–7·50
Pump	0·25–30·00
Timber sawbench	0·75–7·50
Concrete saw	0·75–2·25
Floor grinder and polisher	0·375–2·25

through transformers from a higher voltage intake. In either case, the contractor's responsibility begins at the entry of the 415 V supply to the site.

Site programme for the contractor

1 Arrange pre-contract meeting between responsible executives associated with the work.

2 Decide on electrical requirements of heavy plant, eg tower crane, hoist, compressor, concrete mixer. Prepare layout drawing showing siting and total electrical load. Decide position and layout of site offices, stores, canteen etc.

3 Apply to Area Board for supply. This might range from 10 to 300 kVA.

4 Provide electrical distribution equipment as follows:

415 V, 3-phase, 50 cycle, for fixed plant and movable plant fed by trailing cable.

240 V, single-phase, 50 cycle, for site offices and floodlighting.

110 V, single and three phase, 50 cycle, for local site lighting, portable tools and portable handlamps for general use.

50 V or 25 V single-phase, 50 cycle, for portable hand-lamps for use in confined and damp situations.

Carefully site these installations to minimise interference with construction work as it proceeds.

5 Following the provision of site staff accommodation, eg offices, canteen, toilets, stores, ensure that a satisfactory semi-permanent electrical installation is provided.

6 Connect heavy plant, eg tower crane, hoists, concrete mixers, compressors, to the distribution centre indicated at 4 above.

7 Install floodlighting, if required.

8 As building progresses, a responsible electrical engineer should advise on reduced voltage distribution for local site lighting and for the use of portable tools.

9 As the work pattern changes, a periodic review of electrical load deployment should be made. If this is done either monthly, or preferably weekly, additional ordering of equipment can often be avoided by re-deployment of existing distribution equipment. Any equipment not in full use should be returned to store.

The equipment should be robust, well designed and reusable from site to site. The equipment described here has overall advantages of safety, efficiency and a high recovery value at the end of a contract if returned in good condition.

Unless sub-contractors specialising in this type of work are employed, the main contractor should employ staff to carry out the whole of the installation work or at least a nucleus to supervise it and satisfy themselves on the type of equipment, cabling, etc, to be used. Merely to write a specification and hand it to an electrical sub-contractor is not enough.

Distribution equipment

Electricity supplies should be distributed to the building site through distribution units meeting the requirements of BS 4363:1968. The units comprise the following:

Supply incoming unit (SIU) A unit to house the Supply Undertaking's incoming cable, service fuses, neutral link, current transformers and metering equipment and with provision for one outgoing circuit of 300 A, 200 A or 100 A maximum current controlled by switch and fuse or circuit-breaker.

Main distribution unit (MDU) Equipment for control and distribution at voltages up to 415 V three-phase and 240 V single-phase and fitted with 300 A, 200 A or 100 A isolator, lockable in the OFF position. The available outgoing supplies are controlled by moulded case circuit-breakers (MCCB).

Supply incoming and distribution unit (SIDU) The two foregoing units may be combined in a single unit.

Transformer unit (TU) A unit incorporating a transformer and with provision to distribute electricity at reduced voltage (usually 110 V); available for 240 V single-phase input (TU/1), or 415 V three-phase input (TU/3), or both (TU/1/3).

Outlet unit (OU) Facilities for the control, protection (by miniature circuit breakers) and connection of final sub-circuits, fed from a 32 A supply and operating at 110 V. The single-phase outlet unit (OU/1) has up to eight 16 A double-pole socket outlets; to either, a 32 A socket-outlet may be added but this is not controlled by a circuit-breaker.

Extension outlet unit (EOU) An extension outlet unit is similar to an outlet unit but is fed from a 16 A supply and does not incorporate protection by circuit-breakers. The single-phase extension outlet (EOU/1) has up to four 16 A double-pole socket-outlets and the three-phase unit (EOU/3) has two triple-pole socket-outlets.

Earth monitor unit (EMU) Flexible cables used to supply power at mains voltage to movable plant incorporate a separate pilot conductor in addition to the main earth continuity conductor. A very low voltage current passes between the portable equipment and a fixed monitoring device (EMU) through the pilot conductor and the earth continuity conductor. Failure of the latter will interrupt the current flow; the

Fig. 1 Units for the distribution of electricity on sites
(Photo by courtesy of Wysepower Ltd.)

fixed monitoring device will detect this and automatically isolate the circuit.

Earth leakage circuit breaker A circuit-breaker with an operating coil which trips the breaker when the current due to earth leakage exceeds a predetermined safe value (30 mA is normally recommended).

Plugs and socket-outlets

The identifying colour for plugs and socket-outlets used in 110 V circuits is *yellow*. Colour identification of accessories operating at other voltages should be as shown in Table 2. Accessories should meet the requirements of BS 4343.1968 *Industrial plugs, socket-outlets and couplers for AC and DC supplies* which is based on International Standard CEE 17.

Cabling

For semi-permanent parts of the installation, eg site offices and ancillary buildings, the IEE *Regulations for the electrical equipment of buildings* state all the necessary requirements.

Cables should have a metal sheath and/or armour,

which must be continuously earthed in addition to the earth core of the cable. The earthed sheath or armour must not be used as the sole earth conductor. Except where the risk of mechanical damage is slight, the cables should have an oversheath of pvc or oil-resisting and flame-retardant compound. If the voltage applied to a cable will not normally exceed 65 V, it may be of a type insulated and sheathed with a general purpose or heat-resisting elastomer.

Cables should be fixed on site so as to be clear of constructional operations and not to be a hazard to operatives. They should be kept clear of passageways, walkways, ladders, stairs and the like and should be at least 150mm clear of piped services such as steam, gas and water. Where they pass under roadways and access routes for transport, they should be laid in ducts at a minimum depth of 0·6m with a marker at each end of the crossing.

The use of overhead cables is deprecated but, if used, they should be fixed at a minimum height of 5·8m, or 5·2m in areas where motor transport and mobile plant are prohibited. They should be marked conspicuously either by yellow and black binding tape or with freely-moving fabric or plastics strips.

Maintenance

Site work is, of necessity, in a constant state of change and because of this the associated electrical installation is subject to risk of damage or misuse. Testing, strict maintenance and frequent checking of control apparatus and the wiring distribution system, by a competent person, are essential to promote safety and efficient operation.

Table 2 Identification colours

Operating voltage	Colour
25	violet
50	white
220–240	blue
380–415	red
500–650	black

Layout

Fig 2 shows a typical site layout using the distribution units specified in BS 4363. Fig 3 shows in greater detail how they could be used to provide supplies to a multi-storey building during construction. General floor lighting, access and security lighting are connected direct to transformer units. 1·5 to 3·0 kVA heavy portable tools can be connected through a 32 A socket-outlet on an outlet unit. In a layout such as Fig 3, it may be convenient to combine the supply incoming unit (SIU) and the main distribution unit (MDU) in a supply incoming and distribution unit (SIDU).

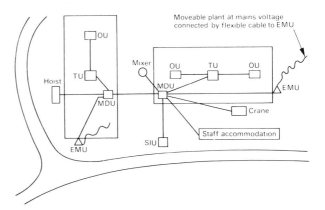

Fig. 2 Typical site arrangement

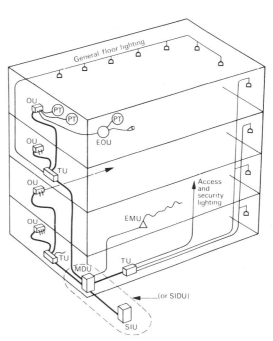

Fig. 3 Typical arrangement of distribution units

Further information

For further details, including references to the various sets of regulations which must be complied with, see BS 4363:1968 *Distribution units for electricity supplies for construction and building sites* and BS CP 1017:1969 *Distributing electricity on construction and building sites*.

Key

MDU	Main distribution unit	**EMU**	Earth monitor unit
SIU	Supply incoming unit	**OU**	Outlet unit
SIDU	Supply incoming and distribution unit	**EOU**	Extension outlet unit
TU	Transformer unit	**PT**	Portable tool (up to 1500 W)

6 General

Drying out buildings

The aims in drying out a building are to bring it to a fit state to receive internal finishes and decoration and to avoid damage or disfigurement of work already done. This digest describes some ways of drying out and also methods of testing the condition of walls, floors and joinery. Although economic considerations will usually call for early completion and, hence, rapid drying, excessively fast drying will cause troubles.

A great deal of water is used in the construction of a building, mainly as mixing water for concrete, plaster and mortar. More is all too often introduced with materials taken from stacks; for example, of bricks, blocks and timber, that have not been protected from rain. Further rain is likely to fall on partially completed work, before the building is roofed in or otherwise made weathertight. Although in favourable conditions the drying process begins as soon as the wet work is placed, there may remain some thousands of litres of water to be dried after completion of the structure.

The heat taken to evaporate five litres of water, even if the temperature of the building never rises above 15°C, is at least 3·6 kWh. So a lot of heat and a lot of time is required for drying out.

Certainly the practical problems presented by drying-out set a premium on protection from the weather both of materials on site and of partially completed work.

The main object in drying out a building is to bring it to a fit state to receive internal finishes and decoration. This should avoid troubles such as expansion of woodwork, particularly floors lifting in compression, loss of adhesion of floor finishes, and paint failures on plaster or wood surfaces. A damp building might also menace the health of its occupants and lead to the development of unsightly mould growths.

Unfortunately, excessively rapid drying can produce its own crop of failures of other kinds so that a balance must be struck between speed and caution, (*see* 'Cautions' on page 3).

The drying out of porous materials, such as brick and concrete, takes place in three distinct phases:

First, all the free water evaporates from the surface; this occurs quite quickly.

Next, the water from the larger pores is lost. This takes longer, partly because there is more water within the material than on the surface but also because it takes time for the water vapour to work its way through the labyrinth of pores to the surface of the material.

Finally, water is lost from the fine pores or cells. This is a very slow process that will continue possibly for several years. Absolute dryness is never attained in actual buildings. Brickwork protected from rain might eventually dry to about one per cent moisture content, concrete about three per cent; brickwork and concrete exposed to rain, about five per cent (all by volume). The equilibrium moisture content of wood varies widely according to its position within the building and the standard of heating: it could be as low as six per cent for wood flooring on a heated sub-floor, or as high as 20 per cent, or more, in some roof timbers, expressed as a percentage of the dry weight.

Drying out is concerned mainly with the second phase.

Two important points to remember are that evaporation inevitably cools the material that loses the water and that the rate of evaporation reduces as the temperature falls.

In a laboratory it is possible to evaporate water so quickly from a brick that the water still contained in it freezes and evaporation comes to a virtual halt until the brick warms up again.

It is not uncommon on walking into a finished building to find that it is noticeably cooler than the air outside. This is the result of the materials drying out, and is, incidentally, the principle on which the old-fashioned earthenware butter-cooler worked.

How to dry out the building

The following alternatives are available:

Use natural ventilation by keeping windows open

Natural ventilation will remove the moisture-laden air from a building replacing it with drier air (except in unfavourable weather conditions) from outside which, in turn, will pick up water vapour and remove it from the building. If the windows are not kept open, water will still evaporate from the damp materials but will accumulate until it condenses not only on windows but also on walls and other surfaces. This often leads to the development of unsightly mould growths, which will incur extra cost in cleaning and, in some instances, redecoration.

Use heaters and keep the windows open

Heat not only increases the rate of evaporation from surfaces but also increases the quantity of water vapour that the air will support. Thus, with the same degree of ventilation, more water vapour will be carried to the outside when heaters are used than when they are not. If the windows are closed in an attempt to reduce the cost by conserving heat, this will build up moisture and the effects will be even more severe than if an unheated room is shut up. Flueless heaters burning gas or oil, a combustion product of which is water, should not be used for drying.

Use dehumidifiers and keep the windows closed

Dehumidifiers (*see* page 3), by removing water from the air, induce the evaporation of water into the air from the surfaces of the construction materials, but as a consequence, the materials cool. The refrigeration types of dehumidifier are very inefficient at low temperatures and therefore it is advisable to provide heaters as well; these should be of the 'dry' type.

Equilibrium vapour pressures within a building can easily be achieved by leaving internal doors open; moving the dehumidifiers from room to room is not necessary. Windows must be kept closed or the equipment will endeavour to dry the outside air as well.

Dehumidifiers remove water more quickly from thin timber sections than from thick masonry walls, and some distortion of the lighter units can therefore occur.

How to tell when a building is dry

Timber

The moisture content of timber can be tested with an electrical moisture meter (*see* page 3) to give immediate results that are accurate enough for practical purposes over the range of moisture contents likely to be encountered in normal usage.

Digest 106 *Painting woodwork* suggests that timber should be as near to its eventual moisture content as possible when painted, ie 10–12 per cent for interior joinery.

Floor screeds

Floor screeds can be tested with a hygrometer, as described on page 4. When the hygrometer readings fall to within the range 75 to 80 per cent, the sub-floor can be assumed to be dry enough for flooring to be laid.

An electrical moisture meter, as described for use on timber, can also be used but it should be noted that although a high reading indicates a wet screed, a low one does not necessarily indicate more than surface dryness: below the surface layer, the screed may still be wet. A more reliable indication can be obtained by taking a reading of the screed surface layer, covering the area with impervious sheeting about one metre square for 24 hours and then taking a second reading on the same spot. It is likely that this will be significantly higher than the earlier reading and, if so, it is the reading on which to decide subsequent action. No actual values can be recommended here because readings for this purpose will usually be taken on an arbitrary scale of numbers, the interpretation of which is explained in the instrument manufacturer's instructions. The presence of contaminating salts in the screed, either as deliberate additions of, say, calcium chloride or by accident, may give misleading results. The use of anhydrous copper sulphate (which turns blue in the presence of moisture) or of anhydrous calcium chloride (which liquifies) is not recommended.

Walls

As with concrete screeds, the surface condition of plaster or brickwork is not a reliable guide to the moisture content of the material behind the surface. It may sometimes be possible to estimate moisture content and suitability for painting from a knowledge of the weather conditions since the wall was built, but it is safer to measure the moisture content. The principal methods available are:

Electrical moisture meters as described for use with timber may be used with probes forced into the wall. To avoid misleading results, it is better to cover an area of wall, not less than 300 mm × 300 mm, with a piece of glass, metal or polythene sheet, for several hours (preferably at least overnight) before taking a reading in the centre of the area covered.

Coloured indicator papers. These are fixed to the surface and change colour according to the amount of moisture present. Again, conclusions may be misleading if based on a small area of surface and overnight covering, as for electrical moisture meters, is advised.

Hygrometers. The form of instrument described for use on floors (*see* page 4) is equally suitable for use on wall surfaces, using a suitable form of prop : overnight readings are advised.

Digest 55 *Painting walls : 1* includes recommendations for treatments appropriate to various moisture contents.

Cautions

It is not possible to tell that a material is dry enough to paint or otherwise finish just by looking at it or touching it. Nor is the length of time since building any reliable guide, although it is true to say 'the longer the better'. A traditional rule of thumb has been 'one month per inch of thickness to be dried in *good* drying conditions'. Even so, experience shows that this is often not enough. Up to the end of the 1930s, new houses were often left for three months before decorating. But if it is necessary to decorate earlier, use a porous, inexpensive finish such as emulsion paint and accept that early redecoration might be needed.

Because the digest is about drying out, it has discussed methods of speeding up the process. Drying out should not, however, be accelerated too much by the use of heaters or dehumidifiers, otherwise excessive cracking of screeds and/or plaster finishes may occur. If woodwork is already fitted, checking, splitting and opening of joints may be encouraged by too rapid drying. After the building is occupied, heating should not immediately be run at its maximum level to complete drying as this also could result in excessive shrinkage and distortion of wood.

Notes on some of the equipment referred to in the digest

Dehumidifiers

There are two types of dehumidifier, one uses a desiccant to extract the moisture, the other condenses it by refrigeration coils. They can extract as much as 6 litres of water per hour, depending on the moisture content of the structure and the air. Their effectiveness falls at low temperatures, because cold air carries less moisture from the structure than warm air, and some heating is therefore recommended.

Chemical desiccant dehumidifiers rely on the removal of moisture from the air by the hygroscopic properties of various chemicals. They are of two types : (a) in which the chemical dissolves in the water which it absorbs and the solution is discarded, and (b) in which a paper impregnated with a hygroscopic chemical holds water until the impregnated paper is subsequently reactivated by heat and reused.

They are thus cheap in capital and operating costs and they work at low temperatures.

Refrigeration dehumidifiers act by drawing a stream of moist air over a series of artificially cooled coils. Water vapour in the air is cooled to its dewpoint and condenses on the coils, whence it is collected and removed. They recover the latent heat of evaporation in the vapour and so deliver slightly warmed dry air back into the room. Some types have a system of automatic defrosting.

Moisture meters for wood

Resistance type meters are the most popular. They measure the electrical resistance of the wood, which varies with changes in moisture content, and are accurate to about ± 2 per cent over a range of about 7–25 per cent moisture content. This meets the normal requirements for wood in buildings.

Measurements are taken by inserting a pair of probes into the wood and reading the electrical resistance between them on a scale calibrated to show the equivalent moisture content. The electrical path measured is only the distance between the probes, usually about 25 mm and to a depth of 6 mm, though some have longer probes that can be driven more deeply. It is therefore important to try to obtain representative values by taking readings at several points on each piece, avoiding any areas of local wetting or visible irregularities such as pitch pockets, knots, stains, etc. Chemically treated wood or wood contaminated, for example, by sea-water, is unlikely to give reliable readings.

The meters are calibrated for a stated temperature and species of timber. Corrections must be made for other temperatures and species, though for periodical measurement of the progress of drying it might be possible to ignore the correction for species.

The instructions of the instrument manufacturer should always be followed : these normally include correction tables for temperature and species.

Dielectric-type meters are based on the dielectric constant of wood. Their use is not restricted to any range of moisture contents but they are generally less accurate than resistance-type meters. Instead of correcting for species a more complicated and careful calibration has to be made for the specific gravity of the wood, on which its dielectric properties partly depend.

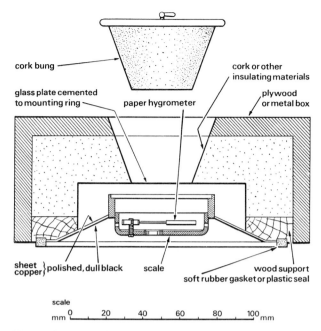

cork bung

glass plate cemented to mounting ring

paper hygrometer

cork or other insulating materials

plywood or metal box

sheet copper} polished, dull black   scale

wood support

soft rubber gasket or plastic seal

scale

mm 0   20   40   60   80   100 mm

Fig. 1   Apparatus for measuring dampness

Hygrometer

A simple form of instrument measures the relative humidity of a pocket of air in equilibrium with the sub-floor (Fig 1). It consists of a paper hygrometer in a vapour-tight mounting housed in a well insulated box. Provision is made for sealing the edges of the instrument against the surface to be tested—Plasticine is a convenient material for this purpose—and for reading the hygrometer scale whilst the instrument is in position on the floor.

After the instrument has been sealed firmly to the floor surface, a period of not less than four hours should be allowed to elapse for the entrapped air to reach moisture equilibrium with the concrete base before the reading is taken. A longer period is preferable and can sometimes be obtained conveniently by placing the instrument in position overnight and taking the reading in the morning. If readings are to be taken at several points with only one instrument available, subsequent positions may be covered by impervious mats (bitumen felt, polythene sheeting, etc) about one metre square, laid down when the instrument is placed in its first position, to speed up the later readings.

If there is difficulty in obtaining the instrument in the form illustrated, it can be made locally using a paper hygrometer which is obtainable from suppliers of laboratory apparatus; the shape and dimensions of the housing are not critical but the principles of thermal insulation and vapour barrier (the copper sheet) should be followed.

Condensation

Principles involved in condensation; conditions producing condensation—atmospheric conditions and artificial influences. The behaviour of absorbent materials and surfaces. An explanation of interstitial condensation. Designing to avoid condensation—characteristics of the building fabric—characteristics of the environment. Estimating condensation risk—a worked example. Lightweight sheeted roofs.

Principles involved in condensation

The amount of water vapour that air can contain is limited and when this limit is reached the air is said to be saturated. The saturation point varies with temperature—the higher the temperature of the air, the greater the weight of water vapour it can contain. Water vapour is a gas, and in a mixture of gases, such as when present in the air, it contributes to the total vapour pressure exerted by the mixture. The ratio of the vapour pressure of any mixture of water vapour and air to the vapour pressure of a saturated mixture at the same temperature is the *relative humidity* (RH), which is expressed as a percentage. Alternatively, relative humidity can be regarded as the amount of water vapour in the air expressed as a percentage of the amount that would saturate it at the same temperature.

In conditions of, for example, 20°C and 80% RH, all the moisture can be held in the air. If more water vapour is introduced into the air and the temperature remains constant, the relative humidity will increase; saturation point (100% RH) may be reached and thereafter any further vapour will be deposited as condensation. If on the other hand the amount of water vapour remains constant but the temperature falls, because the colder air can support less moisture, the RH will rise until at about 15°C it is 100% and any further cooling will cause water to condense. This is the *dew-point* of that air which at 20°C had an RH of 80%.

Conditions producing condensation

It has shown that changes in temperature or in moisture content can cause condensation to occur. These changes can occur naturally—by changes in atmospheric conditions, or artificially—by living habits or industrial processes.

Atmospheric conditions When warm damp weather follows a period of cold, the fabric of a heavy structure which has not been fully heated will not warm up immediately but may remain comparatively cold for several hours or, if the walls are very thick, for a day or more. When the warm, moist, incoming air comes into contact with cold wall surfaces which are below its dew-point, water will condense upon them, but as the walls warm up and eventually exceed the dew-point, condensation ceases and the condensed moisture evaporates. A building of light construction will warm more rapidly and is less likely to suffer condensation from this cause.

A solid floor, with a non-insulating finish, has a surface that is slow to warm, and if there is a rise in temperature and humidity of the air above, it may suffer condensation for several hours. In general, the bigger the heat capacity of the structure, the longer will condensation persist on its surface in adverse conditions.

Artificial influences The humidity inside an occupied building is usually higher than outside. People themselves and many of their activities increase the amount of moisture in the air. Sedentary persons breathe out more than a litre of water, as vapour, in twenty-four hours; physical exertion may raise this to four times the rate. Moisture vapour is released by cooking, by clothes-washing and drying and by the combustion of oil or gas; a litre of oil burnt produces in vapour form the equivalent of about a litre of water and if burnt in a flueless appliance this vapour is emitted into the air within the building.

Many industrial processes require high humidities and temperatures and some release large quantities of steam. The risk of condensation is great when the

RH and temperature of the air remain above 60% and 20°C for long periods.

Condensation, particularly in dwellings, does not necessarily occur in the room where the water vapour is produced. A kitchen or bathroom in which vapour is produced may be warm enough to remain free from condensation except perhaps on cold, single-glazed windows, cold-water pipes and other cold surfaces. But if this water vapour is allowed to diffuse through the dwelling into cold parts such as the stair-well and unheated bedrooms, condensation will occur on the cold surfaces of those rooms, which may be remote from the source of the moisture. Soft furnishings, including bedding, and clothing may become damp because of this, especially as some of these materials are slightly hygroscopic.

Removal of the moisture-laden air from the building, from a point near to the source of the moisture, will greatly reduce the likelihood of condensation.

Water absorbed during construction During its early life, a building may be prone to condensation from the evaporation of water which entered the structure during construction, either as mixing water for the concrete, mortar and plaster or by exposure to the weather before the roof was completed. Much of this moisture evaporates into the internal air in the building and then condenses in the colder regions, usually at night. The amount of water can be as much as 4000 kg, the drying period may be as long as a year.

Some benefit can be gained by leaving all internal doors open and the upper storey windows ajar to facilitate the drying-out process; incoming occupants should be warned of the drying-out period.

Absorbent surfaces and materials

Temporary or intermittent condensation which is clearly visible on a non-absorbent surface may pass unnoticed on an absorbent surface or material. Condensed water can be absorbed and held until conditions change and allow it to dry out, but condensation can only be accommodated in this way if the periods of condensation are short enough and drying periods long enough to avoid complete saturation of the absorbent material. This is the principle on which anti-condensation paint works.

Fig 1 Temperature conditions in wall which may lead to interstitial condensation

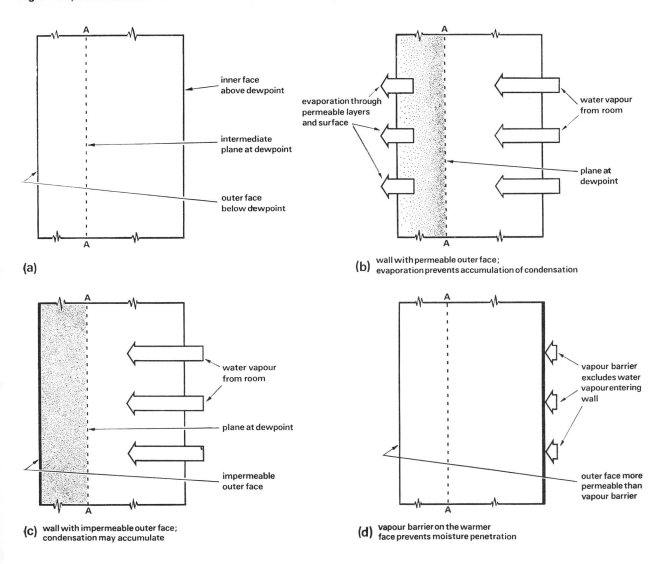

(a)

(b) wall with permeable outer face; evaporation prevents accumulation of condensation

(c) wall with impermeable outer face; condensation may accumulate

(d) vapour barrier on the warmer face prevents moisture penetration

Interstitial condensation

When the material of a wall, roof or similar building element is permeable to water vapour—and this applies to nearly all building materials—a dew-point temperature is associated with each point within the material. In the same way that a temperature gradient exists through a structure, depending on the thermal properties of the component materials, so a dew-point gradient depending on their water vapour diffusion properties exists also. If at any point the actual temperature is below the dew-point, then condensation will occur at that point within the material or structure. For example, the temperature through the thickness of a wall may vary from the inner face being above dew-point to the outer face being below dew-point; at some intermediate position the temperature will then be equal to the dew-point and condensation will begin at this plane (Fig 1a).

The exact processes taking place are complex and are further complicated by the fact that the condensed water changes both the thermal and vapour-transmission properties of porous materials, but the simple concept above is adequate for determining situations in which the risk of condensation trouble is unacceptably high.

If the outer portion of the wall is permeable to moisture, or if ventilation is provided behind impermeable wall or roof claddings, condensation will not be troublesome because the moisture can evaporate gradually to the outside air (Fig 1b).

If the outer surface is impermeable, the condensed moisture tends to accumulate in the wall and may ultimately saturate the material (Fig 1c). The situation will be most severe when the humidity of the indoor air is high.

A vapour barrier on the inner face of the wall (on the potentially warm side of any layer of insulating material) will prevent the passage of water vapour into the wall but only if it is undamaged and continuous. If the outer face of the wall is more permeable than the inner vapour barrier, any moisture contained in the wall can escape to the outside air (Fig 1d). If, however, the outer face of the wall has an impermeable cladding, or if the cladding is of organic material that would suffer in prolonged damp conditions, a ventilated cavity should be formed between the cladding and the wall so that any moisture evaporating from the wall surface is removed.

In some forms of construction, for example, behind timber cladding or tile hanging, a moisture barrier may be needed near to the cold outer face of the wall, to exclude wind-driven rain or snow: a 'breather' type of membrane, i.e. one that will bar the passage of liquids but will transmit vapour, is then required for this position. It should be used in conjunction with a vapour barrier on the warm side of the construction so that less vapour can enter the wall from the warm side than can escape by the breather membrane and there will be no accumulation of condensed moisture on the inner face of the breather membrane.

Designing to avoid condensation

The chart (Fig 2) shows the interdependence of relative humidity, on the left-hand vertical scale, dry-bulb temperature, on the horizontal scale, and the concentration of moisture in the gaseous mixture on the right-hand vertical scale, expressed for this purpose as mixing ratio, the mass of water vapour per unit mass of dry air.

An example of this relationship is shown by the reference points A–D, marked on the chart, which represent the following conditions:

At A, the outdoor dry-bulb temperature is 0°C, the mixing ratio of the air is 3·4 g/kg of dry air, which gives an RH of 90%.

If this air is warmed to 20°C and the mixing ratio remains the same at 3·4 g/kg, the RH will become 23% (point B).

If excess moisture amounting to 7 g/kg is now introduced as a result of activities within the building, at the same temperature the RH will increase to 70% (point C).

D shows the dew-point temperature of the resulting air/moisture mixture, from which it follows that condensation will not occur if the adjoining parts of the building fabric to which the air has access are kept above 15°C.

Two additional scales, not so far mentioned, have rather different uses. The scale of partial pressure of water vapour is directly related to that of mixing ratio, and it may be used, as explained later, to estimate the rate of diffusion of water vapour through the structural fabric. The scale of wet-bulb temperature denotes instrumental readings that are commonly taken with a sling psychrometer. Simultaneous wet-bulb and dry-bulb readings, when transferred to the charts, define the physical properties of an air sample at the conjunction of the respective oblique and vertical temperature lines.

Characteristics of the exposed building fabric

In all buildings heated and occupied during the winter, it may be assumed that the air temperature and water vapour pressure indoors will be in excess of those outside. As a result, heat and water vapour will attempt to flow outward through the fabric in an effort to restore the balance. Their progress is determined by the precise construction through which they must pass, and the result is a characteristic distribution of temperature and vapour pressure throughout the exposed structure.

The relationships governing this distribution are

depicted in Figs 3 and 4. The gradients shown are determined by the total differences of temperature and water vapour pressure across the structure and by the succession of resistances to flow that must be overcome. Table 1 gives a selection of typical values of thermal and water vapour resistance. The estimating procedure described later makes direct use of vapour *resistivity*, for which values may be obtained from the table, but other sources will be found to present figures for vapour permeability, or *diffusivity*, which is the reciprocal of the resistivity.

In Fig 3 it will be seen that the boundary surfaces of structures offer resistance to heat flow. Unless the structure is independently heated, the internal surface is at a temperature lower than that of the indoor air, and it will consequently cool the layer of air in contact with it. How far the local air temperature is

depressed depends on what proportion of the structure's total thermal resistance is contributed by the internal surface. Figure 5 shows the appropriate depressions for a range of temperature differences and comparative levels of structural insulation. It is

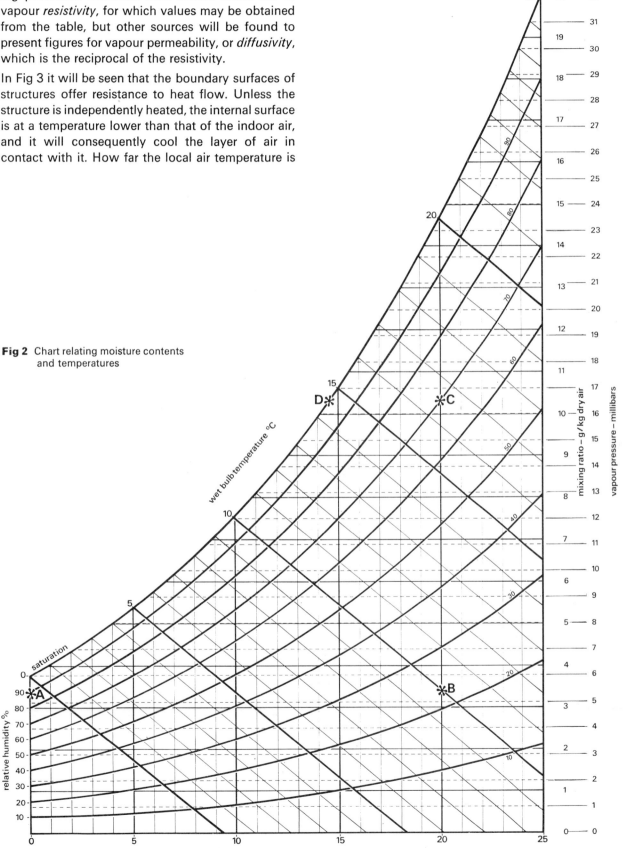

Fig 2 Chart relating moisture contents and temperatures

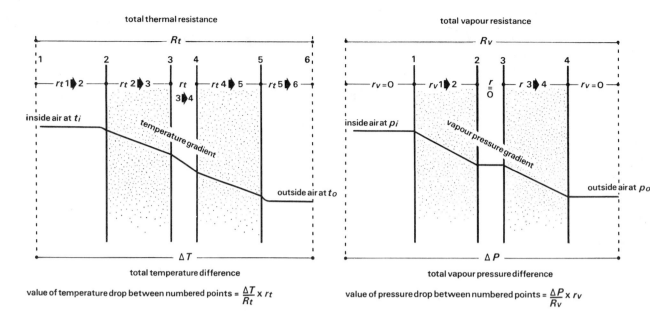

value of temperature drop between numbered points = $\dfrac{\Delta T}{R_t} \times r_t$

value of pressure drop between numbered points = $\dfrac{\Delta P}{R_v} \times r_v$

Fig 3 Temperature gradient through a structure

Fig 4 Vapour pressure gradient through a structure

applicable only to steady state conditions. In buildings of high thermal capacity construction, heated intermittently, the depression of temperature at the internal surface will often be greater, thus increasing the likelihood of condensation. Relating Fig 5 with the earlier example of outdoor air at 0°C, indoor air at 20°C and dew-point at 15°C, the maximum allowable drop of 5°C in the inside surface temperature will be seen to require a structural U-value not greater than 2·1 W/m² °C.

In the gradient of Fig 4, no vapour pressure drop

occurs at the internal surface. Furthermore, it is practically possible to maintain the total indoor/outdoor pressure difference across the internal surface if the latter can be made impervious to water vapour. In this event the sole provision against condensation is that the surface should remain above the internal dew-point temperature. Not all surfaces can be so chosen to stop the passage of moisture vapour, and the more permeable they are, the greater is the chance that, at some internal part of the structure, a sufficiently low temperature will be

Fig 5

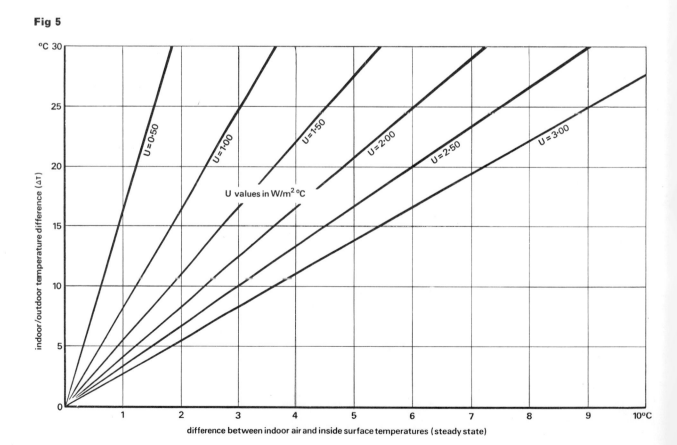

reached to cool the vapour below its dew-point.

The respective temperature and pressure gradients, when charted, can be used to identify the point where condensation might occur in given circumstances, as demonstrated in the worked example later in this digest.

Characteristics of the environment

Except when a warm front brings moist air to follow a cold dry spell, the risk of condensation through natural ventilation in a heated building is slight. It is almost wholly dependent on the amount of excess moisture introduced by the occupants and their activities.

At normal ventilation rates, the gain by the air of body moisture from persons not engaged in physical exertion is roughly 45 g/person in one hour. This results in the indoor atmosphere having an excess moisture content over outdoor air of some 1·7 g of water vapour per kg of dry air. Provided the ventilation rates were properly controlled, this would be a suitable design assumption for shops, offices, classrooms, public meeting-places and dry industrial premises. For dwellings, taking account of the moisture produced by cooking and bathing and the likelihood of restricted ventilation in cold weather, a safer design value for moisture excess might be 34 g/kg. Catering establishments and industrial workshops requiring humid atmospheres or using wet processes may well contribute 68 g/kg or more to the internal air. In naturally ventilated premises such design values may be added to the assumed mixing ratio of the outdoor air.

Table 1. Typical values of heat and vapour resistance

	Thermal resistance (r_t) m² °C/W
Surfaces	
Wall surface—inside	0·12
—outside	0·05
Roof (or ceiling) surface—	
inside	0·11
outside	0·04
Internal airspace	0·18

	Vapour resistance (r_v) MN s/g
Membranes	
Average gloss paint film	7·5–40
Polythene sheet (0·06 mm)	110 – 120
Aluminium foil	4000

	Thermal resistivity m °C/W	Vapour resistivity* MN s/g m
Materials		
Brickwork	0·7–1·4	25–100
Concrete	0·7	30–100
Rendering	0·8	100
Plaster	2	60
Timber	7	45–75
Plywood	7	1500–6000
Fibre building board	15–19	15–60
Hardboard	7	450–750
Plasterboard	6	45–60
Compressed strawboard	10–12	45–75
Wood–wool slab	9	15–40
Expanded polystyrene	30	100–600
Foamed urea-formaldehyde	26	20–30
Foamed polyurethane (open or closed cell)	40–50	30–1000
Expanded ebonite	34	11,000–60,000

*Resistivity—1/diffusivity

Fig 6

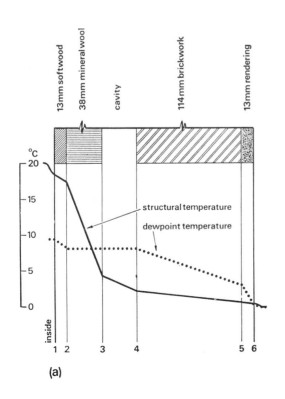

(a)

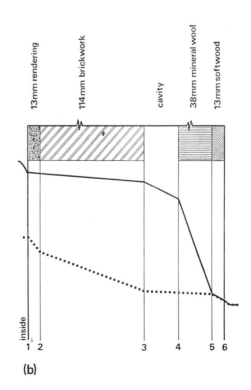

(b)

Estimating condensation risk

The principles used in this digest to predict the likelihood of condensation, and to design so as to avoid it, may be applied to wall, floor or roof constructions, but lightweight sheeted roofs present special problems, see page 183.

Figs 6a–6d show schematically some variations of a wall construction. The four illustrations are drawn with the thicknesses of the components to scale, the boundaries of the materials are numbered in sequence and the temperature difference across the structure is set up on an adjacent vertical scale. Using Fig 6a as an example, the design assumptions and the estimating procedures are then as follows:

1. Indoor/outdoor air temperature difference, ΔT, assumed to be:

$$20 - 0 = 20°C$$

2. Thermal resistance
(value from Table 1 × thickness of component):

Inside air to point 1		0·12
1–2 13 mm softwood		
6·93×0·013		= 0·09
2–3 38 mm mineral wool		
27·72×0·038		= 1·05
3–4 cavity		0·18
4–5 114 mm brickwork		
0·9×0·114		= 0·10
5–6 13 mm rendering		
0·83×0·013		= 0·01
6—outside air		0·05

$$R_t = 1·60$$

3. Temperature drop between points:

$$= \frac{\Delta T}{R_t} \times r_t = \frac{20}{1·6} \times r_t = 12·5 \times r_t \ (°C \text{ point to point})$$

Inside air	=	20°C
Inside air—point 1	= 12·5×0·12 =	1·5
	drop to	18·5
1–2	= 12·5×0·09 =	1·12
	drop to =	17·38
2–3	= 12·5×1·05 =	13·12
	drop to	4·26
3–4	= 12·5×0·18 =	2·25
	drop to	2·01
4–5	= 12·5×0·10 =	1·25
	drop to	0·76
5–6	= 12·5×0·01 =	0·12
	drop to	0·64
6—outside air	= 12·5×0·05 ×	0·63
		0·01 *
Outside air		0°C

*Error due to approximation

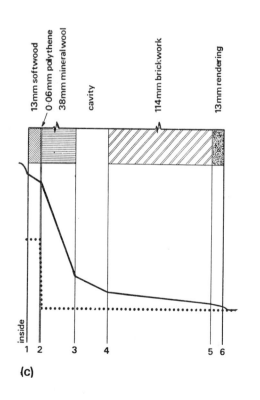

(c)

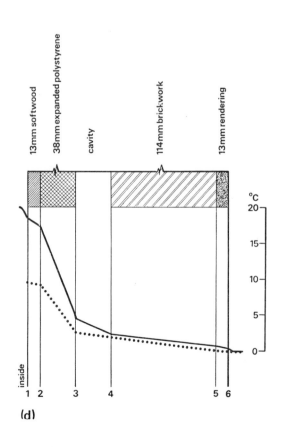

(d)

4. The profile of the temperature drop is plotted against the vertical scale.

5. From Fig 2, using the design assumption that the outside air is at 0°C and is saturated at a mixing ratio of 3·8 g/kg, the outdoor vapour pressure can be read (on the right-hand scale) as 6 mb; if a moisture vapour excess of 3·4 g/kg is contributed by activities indoors, inspection of the two right-hand scales of Fig 2 shows that the total moisture content of 7·2 g/kg gives an indoor vapour pressure of 11·4 mb. The indoor/outdoor vapour pressure difference, ΔP, is therefore $11·4 - 6·0 = 5·4$ mb.

6. Vapour resistance
(value from Table 1 × thickness of component):

$$
\begin{aligned}
\text{point } 1\text{--}2 &= 60 \times 0·013 &&= 0·78 \\
2\text{--}3 \text{ (virtually nil)} &&&= 0·0 \\
3\text{--}4 \text{ (nil)} &&&= 0·0 \\
4\text{--}5 &= 25 \times 0·114 &&= 2·85 \\
5\text{--}6 &= 100 \times 0·013 &&= 1·3 \\
&&& \overline{} \\
R^v &&&= 4·93
\end{aligned}
$$

7. Pressure drop between points:

$$\frac{\Delta P}{R_v} \times r_v = \frac{5·4}{4·93} \times r_v = 1·10 r_v \text{ (mb point to point)}$$

		Corresponding dew-point temp. from Fig 2
Indoor vapour pressure	11·4	9·4°C
point 1–2 1·10×0·78 =	0·86	
	‾‾‾	
drop to	10·54	8·0°C
2–3	= 0·0	
	‾‾‾	
	10·54	8·0°C
3–4	= 0·0	
	‾‾‾	
	10·54	8·0C
4–5 1·10×2·85 =	3·13	
	‾‾‾	
drop to	7·41	3·0°C
5–6 1·10×1·30	1·43	
	‾‾‾	
drop to	5·98	0·0°C

8. Profile of dew-point temperature: Reference to Fig 2 will show the respective dew-point temperatures for the vapour pressures at points 1, 2, 5 and 6 from which the dew-point profile may be constructed.

9. Estimation of condensation risk: At any point where the computed temperature is lower than the computed dew-point temperature, condensation can occur in the conditions assumed. In the worked example, liquid may form in a position where, clearly, it can reduce the effectiveness of insulation and it is likely also to put the nearby timber at risk of rot. As an illustration of the effect that structural detailing may have, Fig. 6b shows the construction reversed and free from risk in the same surrounding conditions. Slight modifications, shown in 6c and 6d, are sufficient, however, to limit the potential risk by using materials that modify the vapour pressure gradient.

Some calculated risk may be accepted, as for example when condensed moisture can do no damage to the weather side of a structure and can be prevented from soaking back. In a great deal of masonry construction it is probable that condensate is soaked up harmlessly until it has an opportunity to distil outwards or evaporate from the surfaces during favourable periods.

Where there is an estimated likelihood of intra-structural condensation the greatest risks arise if vegetable-based materials are exposed to the consequent dampness, particularly where the weather side of the construction is highly resistant to vapour flow. Then, ventilated cavities may offer some help and their value increases as the vapour resistance of the warm-side materials is improved.

Lightweight sheeted roofs

The lowest temperature to which a roof cladding falls occurs in winter, on calm frosty nights when the sky is unclouded. The cooling effect of the low outside temperature is then intensified by the roof radiating heat to the sky, which in these conditions has an effective temperature of about −45°C for radiation from the earth.

With roof cladding of low thermal capacity, e.g. single-skin metal or asbestos-cement sheeting, the change in temperature of the inner surface as a result of a sudden drop in the effective outdoor temperature takes place rapidly, and the cladding may be about 5°C below the outdoor temperature for several hours. The use of vapour barriers, ventilation and insulation to combat the high risk of condensation is discussed in National Building Studies Research Paper 23, A. W. Pratt. *Condensation in sheeted roofs.* HMSO 1958 (*now out of print*).

Further reading
Digest 108, Standardised U-values.

BRS Current Papers Design Series 24, E. F. Ball. Condensation in large panel construction. BRS 1964.

Condensation in dwellings, Part 1: A design guide, £1.00; Part 2: Remedial measures, £1.25. HMSO.

Refuse handling: 1

Local authorities spend about £60 million per year collecting more than fifteen million tonnes of house and trade refuse, which costs a further £20 million for disposal; about twenty million tonnes of commercial and industrial wastes, excluding mining wastes and power station ash, are dealt with by refuse disposal contractors. By 1980, house and trade refuse may exceed twenty million tonnes, with commercial and industrial waste reaching forty million tonnes. The effect of steeply rising costs on an already expensive service provides a strong incentive to local authorities and to commercial and industrial organisations to seek improved methods. This digest and the next examine current methods of dealing with refuse and consider the impact of new equipment and handling systems.

The legal position

The London Government Act of 1963 gave the Inner London Boroughs the duty to collect house refuse and trade refuse (the latter if requested by an occupier of premises) and made the Greater London Council responsible for its disposal. Since 1 April 1974, the new district authorities in England are responsible for the collection of house and trade refuse and the new counties are responsible for its disposal. In Wales, the new districts are responsible for collection and disposal. Legislation is proposed to transfer to the new authorities in Scotland, as from 16 May 1975, the responsibility for collection and disposal.

The Government is at present considering proposals for the introduction of legislation which will extend and develop existing statutory controls over waste collection and disposal. It is proposed to define household, trade and industrial wastes according to the kinds of premises from which the waste originates and to redefine the powers and duties of local authorities in respect of each category of waste. District authorities would be under a statutory obliga-

tion to collect all household waste free of charge (except perhaps for items such as large quantities of organic garden refuse and builders' rubbish) and to collect trade refuse at the request of occupiers but to charge for this unless there are good reasons for not doing so. They would also have a discretionary power to collect industrial waste but would have to make a charge for this.

Yield and composition

Domestic refuse Table 1 shows how domestic refuse has changed during recent years and how it is likely to change by the end of the century. The total volume has tended to rise, though the bulk density has fallen. There is a marked similarity between refuse in the USA in 1966 and that expected in this country by 1980, probably an indication of the effect of rising standards of living. Informed estimates suggest that the increase in the plastics content of refuse will be quite moderate, from 1·2 per cent (by weight) in 1967 to 4 or 5 per cent in 1980; this would change considerably if plastics were to replace glass milk bottles.

Table 1 Domestic refuse: composition and yield (percentages are by weight)

	Fine dust or cinder	Vegetable, bone, etc	Paper & rag	Metal & glass	Misc (incl plastics)	Av. household/week weight	volume	density
	%	%	%	%	%	kg	m³	kg/m³
1935	57	14	16	7	6	17·0	0·058	291
1963	40	14	25	16	5	14·1	0·071	200
1967	31	16	31	16	6	12·9	0·081	161
1968	22	18	39	18	3	13·2	0·084	158
1969	17	20	40	20	3	12·7	0·089	143
1970	15	25	39	18	3	13·5	0·092	146
1972	21	19	33	19	8	11·8	0·077	153
1973	19	18	36	19	8	11·5	0·076	152
1966*	10	24	43	14	9	19·1	—	—
1975†	18	13	53	12	4	16·8	0·13	130
1980†	12	17	46	18	7	14·5	0·12	120

* typical town in USA
† estimated

Table 2 Commercial refuse

	Typical bulk density	Multiple stores	Depart-mental stores	Super-markets	Hotels	Offices
	kg/m³	%	%	%	%	%
Folded newspaper; cardboard packed or baled	500	81	65	50	8	80
Loosely crumpled paper; office stationery	50					
Wastepaper (loose in sacks)	20					
Mixed general refuse, similar to domestic (no solid fuel residues)	150	13	31	40	55	16
Separated food wastes: —uncompacted vegetable waste	200	4	2	—	33	4
—well-compacted, moist pig swill	650					
Salvaged bones and fat	600	2	2	10	—	—
Empty bottles	300	—	—	—	4	—
Yield per week, kilogrammes		1·0	0·54	1·8/5·8*	3·0 per head, staff and residents	1·68 per employee
		←per m² of sales area→				
		6-day week			7-day week	5-day week

* Supermarkets, judged by their output of refuse, fall into two categories.

Table 3 Hospital waste

Function	Output/bed
	kg/day
Long stay	0·2
Mental	0·7
General	1·9
Maternity	3·8

Table 4 Resort waste (US data)

Situation	Output	Unit
	kg/day	
Campground	0·57±0·04	camper
Rented cabin with kitchen	0·66±0·14	occupant
Restaurant	0·32±0·18	main meal
Residence	0·97±0·24	occupant
Observation site	0·02±0·01	vehicle
Swimming beach	0·18±0·01	swimmer

The GLC[1] has suggested that the total volume of refuse may double in the next twenty years and its weight increase by 70 per cent. The Working Party on Refuse Disposal[2] thought that where open fires were still in use, shown by an ash content exceeding 30 per cent, no weight increase would occur. Where the ash content is already at or below 20 per cent, the increase would be about one per cent per year over the next decade. Taking this weight increase into account, the volume increase could reach 44 per cent by 1980.

Fig 1 shows a seasonal variation in the output of refuse (by weight) from centrally heated multi-storey flats. The variation is greater in dwellings with open fires.

Commercial and institutional buildings The greater range of building types and activities makes it difficult to forecast refuse yield and composition in these classes of buildings. Table 2 is based on the results of a small-scale survey made in 1969/70 by BRS[3] and gives the quantities of refuse produced in some typical commercial buildings, its composition and an indication of the bulk densities of the constituents. Paper forms up to 80 per cent of the output from offices and shops, one of the principal sources of waste paper for recycling.

Table 3 summarises information on hospital waste, obtained by the Department of Health and Social Security in 1969[4]. It has been estimated that this type of refuse increases at a rate of 10 per cent per annum.

Table 4 is based on American data of 1973 for resort areas which might be expected to be comparable with similar activities in this country.

Equipment currently available

1 Storage containers (CP 306)[5]

(a) Dustbins, paper and plastics sacks, up to 0·11 m³ capacity; for domestic and similar use, handled manually.

(b) Metal containers not exceeding 1 m³ (straight-sided circular 'Paladin' on castors or feet); used in conjunction with chutes, and in commercial areas. These require collection vehicles equipped to lift and tilt.

(c) Bulk containers: up to 9 m³—for use with skip lift vehicles
up to 23 m³—for use with lift hoist vehicles
up to 30 m³—as trailers for articulated vehicles

2 Incinerators

(a) Small units, gas or electrically heated, for immediate disposal of soiled dressings and similar items.

(b) Chute-fed incinerators—capable of handling refuse from large residential buildings, and all hospital wastes.

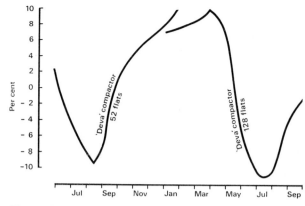

Fig 1   Seasonal variation in refuse output

(c) Municipal plants, handling many tonnes per hour.

Incineration can add to the total volume of air pollution and, furthermore, some constituents of refuse, eg pvc and other plastics, produce toxic gases when burnt. Tests of two incinerators in blocks of multi-storey flats[6] have shown flue-gas quality to be within acceptable limits, but these were new plants correctly adjusted. A badly maintained incinerator could pollute the air. Municipal installations are big enough to justify the provision of elaborate plant for treating effluent gas.

3 Compaction equipment Volume reduction of up to 8 to 1 is possible, but 2–4 to 1 is more usual. Compactors used with:

(a) Disposable containers (paper or plastics sacks, cardboard cartons) require small floor space and low energy input (sometimes only a standard 13A single-phase power supply); may be used by unskilled labour. Suitable for shops, restaurants, hospitals etc and some are intended for domestic kitchens. Chute-fed machines are suitable for multi-storey flats[6].

(b) Interchangeable containers: sizes from chute-fed machines using 3 m³ containers, up to those with a feed-hopper capacity of 10 m³ packing into 30 m³ containers; particularly suitable for handling low density commercial and industrial wastes.

4 Crushing and comminution The size of individual pieces of refuse is reduced by crushing or cutting to produce material that is more readily compacted when tipped, is less unsightly and is unattractive to flies and rodents.

(a) Small bottle crushers, for use in restaurants etc where large quantities of non-returnable bottles have to be disposed of.

(b) Machines capable of crushing or cutting tin cans of up to about 20 litres capacity; some of these can deal with virtually all domestic refuse, except heavy metal objects.

(c) Large machines capable of dealing with domestic refuse on a municipal scale or which have been developed for special purposes, eg tyre shredders.

5 Extruders These crush and compact simultaneously by forcing refuse through a tapered tube or against a spring-loaded plate; the forces are applied by hydraulic ram or Archimedian screw. Volume reduction as high as 15 to 1 is obtainable.

6 Wet processors Mechanical treatment can be assisted by the addition of water; the resultant mixture can then flow by gravity, or be pumped, through pipes to a point for further treatment.

(a) Kitchen waste disposers, using 6–7 litres/minute of water, can handle about 15 per cent of domestic refuse.

(b) Larger machines can deal with catering wastes, hospital dressings, disposable bed-pans etc.

(c) Some machines are designed to deal exclusively with waste paper, but will macerate other organic wastes. The resultant pulp is flushed into the sewer or dewatered for disposal by other means.

(d) A group of machines related to paper-making; a rough-surfaced plate rotating at the bottom of a water-filled circular tank deals with pulpable materials and a cutting blade reduces plastics materials. Light-gauge metals are 'balled' and thrown, with any other very hard materials, by centrifugal force into a reject container. A slurry of pulped material is continuously extracted and may be pumped to a considerable distance, if required, for dewatering and recirculation of the clarified water.

7 Transport systems

(a) Vertical transport by chutes (CP 306[5] and BS 1703[7]) is widely used in multi-storey flats and should be considered for use in shopping and commercial areas.

(b) The Garchey system uses trapped waste water to transport refuse from a receiver fitted beneath the kitchen sink through a 150 mm dia vertical stack pipe to a holding tank for removal by tanker. One vertical pipe will serve a tier of dwellings in a typical multi-storey block; in some larger installations several holding tanks can be evacuated through a network of underground pipes to a central tank. The stack pipe size imposes a limit on the handling capability and 40–50 per cent of normal domestic refuse will need a separate manual collection.

Trends

At the present state of technology, either pneumatic or hydraulic pipeline systems[8] offer the best means of mechanising refuse handling. Pneumatic systems in which crude refuse is sucked along 500 mm bore pipes, or where refuse is first pulverised and then sucked along 200 mm bore, or smaller, pipes are both in use, transporting over distances up to 1 km. These systems can provide an efficient service to a relatively small number of inlet points, and are thus particularly suitable for use in conjunction with refuse chutes. The capital outlay is, however, greater at present than most property owners could afford.

Higher solids loadings, smaller pipes and low transport speeds are possible in hydraulic systems with consequently reduced power requirements and bulk of central plant. They would, however, increase the loads on sewage works for treating heavily contaminated water and the problems of disposing of large quantities of wet refuse sludge would be formidable. There is also a strong probability of rapid corrosion of metals by the slurry. In spite of problems such as these, pipeline systems are likely to be used more frequently in large, compact, residential and commercial developments and for long distance transport between centres of population and disposal areas.

References

1 *London's refuse* Greater London Council: London 1969

2 *Refuse disposal* Report of the Working Party on Refuse Disposal: Department of the Environment: HMSO: London 1971

3 *Non-domestic refuse: an investigation by BRS* 'Surveyor' 23 Apr 1971 pp 23–24

4 *Hospital waste* Department of Health and Social Security; Technical Information June 1969

5 British Standard Code of Practice CP 306: Part 1: 1972 *The storage and on-site treatment of refuse from buildings*

6 Sexton D E and Smith J T *Studies of refuse compaction and incineration in multi-storey flats* 'Public cleansing' 62(12) 1972, pp 604–623

7 British Standard BS 1703 *Refuse chutes:*
              Part 1: 1967 *Hoppers*
              Part 2: 1968 *Chutes*

8 Courtney R G and Sexton D E *Refuse collection from houses and flats by pipeline* Building Research Establishment Current Paper CP 4/73

Refuse handling: 2

This digest concludes the brief survey of refuse systems begun in the previous digest (No 168)

Digest No 116 'Domestic Refuse' should now be withdrawn.

(The numbering of figures, tables and references is continued on from Digest 168)

Systems for handling refuse

There is a well-defined chain of events in the refuse system: generation—storage—on-site treatment—collection—intermediate processing—disposal. The design of a refuse handling system involves selecting for each of these links a solution that is compatible with the neighbouring stages and acceptable to the parties involved—the waste producer, the local authority and/or the treatment and disposal contractor—between whom collaboration early in the design stage is essential.

The changing composition of refuse, particularly the increasing volume of lower density material, has compelled many local authorities to acquire compacting collection vehicles and is tending to make on-site volume reduction almost essential. The use of compactors is increasing rapidly and some vehicles suitable for sack and bin collections are now equipped with auxiliary lifting devices for emptying packed containers. Pending the perfection of pipeline transport systems, compaction may be expected to play an important part in reducing calls on transport and disposal facilities. Compactors will not become commonplace in individual dwellings in the near future. Existing collection methods should be able to cope with outputs up to twice their present level if rigid dustbins are replaced by sack systems. If present trends are maintained, more than two sacks will be required after 1980 when in-kitchen volume reduction will become essential if collection frequency is not increased since one man will no longer be able to carry all the refuse in one journey between the storage point and collection vehicle. This condition has been reached in the USA where daily output is now about 1·2 kg per person, producing about 0·25 m³ weekly for an average household of three persons.

There are two scales of refuse production in shopping and commercial areas: small units with outputs of the same order as domestic premises; and office buildings and large stores producing up to 1·5 tonnes (14 m³) per day, five days a week. This situation is usually dealt with by frequent collections by local authority vehicles for small premises and those using 'Paladins' or similar containers, and a secondary system based on container interchange for the larger premises. A single system, possibly backed up by special collections for materials for recycling, would be preferable. Compaction offers an acceptable solution. A central compactor, fed manually or by some form of conveyor system, could replace multiple collections by a number of different vehicles with consequent savings in cost, reductions in traffic and improvements in amenity.

Any system must satisfy the following general requirements:

Safety On-site treatment plant uses large machines and high power to compact or disintegrate refuse and there is thus danger of injury to careless operators and unauthorised persons, particularly children, who might be attracted by moving machinery. All equipment should be guarded and provided with devices to prevent unauthorised operation. There is also a fire risk where refuse is accumulated in large quantities, for example in shopping centres. Professional advice should be sought on the design of refuse stores and processing rooms. Both fixed and portable fire-fighting equipment must be provided.

Hygiene Crude refuse provides food and shelter for vermin and is an ideal breeding medium for flies. During summer conditions, removal of refuse at intervals not greater than one week breaks the life cycle of blow-flies and prevents infestation building up. Putrescible waste becomes obnoxious in a much shorter time and food wastes must be cleared frequently. Food waste grinders are recommended for food preparation rooms in areas where disposal into the sewage system is permitted.

Refuse containers must be kept clean; the Working Party on Storage and Collection[9] recommended monthly cleaning, an ideal rarely attained. Disposable containers overcome this problem. Containers for compacted refuse are to some extent self-cleansing and since they are usually removed from the site at each emptying it is not difficult to route them through a cleansing process before putting them back into service. Unfortunately, 'Paladin' containers emptied on site are commonest and many of these are seldom, if ever, cleaned. A system in which the containers are removed to a central cleaning plant and replaced by clean containers is really necessary; this costs about £1 per treatment but no completely effective cleansing method has yet been developed. Experiments to develop a mobile cleaning machine are in progress.[10]

Access and container handling Adequate access for collectors and vehicles is essential. CP 306 suggests that collectors should not be required to carry dustbins more than 25 m. In some areas where the pattern of development limits road access, eg 'Radburn' layouts, sack storage systems with auxiliary collection by electric trolley using footpaths have been used. Containers of 1 to 2 m^3 capacity, widely used beneath chutes and in shops and commercial buildings, are usually fitted with castors and may be manhandled over distances up to 9 m over level ground; direct access for a collection vehicle equipped with the appropriate lifting gear is preferable.

Larger containers are usually lifted direct from site by a special vehicle which needs room to manoeuvre and considerable headroom. The latter point is particularly important when containers are sited within buildings. A recent installation in Western Germany uses a low trolley to carry 11 m^3 containers which are moved between compactor and pick-up point with a small, general-purpose electric truck.

Environmental considerations

The improvements in hygiene and amenity that can be obtained from volume and size reduction of refuse on the one hand, and the undesirable aspects of on-site incineration on the other, have already been mentioned. Recycling of certain materials can reduce demands on disposal facilities and conserve valuable resources, but the case for reclamation must be considered on both environmental and economic grounds. The environmental case is that the Earth's resources are finite and should be used responsibly, making every effort to conserve irreplaceable materials, and that pollution should be avoided. Economics, however, suggests that conservation is only worth while if it saves money and, unfortunately, it often costs more to reclaim materials than to use new, virgin, material. Paper is perhaps the exception, being easily separated at the beginning of the refuse chain. BRS studies[3] showed that about 80 per cent of the paper disposed of in office buildings was reclaimed. Other published figures show that about 30 per cent of the annual production of paper is reclaimed, a quarter of this by local authorities. Further large quantities could be reclaimed from municipal waste but the major problem is separation from the bulk refuse. Householders co-operated in the separation of reclaimable materials in a time of grave national emergency thirty years ago but this is not now generally practised except perhaps for clean newsprint and clean waste paper. Manual separation on picking belts in municipal plants is not acceptable. The problems of separation are being studied in a number of research establishments and the development of effective mechanical methods could extend the scope of economic reclamation.

Costs

The traditional refuse handling system, manual emptying of dustbins to a collection vehicle, still serves the majority of dwellings and small commercial premises at a cost of about £3 for a dustbin with a life of 6–7 years and collection costs of the order of £1 per resident per annum. Where sacks are used as liners to bins, a more acceptable 'dustless' service can be achieved; if there is occupier co-operation in placing sacks at the front of the house on the day of collection, present evidence suggests that the cost is no greater than the traditional collection and return of bin. No other system can yet compete with these costs for single dwellings but the larger scale of operations in flats and commercial buildings can justify the higher costs of more sophisticated equipment.

With a wide range of handling methods available, which may be combined in many different ways, it is difficult to present reliable cost data. Operating and maintenance costs will not be available for some of the newer systems until further experience has been gained. In the following paragraphs, budget costs are given for elements which are common to several systems.

1 **Containers** Fig 2 compares the costs of different types of containers for storing compacted refuse. Table 5 includes hire charges for interchangeable containers as this is more common than outright purchase where a refuse disposal contractor is employed. An allowance of 2·6 p/m^3 of uncompacted refuse is added to the costs of interchangeable containers to cover cleaning.

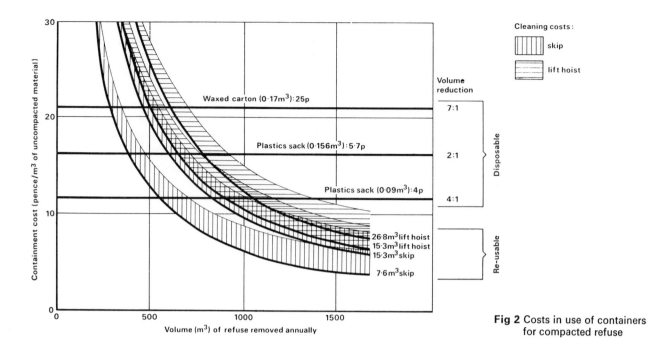

Fig 2 Costs in use of containers for compacted refuse

2 On-site processing The plant for these processes is usually made up of a number of separate units which may be arranged in various ways. For instance, the costs of a chute-fed compactor installation in a block of multi-storey flats would include sacks, the compactor and chute, and refuse chamber. The same compactor could be used in a different type of chamber with no chute feed in a low-rise development.

The simplest system using a chute with storage containers in a block of flats costs up to £50 per dwelling: the BRE survey[6] produced the data in Table 6 for processing equipment to replace the storage containers. A carton packer for a similar duty would cost about £1575 and a static packer filling 8 m³ interchangeable containers about £1100. This static packer could also be used with direct hopper feeding in a shop or commercial building. A large static packer filling 30 m³ containers would cost £5000.

A domestic sink grinder installation costs £60 and a large machine for dealing with catering wastes in a large restaurant or institutional kitchen about £1750.

Simple bottle and can crushers cost £130 and a crusher capable of dealing with 0·5 tonne/hour of domestic type refuse, discharging into sacks or bins, costs £1150.

Capital costs for a Garchey installation in multi-storey flats are about £150 per dwelling with running costs of £15 per annum whilst a pneumatic pipeline system costs £200–£400 per dwelling to install at its present stage of development.

For detailed costings, consultation with manufacturers at an early stage in design is recommended.

3 Labour Most of the systems referred to in Table 6 require some day-to-day maintenance and servicing. Static packers require attention during operation; this might range from a few minutes a day to full-time. The following observations made during the BRS appraisal of compactors and incinerators in multi-storey flats,[6] give some idea of the amount of work entailed:

(a) **Sack-packing compactor** (Deva) Remove filled sacks and replace; clean machine and refuse chamber; oil machine, service air-compressor.

50 flats/machine	2 hour/week
128 flats/machine	3 hour/week

Table 5 Cost of refuse containers

Container	Capacity	Cost
	m³	£
Dustbin	0·09	3
Sack, plastics	0·09	0·04
	0·156	0·057
Waxed carton	0·17	0·25
Paladin	1·0	30
Interchangeable steel	7·5–30	26–39/m³ (purchase)
		0·2–0·3m³ (per week hire)

Table 6 Machine costs

Machine	Capital cost	Capital per dwelling	Annual costs	Running and maintenance
	£	£	£	
Sack-packing compactor	1300	10–26	300–450	35% of annual costs
Incinerator (including flue)	4000	50–80	1000	50% of annual costs

(b) **Incinerator** (Refumatic, Cintafuse) Clean grate, remove residues, charge oversize items, day-to-day maintenance: 7 hour/week

Work at (a) and (b) is carried out by the caretaker of the building.

4 Storage and processing space The space required for dealing with refuse ranges from a hard-standing for a dustbin beside the kitchen door to areas in large commercial buildings which have to accommodate processing machines and storage of refuse for disposal and reclaimed materials. The value of space saved can offset the higher cost of more appropriate plant in some commercial and industrial premises.

References (see also Digest 168)

9 *Refuse storage and collection* Report of the Working Party on Refuse Storage and Collection; Ministry of Housing and Local Government; HMSO: London 1967.

10 *First Report of the Standing Committee on Research into Refuse Collection Storage and Disposal* Department of the Environment; HMSO: London 1973.

Honeycomb fire dampers

Intumescent paints, that swell to many times their original volume when heated, have been used for many years on flammable wallboards to prevent the surface spread of flame. A honeycomb base structure can provide sufficient surface area to support the weight of intumescent paint needed to close its cells. A honeycomb fire damper will permit an almost unrestricted flow of air through a duct or a transfer grille in a wall or door but will effectively close when subjected to the heat of a fire and will provide good insulation downstream from the fire.

The Building (First Amendment) Regulations 1973 require that where a ventilation duct passes through the wall or floor of a compartment an automatic fire shutter must be fitted within the duct (Regulation E9–(1)(c). The fire resistance of the shutter must comply with the standards laid down for the compartment wall or floor (Regulation E5).

If the ventilation duct lies within a protected shaft, or if the shaft itself serves as a ventilating duct, then the duct must be fitted internally with automatic fire shutters to reduce the risk of fire spreading from one compartment to another (Regulation E10(9a)).

Mechanical fire dampers in various forms actuated by fusible-link have been available for many years and they operate rapidly once a fire occurs. Any mechanical device, however, requires careful maintenance if it is to give the standard of reliability that should be associated with fire protection devices. In common with other safety equipment, fire dampers must remain effective after many years often in inaccessible places and in conditions inimical to successful mechanical operation. With this in mind, the honeycomb fire damper was developed.

Intumescent paints
The honeycomb fire damper depends for its action on the property of some paints to swell to many times their original volume when heated. These paints were first produced commercially about a quarter of a century ago to prevent the spread of flame over flammable wall boards and for this they proved to be extremely valuable.

When heated, the paint film froths into a dough-like mass which, as the heating progresses, sets into a carbonaceous meringue having a volume which can be up to several hundred times the original volume of the paint film. The froth at once seals and insulates the surface that is being protected. The insulating properties of these paints naturally attracted attention in the hope that they might be used to increase

the fire resistance of structures as well as reducing the surface spread of flame, but it was found not to be possible to produce a heavy enough paint film on the surface for the required insulating film; increasing the paint film thickness resulted in the paint layer falling slowly away from the surface as the film expanded and became mastic.

In 1970 experiments were started on the use of intumescent paints applied to honeycomb structures as the use of a honeycomb both increased the surface area available to carry paint at the same time that it provided a matrix to support the expanded film of paint. The applications of this work are the subject of patent applications; the dampers are being manufactured under licence by:

Dufaylite Developments Ltd, Cromwell Road, St Neots, Huntingdonshire PE19 1QW.

Painted honeycombs
It is a matter of simple geometry to show that for a honeycomb having a cell size d (measured across opposite flats) and a thickness t, the internal area covered by paint is $4\,t/d$ that of the plane area of the face of the honeycomb. For honeycombs as thick as the cell size the gain in paint weight is x 4 and for a cell size of 10 mm and a thickness of 50 mm the gain in paint weight is x 20.

It does not appear to matter whether the base material of the honeycomb is metal or paper because the honeycomb material is protected by the expanded paint. In fact, honeycombs of paper which are considerably cheaper than those of metal give a slightly higher fire resistance no doubt because of the higher thermal conductivity of metal.

Fire dampers
A honeycomb fire damper when open permits the free transmission of air, but once the honeycomb is heated the paint swells to fill the cells of the honeycomb and form a seal. Two types of damper have been tested in a vertical position for fire resistance according to the time/temperature curve specified

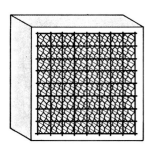
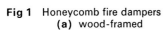
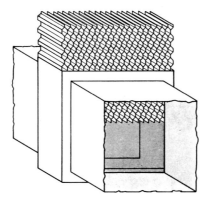

Fig 1 Honeycomb fire dampers
(a) wood-framed   (b) metal envelope

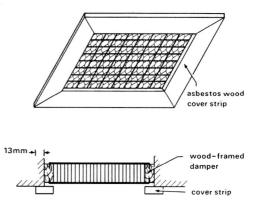

Fig 2 Damper fixed in ceiling

in BS 476 : Pt 8, one framed in wood and another in which the honeycomb is slipped into a metal holder.

Wood framed dampers Dampers framed in wood coated with intumescent paint (Fig 1a) were tested in sizes 228 mm square and 610 mm square; the honeycomb was protected on each face by a welded wire grille of 25 mm square mesh and the honeycomb was supported by metal ties between the grilles at approximately every 125 mm. The honeycomb of the 610 mm square damper was protected along its upper edge by an asbestos board cover extending for 12 mm into the honeycomb space. This type of damper is easy to construct and can readily be screwed into the wall of a duct.

Metal envelope dampers In this type of damper (Fig 1b) a metal slide, to contain the honeycomb, forms part of the duct wall. The honeycomb occupies the whole of the interior of the duct and there is no obscuration by framing. After a fire, the honeycomb can be readily withdrawn and replaced.

In a 50 mm thick damper with a 10 mm cell size (typical of a commercial product) the sequence of events during heating is as follows. The paint first swells on the furnace side of the damper and completely closes it in just over one minute. The paint then progressively swells throughout the thickness of the damper until at the end of 30 minutes the cells are blocked throughout the full thickness of the damper. Meanwhile the paint on the furnace side begins to become carbonised and attrition takes place until fire eventually penetrates along the upper edge of the damper.

The relationship between the weight of paint and the fire resistance is approximately T (minutes) = $13 W$ (kg/m^2). Hence, a fire resistance of one hour would be given by a paint weight of about 5 kg/m^2.

From this sequence of events it would be expected that the fire resistance of fire dampers would be additive and consequently dampers could be placed in cascade. This was tested by placing two one-hour dampers together and the fire resistance was found in fact to be rather more than two hours.

Practical fire tests
In addition to the BS 476 : Pt 8 fire resistance tests, a number of practical tests have been carried out in fire chambers 2·75 m x 3 m x 2·45 m high, carrying a fire load of 24·4 kg/m^2 composed of cribs of 25 mm and 40 mm wooden sections. The window areas of the compartment were such as to give severe fire conditions and the temperature of the fire at its peak was between 900 and 1000°C. The fire test was terminated at about 45 min as by this time the fuel had been consumed. The fire damper was built into one wall of the compartment and the combustion products of the fire were drawn through the damper at different speeds in the various tests.

Two proprietary paints were used in the tests which were made with honeycombs 50 mm thick with a cell size of 10 mm.

After the fire was lit the temperature in the compartment rose steadily to about 250° C when the damper closed. After this, the temperature behind the damper fell to below 100° C and remained at this level until the fuel was consumed. The paints differed in their performance in these tests. The better one formed a seal with the products of combustion being drawn through it at 10 m/s; it failed only towards the end of a test with a velocity of 15 m/s. The other paint was satisfactory only up to an air speed of 5 m/s.

Horizontal operation
The honeycomb apertures close equally well whether the damper is in the horizontal or the vertical position, but care must be taken to support the honeycomb of horizontal dampers adequately otherwise when the paint becomes mastic the honeycomb will sag under its own weight and gaps may be formed at the sides. Proper support can be achieved in the manner shown in Fig 2.

Toxic considerations
In some of the fire resistance tests a search was made for toxic products resulting from the thermal degradation of the damper material. As this consists of carbo-hydrates and ammonium phosphate it would be expected that ammonia and water would be evolved on heating and these products were found though not in quantities that would be

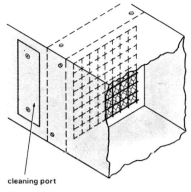

Fig 3 Wood-framed damper in duct

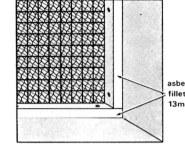

Fig 4 Damper in fire-resisting wall

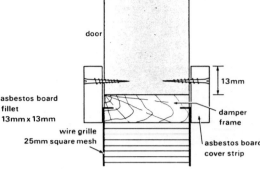

Fig 5 Fixing damper to door

biologically significant. A search was also made for hydrogen cyanide but none was found and tests for phenol revealed only a trace.

It would, of course, be surprising if significant levels of toxic products were found as the paints have been in use as flame-retardant treatments for the last 25 years.

Life of honeycombs

The chemical compounds used in the manufacture of honeycombs are stable and not likely to deteriorate with time. Tests at the Fire Research Station on painted samples of wallboard exposed on corridors showed no change in their fire performance after four years; a sample withdrawn from a building about to be demolished still performed satisfactorily under test after 15 years.

Resistance to air flow

The pressure drop across a honeycomb 50 mm thick and with a cell size of 10 mm, covered by a wire grille, is 9 N/m²* at an air speed of 5 m/s. This low figure is no doubt due to the fact that the honeycomb apertures are aligned to the direction of the air flow and thus have a straightening effect.

The 'free area' of the honeycomb may be deduced by finding the volume of the material of the honeycomb and comparing this with the overall volume of the honeycomb; this amounts to 88 per cent.

The wood-framed honeycomb offers more resistance to air flow than the honeycomb itself because of the restriction of the frame and wire grille. The extra

* 1 N/m² = 0·102 mm, or 0·004 in., water gauge

Table 1 Pressure loss across framed fire dampers

Air speed	Pressure loss					
	due to honeycomb and wire grille		due to frame		total	
m/s	N/m²	in. wg	N/m²	in. wg	N/m²	in. wg
1·25	1·5	0·006	0·2	0·001	1·7	0·007
2·5	3·4	0·014	0·7	0·003	4·1	0·017
3·75	7	0·03	2	0·008	9	0·04
5·0	9	0·04	3	0·01	12	0·05
6·25	12	0·05	5	0·02	17	0·07
7·0	15	0·06	7	0·03	22	0·09

pressure loss due to the restriction of the frame amounts to 0·2 of the velocity head (assuming a 25 mm wide frame to a 610 mm square damper). The total loss in pressure for various air velocities is given in Table 1.

Although the pressure loss across a honeycomb at various air speeds gives the best indication of its effect on the flow of air in a duct, designers may need to know the 'free area': if allowance is made for the frame as well as the honeycomb, the free area is 74 per cent for this size of honeycomb.

Mechanical strength

The honeycomb when painted has great strength along the axis of the apertures. A paint weight such as would give a fire resistance of one hour would give the honeycomb a crushing strength of 1·9 $\times 10^5$ kg/m².

Dirt, moisture and industrial conditions

The intumescence of the paint is not affected by fluff or oil spray, nor is it affected by an atmosphere containing 5 per cent of sulphur dioxide, irrespective of whether the atmosphere is dry or has a relative humidity of up to 95 per cent at normal room temperature (18° C). The dampers ought not to be used, however, in conditions such that water may condense on them as this could lead to salt migration in the paint and this would interfere with correct closure.

Fixing

In ducts Wooden-framed dampers may be fixed directly into ventilation trunking by screws. Two rows of screws should be used so that the damper remains anchored whichever face is attacked by fire and there should be not less than two screws in each side extending to three-quarters of the thickness of the frame (Fig 3).

Cleaning ports should be provided for the removal of any material that would otherwise obstruct the flow of air. Greasy dirt may be cleaned off the damper with white spirit, the damper being withdrawn from the duct during cleaning to prevent an explosive atmosphere being formed within the duct.

Honeycomb dampers in metal holders are easy to withdraw and clean.

Fig 6 Honeycomb being placed in duct for test

As transfer grilles in walls In some ventilating systems air is required to pass through a fire resisting wall as part of the air distribution system or, in some instances, an air supply is required for combustion, as for example in a boiler room. As the damper may be only 50 mm deep, it can readily be accommodated in the thickness of a wall. The heavy coating of intumescent paint on the frame will expand and help to prevent the fire exploiting the joint between the wall and the damper. To increase the safety margin, a fillet of asbestos wallboard should be fixed to cover the joint between the wall and the frame on both sides of the wall (Fig 4).

As transfer grilles in doors Dampers may be included in fire resisting doors in order to transfer air; they should preferably be placed in the upper part of the door so that they close as early as possible in fire conditions. To avoid exploitation of the joint between the damper and the door by fire, this should be covered with a fillet of asbestos as shown in Fig 5.

Partitions and doors
Although most of the work with honeycombs has related so far to their use as fire dampers, it will be seen that if plane faces are attached to the honeycomb—and the intumescent paint on the honeycombs has been used as an adhesive for this purpose—a lightweight fire-resistant partition is formed. With hardboard faces, the weight of the partition would be about one-third that of the lightest conventional partition of equivalent fire resistance. Tests so far have been limited to small panels, about 1 m square. Obvious applications would be in the larger passenger-carrying aircraft of the future, hovercraft and other high-speed transport systems in which it is necessary to reduce mass as far as possible. Such partitions might also be used in buildings in which movable walls could give flexibility in layout, for example, to meet the need for changes in the home as families grow.

The controlling factor in the fire resistance of a partition or door of this construction is the weight of paint per unit plane area of the surface.

The fire resistance of a partition is related to its thickness since the partition must be thick enough to carry the weight of paint necessary for the required fire resistance without the paint rolling out of the honeycomb as it intumesces. For partitions with hardboard faces, if the paint weight W is expressed in kg/m^2, the fire resistance T (minutes) $= 3(4W+3)$. Thus a paint weight of about 4·3 kg/m^2 would be sufficient for a fire resistance of one hour. For doors, rather less paint is necessary because the insulation criterion is waived and $T = 4(4W+3)$; a paint weight of 3 kg/m^2 would therefore give a fire resistance grading of one hour.

Summary
Intumescent honeycomb fire dampers can provide an effective fire barrier for air speeds up to 10 m/s; they provide insulation so that the temperature downstream from the fire is kept below 100° C and this could be a useful property in some situations. They should not be used in conditions in which they are likely to become wet, either by condensation or other cause.

Fig 7 Dampers after test
    (a) seen from furnace side   **(b)** seen from unexposed side

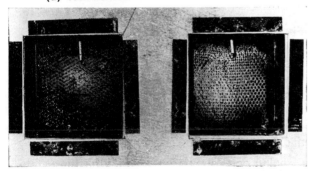

Smoke control in single-storey shopping malls

Should a fire occur in a shop in a covered shopping centre, people escaping along malls may become enveloped in thick smoke. A glass shop front is not a reliable barrier to the flow of hot smoky gases on to the mall because it may break early in the fire.

It is therefore necessary to restrict the travel of smoke either within the shop or within a short length of the mall. The measures necessary to achieve this in single-storey or horizontally-compartmented shops or malls are described in this digest.

When a fire occurs

A column of hot gases and flames rising above a fire draws in or entrains surrounding air (Fig 1). The entrained air becomes contaminated with smoke and other combustion products, which may be toxic, and its temperature is raised. The resulting gases spread out to form a layer beneath the ceiling.

If the fire occurs in an open-fronted shop in a mall, the hot smoky gases will start to flow out beneath the fascia as soon as they reach it, unless there is a system for exhausting the gases from the shop. If the shop has a glass front, it must be assumed that this will break early in a fire to produce, in effect, an open-fronted shop.

The gases will flow out of the upper part of the shop front and rise to the ceiling of the mall while the air to replace them will flow in through the shop front at a lower level.

In the mall, the gases will form a layer beneath the ceiling; this layer will flow faster than the rate of flow from the shop because its volume is increased by the entrainment of air as the gases flow out of the shop front and for some distance along the mall. The layer of hot gases will advance along the mall by the pressures resulting from the buoyancy of the hot gases.

At the same time, air is drawn into the shop on fire (Fig 2) so that there will be a flow of air along the lower part of the mall towards the fire. At first, this air will be free from smoke and the hot smoky layer will probably occupy no more than the upper third of the mall. If the mall is high enough, the smoke layer will be well above people's heads and escape will not be impeded by smoke. But some mixing between the two flows does occur; a relatively stagnant smoky layer forms between the layer of hot gases flowing away from the fire and the return flow of fresh air (Fig 3) and some smoke from this layer mixes into the return air flow which thus gets progressively more smoky. When the hot, smoky upper layer reaches the end of the mall, further mixing occurs. At an open end, the hot gases flow out of the top part and fresh air flows in beneath them, but even a slight breeze will cause sufficient turbulence to mix the smoke with the return air flow, and the mall becomes filled with smoke in the vicinity of the open end— which is usually the exit. At a closed end, or at an obstruction, the layer virtually doubles in depth and mixing of smoke into the return air flow increases along the length of the mall. The mall becomes more thickly filled with smoke moving at pedestrian height from the closed end or obstruction, back towards the fire (Fig 4). This general flow of smoke towards the fire could easily mislead people about the location of the fire and the best direction in which to escape.

Fig 1 Entrainment of air into flames and hot gases rising from fire

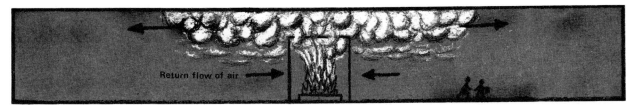

Fig 2 Hot gas layer spreading along the mall

How fast does this happen?

When the fire has attained a moderate size, with flames approaching the ceiling, the smoke will take only a few seconds to fill the space beneath the ceiling of the shop before starting to flow out into the mall. It has been calculated that a fire of 3 m x 3 m in plan area, in a 1000 m³ shop 5 m high, would fill the upper 2 m of the shop with smoke and hot gases in about one minute.

In the mall, the layer of hot gases will advance at a rate of about 1 m/s if the fire is being held in check by sprinklers. If the fire is unchecked, the rate of advance will be higher; in one instance, it is likely that a mall 100 m long became untenable because of heat and smoke in less than one minute.

1 m/s is only a slow walking pace if progress is unimpeded; but congested crowds move even more slowly and the rate of flow on escape routes may be only about 0.5 m/s. Congestion is likely near exits which are frequently at the ends of malls; here, in the absence of smoke-control measures, the smoke becomes thick at ground level before it does so further inside the mall. People may emerge from shops at a time when a layer of smoke is above their heads and the exits are clearly visible, only to find that the end of the mall is filled with smoke before they can reach it.

Design measures

Four measures are needed to cope with this potential situation:

1 Install a sprinkler system in each shop to limit the size of a fire and the amount of smoke produced.
2 Form smoke reservoirs below the ceilings.
3 Extract the smoke from these reservoirs.
4 Provide for the entry of replacement air.

All four of these provisions must be made; none will be effective on its own. The system may be used to confine the smoke to the shop of origin but for this to be effective it is necessary to ensure that the shop front will retain its integrity during a fire, ie glass front not to fall out, self-closing fire doors with adequate seals and, possibly, pressurising the mall relative to the shop. It is therefore more usual to design a system that will limit the spread of fire and smoke in the malls, though it is also desirable to ventilate the smoke from large shops.

To minimise apprehension and reduce the possibility of panic, the bottom of the layer of hot smoky gases in a mall should be as high as possible above people's heads. A height of 2.5 m for the bottom of the gas layer is probably the absolute minimum; dense smoke is generally confined to the upper third of the mall and hence the height of a mall should not be less than about 4 m.

Sprinklers

It is seldom practicable, for economic reasons, to design a smoke-control system for a shopping centre that will exhaust the hot gases from a fire of more than limited size. Reliance will generally be placed on sprinklers to ensure that the fire does not exceed the design size.

Sprinkler discharge in a shop will greatly increase the amount of mixing of smoke and air, so that the shop may fill with swirling smoke down to floor level. The smoke will, however, flow out of the upper part of an open shop front and air will flow in through the lower part in the same manner as with an unsprinklered fire. In operation, sprinklers will not impair natural venting and will increase the effectiveness of a mechanical extraction system.

Sprinklers are not desirable in a mall because of their effect on smoke movement, but if the mall is to

Fig 3 Formation of stagnant layer of smoke

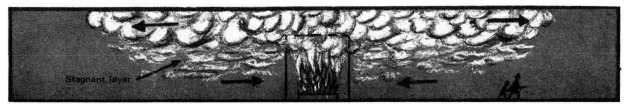

Fig 4 Smoke flow back along the mall

be used for displays or exhibitions which are likely to provide combustible materials, it must have the same measure of protection as the rest of the complex, ie sprinklers must be fitted.

Any smoke-control measure which requires functioning of fans, ventilation, screens, doors, etc should come into operation automatically at the earliest possible stage. By the time a fire has become large enough to actuate a fusible link, it is likely to be producing a very considerable volume of hot smoky gases. Fusible links in the mall are incapable of reacting early enough to a fire in a shop. Smoke detectors are a suitable means of activating smoke-control devices; if the system is intended to restrict gases to the shop of origin, the detectors must generally be situated in the individual shops.

Smoke reservoirs

To help control movement of the layer of gases below the ceiling, smoke reservoirs should be formed to a depth of about one-third of the storey height. If the ceiling of the mall is higher than that of the shops, the difference in levels may be of sufficient depth for these reservoirs (Fig 5a), but if the difference is not sufficient, deep fascias will be needed above the shop fronts (Fig 5b). To limit the distance of smoke travel along the mall, the reservoirs should be no more than about 60 m long and dividing screens might therefore be needed. Within a shop, a reservoir can be formed by the enclosing walls and a deep fascia over the shopfront. The area of a smoke reservoir should not exceed 1000 m² to avoid cooling the layer to an extent that would reduce the efficiency of smoke extraction; screens may be needed to sub-divide the area of a larger shop.

Ceiling screens must be continued up through the space above a suspended ceiling, unless the latter is fire-resisting and sufficiently airtight to prevent smoke leakage. If a suspended ceiling is of an open type of construction, and forms a sufficiently deep space, the smoke reservoirs may be confined entirely to this space and dividing screens need not extend below it.

Smoke extraction from reservoirs

The hot smoky gases must be extracted from the smoke reservoirs either by natural or mechanical ventilation. Extract rates and estimates of the maximum gas temperatures which the extract system must withstand are given in Table 1. These are based on a fire covering a floor area of 3 m x 3 m with a heat output of 5 MW, which is considered reasonable for a fire that is controlled by sprinklers. The estimates neglect the cooling of the hot gases by heat loss to the ceiling and by radiation downwards, so that temperatures a long way from the fire will be somewhat lower. This cooling is advantageous with a mechanical extract system but can reduce the efficiency of a natural system. Temperatures will be higher than those stated if the heat output of the fire exceeds 5 MW, as it may if a sprinkler head fails to open properly.

Table 1 Rates of extract required from smoke reservoirs

Height of lower edge of screen	Mass rate of extract		Max temp of hot gases above ambient		Extract rate (at max temp)	
	shop	mall	shop	mall	shop	mall
m	kg/s	kg/s	°C	°C	m³/s	m³/s
2.5	9	18	560	280	22	29
3	12	24	420	210	24	34
3.5	15	30	340	170	26	39
4	18	36	280	140	29	44
5	25	50	200	100	35	56
6	35	70	150	75	44	73

The areas of vent necessary to exhaust the gases are given in Table 2. These assume that the temperature in the vicinity of the vents is at least 100°C above ambient; in practice, some cooling occurs, particularly with large area reservoirs, and the

Fig 5a

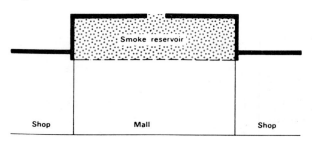

Fig 5b

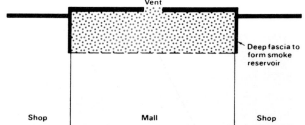

Table 2 Minimum areas of vent (m²) to smoke reservoirs in single storey malls (For smoke reservoirs in shops, halve the areas given below)

Height of lower edge of screen	Height of outlet above lower edge of screen (m)					
m	1.5	2	3	5	8	10
2.5	10	9	7	6	4	4
3	13	12	9	7	6	5
3.5		15	12	9	7	6
4		18	14	11	9	8
5			20	16	12	11
6			26	21	16	15

Note: Generally, areas should be 1½ per cent of floor area in large shops or 3 per cent of floor area in small shops, but not less than the figures given in the above table.

areas should be regarded as minimum ones. Because of the difficulty of estimating the temperature of the gas layer, a rule-of-thumb is frequently adopted; in a single-storey mall where the lower edges of screens are not more than 3.5 m above the floor, the total area of vent should be 3 per cent of the floor area; in large shops, the total area of vent should be 1½ per cent of the floor area.

The area of vent in a single smoke reservoir should be not less than is given in Table 2. This area should be divided between a number of outlets, uniformly distributed, in accordance with Table 3 which is based on a temperature of 100°C for the gas layer. Smoke outlets should be sited at the highest possible position in the reservoir, away from walls and other obstructions to flow.

Fresh air inlets

Inlets must be provided for fresh air to enter the building. Although this is sometimes said to be undesirable because it might promote a faster rate of burning, it is essential if smoke is to be extracted and the escape of people facilitated.

With a natural ventilation system, the area of the inlets should be at least twice the area of the vents in any single smoke reservoir. A smaller inlet area may be sufficient with a mechanical system.

Doors in the ends of malls are suitable inlets if it can be ensured that they will be opened in the event of a fire. Fresh air flowing along the malls from end doors will reduce the likelihood of smoke flowing out of the smoke reservoirs but a disadvantage of using end doors is the mixing that occurs at the end

of the mall with even a slight wind. The effect of this can be minimised by placing a smoke screen several metres inside the mall.

Other possible inlets are windows or ventilators at the rear of shops, or vents in smoke reservoirs remote from the fire, but the method adopted should not induce mixing between the layer of hot smoky gases and the air beneath. A mechanical ventilation system might be capable of supplying at least some of the required air but inlet grilles must be designed and located so that they do not cause the smoke to mix with the fresh air flow. If air is admitted through ceiling grilles, it may be necessary to shut off the mechanical inlet ventilation in the event of fire.

Table 3 Minimum number of outlets from a single smoke reservoir in a mall (For extraction from shops, halve the numbers given below)

Height of lower edge of screen	Depth of reservoir (m)						
m	1	1.5	2	2.5	3	3.5	4
2.5	9	4	2	1	1	1	1
3	12	4	2	2	1	1	1
3.5	15	6	3	2	1	1	1
4	18	7	3	2	2	1	1
5	24	9	4	3	2	2	1
6	34	12	6	4	3	2	1

Advice

It is essential that smoke control measures should be considered at an early stage in the design of the building complex, since they may be difficult to incorporate as an afterthought. Developments of the kind described present particular problems in relation to the Building Regulations and advice should be sought, at the design stage, from the local authorities, the Building Regulations Division of the Department of the Environment, and the local fire authority.

Further reading

HINKLEY P L *Some notes on the control of smoke in enclosed shopping centres* Fire Research Note No 875/1971

SPRATT D and HESELDEN A J M *Efficient extraction of smoke from a thin layer under a ceiling* Fire Research Note No 1001/1974

Fire Prevention Guide No 1 *Fire precautions in town centre redevelopment* Home Office and Scottish Home and Health Department. London HMSO (1972)

Operating costs of services in office buildings

The complexity of engineering services has increased as the standards for the internal environment have been raised and as more measures are taken to combat external noise. This has increased not only the initial capital costs of the installations but also the costs of maintaining and operating them.

A BRS survey has shown that, for a broad assessment, operating costs can be related to floor area. A closer estimate is possible if the plant capacity is known. Maintenance and energy costs are considered separately. The survey results provide a better basis for estimating costs than has previously been available.

The annual expenditure on services maintenance and energy is about equal to the amortised capital cost of the services in office buildings. Current estimates put the cost of owning and operating services in a range from 30 to 45 per cent of the total costs-in-use of a building. This proportion is likely to increase not only because of further improvements in environmental standards but also because external noise levels may require the windows of many buildings to be kept closed, a situation requiring mechanical ventilation and, frequently, air-conditioning.* Services costs are therefore of increasing importance to building owners, but, unfortunately, the designer or client who wishes to estimate their costs-in-use is faced with a lack of systematic data both on maintenance and energy aspects.

The survey of office buildings[1] has shown that, on average, the annual total operating cost for services is nearly £3 per square metre of floor where heating is provided and nearly £4 per square metre with air-conditioning.

The work was done in 25 buildings, 14 of which were air-conditioned. Floor areas ranged from 1000 m² to 100 000 m² and most of the buildings were erected in the last 15 years. Seven of the buildings could not provide time-sheet returns for the resident labour force and so in these cases the labour cost could not be fully partitioned. The survey was limited to the extent that all buildings had oil-fired heating systems, the air-conditioning systems were mostly of the induction type, most used fluorescent artificial lighting with daylight from side windows and there were no deep open-planned buildings.

* The term 'air-conditioning' implies the provision of sufficient refrigeration and heating plant to keep internal temperatures and humidities within the desired range, in addition to a mechanical ventilation system to meet fresh air requirements.

The cost figures quoted in this digest were those which applied in 1968. Since maintenance is mainly labour it should be practicable to up-date these costs by reference to an annual wages index (eg Department of Employment Gazette, HMSO, 50p monthly). Energy costs can be determined from current fuel tariffs. Both sets of data should remain valid whilst there is no major change in the type of equipment used.

Maintenance costs

Maintenance costs are incurred in keeping the plant operational—repairs, cleaning, watchkeeping and supervisory activities. Costs fall into two categories—the direct charge, which covers the labour and materials used on each piece of equipment, and the on-cost which includes supervision, support services such as stores and workshops, sick leave and holiday pay, National Insurance, etc. Typically, on-costs add about 40 per cent to the direct charges.

Costs related to floor area

For budgeting purposes it may often be sufficient to estimate costs from the size of the building. For this purpose the best estimates are given by the index 'functional floor area', which is the combined areas of office space, corridors and toilets; it is usually about 70 per cent of the gross floor area. Figure 1 shows the total direct charges for the survey buildings. Not unexpectedly, costs in the air-conditioned buildings are greater than in those with heating only. When the values are adjusted to include 40 per cent on-costs, the average annual charges for services maintenance, related to functional floor area, are:

heated offices   £1·1 per square metre
air-conditioned offices   £1·6 per square metre

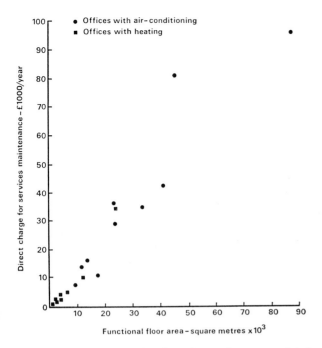

Fig 1 The direct charge for all services maintenance related to functional floor area—the results in this form can be used for broad budgeting purposes.

This method of estimation takes no account of the standards for the internal environment which is a serious limitation when a cost prediction is required for environments much different from those in the survey. This is illustrated by the results for air-conditioned buildings where the wide range of standards for lighting and ventilation causes most of the scatter in the costs.

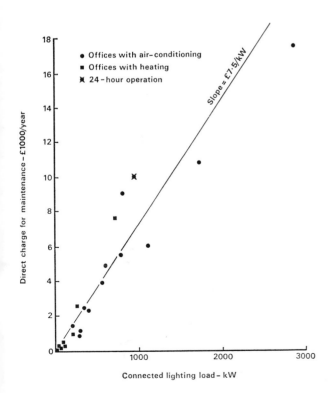

Fig 3 Cost trend—lighting and small power.

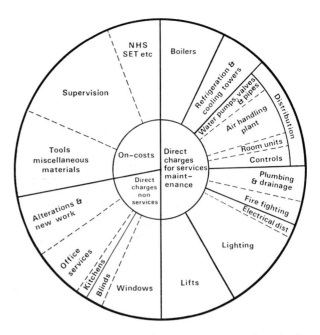

Fig 2 Breakdown of annual maintenance expenditure in air-conditioned buildings.

Costs related to plant duty

Better estimates of maintenance costs are obtained by considering the duty of the installed plant. Figure 2 shows the main areas of expenditure on the maintenance of services in air-conditioned buildings. The groups are described more fully in Table 1.

For each of the groups, the best relationships found so far are with measures of the plant capacity. The survey results are summarised in Table 1 in the form of coefficients derived by assuming straight-line relationships which are close enough for estimating purposes. The coefficients were obtained from cost trends, such as Fig 3, which shows the survey findings for the lighting group. For comparable equipment, there is no difference in costs whether it is installed in heated or air-conditioned buildings.

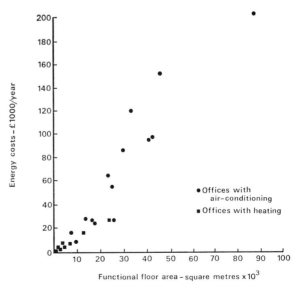

Fig 4 Total annual cost of electricity, fuel-oil, water and gas —this can be used for broad budgeting purposes.

Table 1 Coefficients for annual direct maintenance charge related to plant duty

Group	Constituent Plant	Parameter	Unit	Coefficient £/Unit
Distribution	Fans, pumps, pipes, ductwork, filters, valves, controls and room units for heating and air-conditioning	Shaft power of pumps and fans	kW	25
Lighting	Lighting, small power and main electrical distribution	Connected lighting load	kW	7·5
Lifts	Lifts	Passenger capacity × no. of floors	Pass. floors	2·5
Boiler plant	Boilers, burners, fuel storage, chimneys	Rated heating capacity	kW	0·60
Refrigeration plant	Refrigeration machine and cooling towers	Rated cooling capacity	kW	0·85
Plumbing and drainage	Baths, sinks, WCs, urinals, showers, etc	Number of appliances	Appliance	5·50

An on-cost of 40 per cent must be added to get the total annual cost

Apart from floor area and installed plant capacity, the following factors could also be expected to have a bearing on maintenance costs, although at this time it is not possible to give guidance on their importance : the quality and age of the equipment, the standard of maintenance and the use of resident as opposed to contract labour.

Energy costs

Energy costs are the expenditure on fuel-oil, electricity, gas and water. Of these only water costs are likely to be unrelated to the quantity used ; they are usually assessed as a percentage of the rateable value of the building. The other three costs are based on consumption and tariff structures. As with maintenance costs, floor area gives an approximate guide to energy costs (Fig 4) but this method of estimation is open to the same criticism, that no account is taken of the environment provided in the building. The average expenditure on energy is shown in Fig 5.

Electricity costs

Except for small tenanted areas, the electricity charges in office buildings are based on the 'maximum demand' tariff. This has two main components, a charge for the units of electricity consumed (kWh) and a charge for the highest rate of consumption, the maximum demand (kW). The maximum demand charge is about 40 per cent of the total bill, but this proportion is greater in the smaller buildings.

Electricity consumption

The average annual consumption of electricity is shown in Fig 6, from which it can be seen that buildings with air-conditioning use almost three times as much power as those with heating only. In air-conditioned buildings the typical breakdown is for lighting to use half the energy ; fans and pumps take about one quarter and the remainder is divided

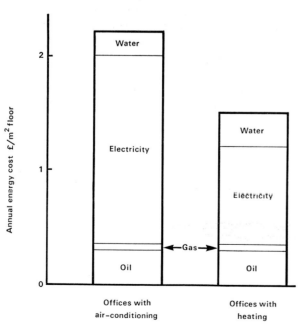

Fig 5   Electricity costs are the largest item in the energy bill.

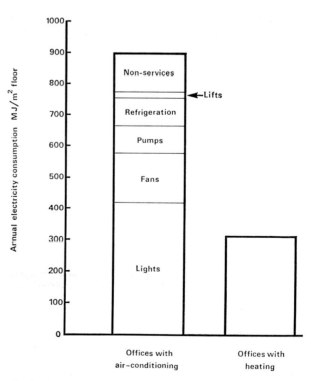

Fig 6   Average annual electricity consumption.

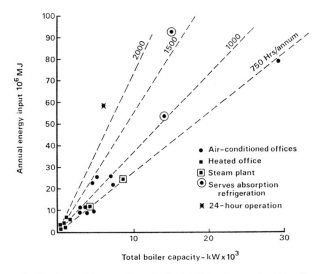

Fig 7 Annual hours of equivalent full-load operation for oil-fired boilers.

between refrigeration and ancillary equipment. Lighting accounts for nearly all the electrical energy in offices that are not air-conditioned. Lift consumption is only about two per cent of the total and can be ignored for most practical purposes.

Energy consumption can be calculated for the main items of plant by using the concept of 'hours of equivalent full-load operation', which is the ratio of the annual energy consumption to the full-load operating power for each item of plant. Values for office buildings are given in Table 2 where it is seen that the annual hours of use of lighting, fans, and pumps exceeds the nominal working year of 2150 hours. For fans and pumps the extra hours are due to the early start-up needed each day to gain control of the thermal environment and to the need for ventilation when the building is occupied, however sparsely, in the evenings. For lighting the increased usage is due to overtime working, security requirements and the use of lights by cleaners. The services are used for longer periods in the prestige air-conditioned buildings.

Table 2 Annual hours of equivalent full-load operation

| Plant | Index | Offices with air-conditioning | | Offices with heating |
		Prestige	Standard	
Refrigeration plant and cooling towers	Full load operating power	1040	1040	—
Fans	Operating load	3800	3000	—
Pumps	Operating load	4000	3000	2460
Lighting	Connected load	3300	2500	2460

The maximum demand for electricity

In heated buildings the maximum demand occurs in the winter months when all the electrical equipment, and particularly the lighting, is operating simultaneously at full load. With air-conditioning the demand is less easy to predict because of the variable nature of the refrigeration load. In winter, with all other plant at full load, refrigeration plant operates typically at about half load, but there is considerable variation in this proportion from building to building. Since the maximum demand levy is usually highest in the winter months there is an advantage in designing to avoid the need for refrigeration at this time. The highest maximum demand occurs in summer, when, in addition to the pumps and fans, refrigeration is at full-load together with between 60 and 70 per cent of the lighting.

Fuel-oil

Full-load operation for boilers ranges from about 750 to 1500 hours annually, with an average value of 1000 hours (see Fig 7). Some of this scatter results from the variety of usage because, in addition to heating the buildings, boilers also supply domestic hot water and sometimes heat for canteens and absorption refrigeration plant. Fuel consumption can be estimated using a method[2] which takes account of the design of the building and of intermittent plant operation; results using this method are in general agreement with the survey figures.

References and further reading
(1) N O Milbank, J P Dowdall and A Slater. Investigation of maintenance and energy costs for services in office buildings. IHVE 1971, 39 (*BRS Current Paper 38/71*)

(2) N S Billington. Estimation of annual fuel consumption. JIHVE 1966

N O Milbank. Energy consumption and cost in two large air-conditioned buildings. IHVE/BRS Symposium: Thermal environment in modern buildings 1968 (*BRS Current Paper 40/68*)

Energy consumption and conservation in buildings

Buildings represent an important part of the UK energy economy. It is estimated that 40 to 50 per cent of the national consumption of primary energy takes place in building services, and that over half this energy is consumed in the domestic sector. By undertaking the technically feasible options it should be possible to achieve an ultimate saving of over 15 per cent of annual consumption of primary energy by measures in building services which would not impair environmental standards. Even undertaking only those measures which use well-established technologies and which are cost effective in the existing housing stock would save about six per cent of the annual UK consumption of primary energy.

Definition of terms

Gigajoule: The unit of energy used in this digest. It equals 10^9 joules and corresponds to about 278 kW hours or 9·5 therms.

Primary or gross energy: The (higher) calorific value of the raw fuel, eg oil, coal, natural gas, nuclear and hydro-electricity, which is input into the UK economy. Both nuclear and hydro-power are used only for the generation of electricity and the convention is adopted that the primary energy attributed to these inputs is equivalent to an efficient coal-burning power station producing the same electrical output.

Delivered or net energy: The energy content of the fuel actually received by the final consumer.

Energy overhead: The difference between the gross energy input to a particular fuel industry and the net energy delivered by that producer.

Useful energy: The energy required to perform a given task. The ratio of useful energy to net energy represents the efficiency of the device employed. The useful energy may not always be well defined particularly when the wasted energy of one process serves to reduce the useful demand of another, eg the fortuitous heat gain from the lighting in an office can reduce the demand for heat from the heating system.

Prepared at Building Research Station, Garston, Watford WD2 7JR
Technical enquiries arising from this Digest should be directed to Building Research Advisory Service at the above address.

Buildings in the national energy economy

The following discussion is based on 1972 figures to avoid any distortions from difficulties of supply in the period which immediately followed that year or the effect of price increases now being passed on to consumers. Primary energy consumption of the UK was 8.9×10^9 GJ and energy delivered to final users was 6.1×10^9 GJ, the difference representing losses in conversion and distribution by the fuel producers (Fig 1). The difference between the gross energy input and net energy output of the electricity industry is almost equal to that delivered to the whole of the industry sector. The relationship between the gross energy used and the net energy delivered for UK fuel industries is given in Table 1.

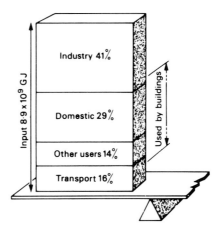

Fig 2 Gross energy used by final users

Table 1 Ratio of gross energy input to net energy delivered to users: UK fuel industries, 1972

Coal	1·03
Oil	1·09
Natural gas	1·07
Town gas	1·42
Electricity	3·82
Other manufactured fuels	1·38

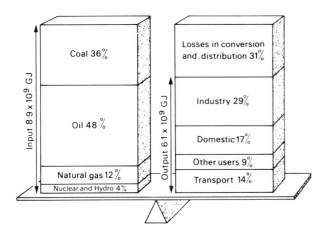

Fig 1 Gross energy input and net energy output of UK economy

It is important to bear in mind that the total amount of energy consumed to perform a particular task depends on the overall efficiency of converting primary energy to useful energy. For example, the high efficiency with which the electric motor converts delivered energy to useful mechanical energy more than recoups its losses in conversion when compared with, say, the internal combustion engine. Figure 2 is derived from Fig 1 by distributing the losses in conversion and distribution amongst the final users in proportion to their use of the various fuels in Table 1.

Nearly all the energy used by the domestic sector is used to provide heating and light within the building. This is also true of a large part of the energy used in the 'other users' sector since this sector also includes schools, hospitals and most offices. There will also be some energy used in the transport sector for building services (associated with railway and bus stations for example) and a good deal in the industry sector. Overall, at least 40 per cent and possibly 50 per cent of the national primary energy input can be associated with building services.

Energy use in the domestic sector

There were about 19 million domestic households in 1972, so the average net energy consumption in the home was about 81 GJ. This provided space heating, water heating, cooking, lighting and other sundry usages. Estimates based on field studies give the approximate usages shown in Table 2.

Table 2 Average annual domestic energy consumptions (GJ)

	Net	Gross	Useful (approx)
Space heating	52	74	30
Water heating	17	30	9
Cooking	7	16	5
Lighting, TV, etc	5	18	5
	81	138	49

An important feature of the UK domestic sector is the relatively large usage of electricity. It is about twice that *per capita* of the original six EEC member countries. Belgium and Holland provide particularly good examples as they share with the UK a maritime climate. The domestic sector *per capita* consumption of electricity in the UK is about $2\frac{1}{2}$ times that of Belgium and twice that of the Netherlands.

The Family Expenditure Survey carried out in 1972 can be used to estimate the consumption of fuel for groups of income in the range 0–£10 per week to greater than £80 per week. The result is shown in Fig 3. The average energy consumption obtained from the survey was 82 GJ, in good agreement with Table 2, derived from other sources. It can be seen

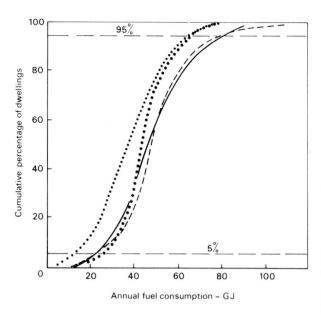

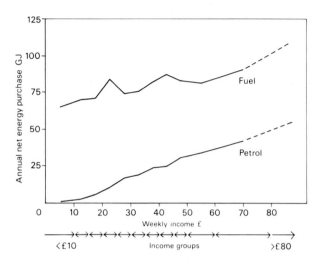

Fig 3 Estimated annual net energy purchases per household based on 1972 Family Expenditure Survey

Fig 4 Distribution of annual fuel usage in four local authority housing estates of 100–200 houses. Average annual fuel consumption 42GJ

that consumption rises fairly slowly with income. For an 8-fold increase in income, expenditure on domestic energy increases by only 74 per cent. This compares with an almost 20-fold increase in expenditure on petrol over the same income range. The small effect that income has on the consumption of domestic energy may possibly be attributed to a combination of partially cancelling factors. High income groups will probably have more modern, efficient appliances and better levels of thermal insulation, but there will also be a tendency towards higher comfort levels and larger heated zones. Patterns of occupation will also play a contributory role. Amongst the higher income groups of households there will be a larger fraction of dwellings which are unoccupied during working hours, probably in excess of 60 per cent at the median income group. This falls to as low as five per cent in the lowest income group which is made up predominantly of persons over 65 years of age.

In view of the differing social factors which affect energy consumption, it is not surprising that within any one socio-economic group in apparently identical housing, large differences in energy consumption are found. For example, a factor of five has been observed in a recent study of Scottish local authority housing between the lowest five per cent consumers and the top five per cent (Fig 4).

Variations also arise from differences in built-form, thermal insulation level and type of heating system. These last two factors are illustrated in Fig 5 which shows the total net energy consumption for over 1000 Scottish local authority dwellings monitored by BRE during 1972. When the consumption is corrected by the factors in Table 1 the gas systems have twice the primary energy efficiency of the electric systems. This difference is greater than that shown in the figure for dwellings with different levels of fabric heat loss but with the same type of heating system.

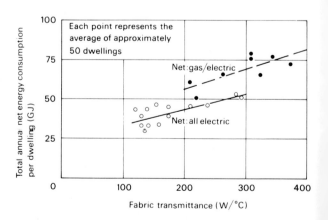

Fig 5 BRE (Scottish Laboratory) results. Total annual net energy consumption per dwelling for gas and electric space heating against fabric transmittance

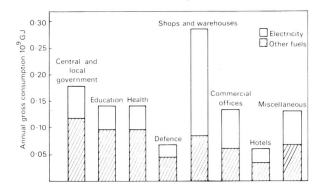

Fig 6 Estimated gross energy consumption in the categories public service and miscellaneous (derived from various tables in UK Energy Statistics 1973)

can represent a dominant heat source to an area and so incur a cooling load even in winter. Figure 7 illustrates the consequences for energy consumption where the gross energy consumption of a shallow, naturally-lit, naturally-ventilated office is compared with that of a deep-plan air-conditioned office lit by traditional fluorescent light fittings to a level of 750 lux.

The naturally-lit, naturally-ventilated office has a primary energy consumption almost one-third less than the air-conditioned, artificially-lit office. The use of more efficient fittings and particularly lighting standards optimised to the tasks to be performed can help to reduce the difference between the two built-forms.

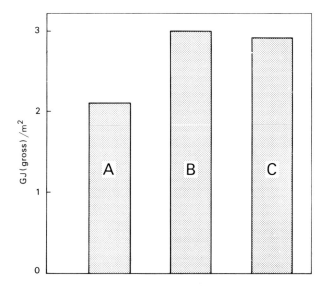

Fig 7 Primary energy consumptions for three built-forms of tall offices
A naturally-ventilated, naturally-lit, shallow plan
B air-conditioned, naturally-lit, shallow plan
C air-conditioned, artificially-lit at 750 lux, deep plan

Energy conservation economics

Most measures which conserve energy consume other scarce resources, with the possible exception of better husbandry. If the optimum use of all resources is to be made, resources devoted to energy conservation must be cost-effectively applied. In principle, this should be checked by comparing the stream of savings resulting from the measure with the rate of return on the same sum of capital placed in the next alternative investment. It is thus not unusual for the cost-effectiveness of an energy conserving measure to differ between consumers because of differences in alternative investment (or loan) opportunities and differences in the time horizons for decisions.

The principles of cost-effectiveness analysis for decisions in the public sector are well established. Both capital expenditure and cost savings are discounted at 10 per cent in real terms to their present value. The measure is cost-effective if the present value of the savings exceeds the present value of the capital costs. The present value of the savings is taken as that sum of money which, when invested at 10 per cent in real terms, would produce the same stream of savings through time. The present value of the capital costs is similarly defined. The present value can be conveniently computed from tables; a few useful values are given in Table 3.

Energy use in non-domestic buildings

There are considerable difficulties in using existing statistics to estimate energy use in non-domestic buildings. Figure 6 gives estimates of the division of the gross energy consumed by the 'Other users' sector against eight categories. The pattern of energy use varies widely between different building types, and from the pattern of use in the domestic sector, lighting plays a much more important role in the non-domestic sector. The lighting of offices and schools alone accounts for about one per cent of the national primary energy consumption. Lighting levels

Table 3 Present value of £1·00 p.a. for n years at 10 per cent discount rate

n	10	15	30	40	60	∞
pv (£)	6·15	7·61	9·43	9·78	9·97	10

Present value of £1·00 paid n years from now at 10 per cent discount rate

n	5	10	15	20	30
pv (£)	0·621	0·386	0·239	0·149	0·06

As far as possible all costs that are incurred during the measure's life, such as maintenance cost, should be included in the analysis. The future cost of energy is a source of considerable uncertainty, particularly when it is projected a considerable distance into the future. This element of uncertainty is largely removed if it is thought that the cost of energy is not likely to fall in real terms over the life time of the measure. Under these conditions, the best use of resources occurs if the measures are implemented as they become cost-effective with the fuel costs prevailing at that time.

The one important exception to this rule occurs when there is some first cost advantage in taking action now; for example, including a measure when the building is constructed. One possible way to meet this case is to test the sensitivity of the analysis to the use of several different assumptions concerning the variation of cost with time (examples are given in Ref 1).

The costs used in a public sector analysis should, strictly speaking, be the cost of the resources employed and should not include elements of tax or subsidy. However, current market prices can in most cases be taken as a good approximation. Many measures also offer a number of benefits (such as the reduced sound transmission through widely-spaced double glazing) but care should be taken in identifying the purpose of the analysis before they are included.

When undertaken by the consumer financing his own conservation measures, the analysis should use the prevailing prices, irrespective of any hidden tax or subsidy element. The consumer's own valuation of any additional benefits resulting from the measure should be included where possible. Many businesses have an internal discount rate as part of their management accounting procedure, so that a present value analysis can be undertaken along the same lines as discussed for the public sector. For the individual consumer the choice of a discount rate is more difficult. If, for example, he finances the measure as part of a mortgage, the mortgage rate would represent the correct discount rate after allowance for tax. The discount rate would be lower, however, if he used his savings. An important consideration for the individual consumer which might not appear in a public sector analysis is a limit on the time in which costs could be recovered. Thus, if the house market is relatively insensitive to the value of the particular measure the house-owner will wish to have recovered his capital (and lost interest) by the time his house is sold. Discussion of the consumer's point of view is given in Ref 1.

The conclusions of a cost-effective analysis can differ depending on whether the analysis is taken from the point of view of the nation, the firm or the consumer. Particular attention is to be paid to the choice of discount rate and to the costs and incidental benefits to be included in the analysis.

Options for energy conservation

In appraising the amount of energy saved by a given measure, it is important to allow for both the efficiency of the energy-consuming appliance and the energy overhead of the relevant fuel industry so that the total primary energy savings can be evaluated. Electricity, for example, has a consistently high efficiency at the point of use, but it is necessary to allow for the efficiency of electricity generation before it is possible to determine the overall efficiency of using an electric appliance compared with one which utilises the direct combustion of fossil fuel. An indication is given of the cost-effectiveness of the following measures as a means of conserving energy from a national point of view based on January 1975 resource costs. In general, a better indication of the use of resources will be obtained by determining the cost-effectiveness of each individual application. This is particularly true in the non-domestic sector where building types vary considerably.

Combined generation of heat and power

District heating schemes seldom offer substantial savings over individual fossil-fuelled appliances. They do, however, offer the opportunity for a combined generation of heat and electrical power. About 10 per cent of the national primary energy consumption could be saved if all buildings could be heated by such a scheme. There are substantial problems in matching the load factors of the heat demand and the electricity demand which can adversely affect both the energy efficiency and economics. One way in which this problem can be overcome is by utilising large-scale thermal storage to decouple the two demands for energy. The economics of such storage needs to be determined in each application.

Heat pumps

These use mechanical energy to transfer heat from a free heat source (commonly the outside air) to the building. It seems technically feasible to design heat pumps which would transfer three times as much heat as mechanical energy consumed. An electrical heat pump thus has a primary energy efficiency better than that of the best modern domestic boilers. The use of heat pumps to supply heat to buildings would save about seven per cent of the national primary energy consumption. Heat pumps at present available for domestic use are not economic at 1975 costs when compared with natural gas central heating systems, but there is considerable scope for cost reductions when the designs are optimised for UK conditions. Their economics would be most favourable in applications where electricity was the only energy source available.

Better primary energy efficiency

Electricity is a high-grade energy source in that it can provide motive power very efficiently even after allowing for power station losses. However, the demand for heating buildings is generally for low-grade heat and this can be provided with a better primary energy efficiency and usually more economically by the direct combustion of fossil fuels. It can be seen from Fig 5 that the net efficiency of gas relative to electricity is 70–80 per cent and the high primary energy efficiency of gas easily recoups this figure in primary energy terms. In primary energy terms, gas is about twice as efficient as electricity for space heating.

Further savings can be made in fossil-fuelled systems by improvement of control and husbandry and in avoiding oversizing at the design stage. Savings in the order of four per cent in the national consumption of primary energy can probably be made by improvements in primary energy efficiencies. The cost-effectiveness of different measures varies. In the domestic sector it would be cost-effective in most cases to replace existing electric space and water heating by direct fired fossil-fuelled appliances.

Thermal insulation

Possible options in the existing stock include roof insulation, cavity fill, double glazing, internal lining and external cladding. The full potential savings from insulation may not be realised, particularly in housing. This is because some consumers, especially those with a poor standard of heating, may take up some of the savings in higher comfort standards. An important additional contributory factor reducing the potential savings is the redistribution of internal temperature following improved insulation. The savings possible from improved thermal insulation amount to some four per cent of the national primary energy consumption.

In the domestic sector it would be cost-effective at January 1975 costs to undertake loft insulation and cavity fill where possible in the housing stock. The most advantageous implementations from a cost-effective point of view are in buildings heated continuously to a high standard, such as hospitals.

Ventilation control and heat recovery

Winter-time control of ventilation to the levels required for safety and comfort could save about one per cent of the national primary energy consumption. Further savings can be achieved by heat recovery from the exhausted air using a heat wheel, heat pump or other heat exchanging devices. Ultimate savings then rise to about two per cent of the national energy savings. The cost-effectiveness for dwellings

at current costs remains to be established. This is particularly so in more recent constructions where the ventilation rates achieved in practice appear to be slightly below one air change per hour.

The most cost-advantageous applications of heat recovery are where the latent heat of water may also be recovered, such as in indoor swimming pools.

Natural energy sources

Solar gain already contributes to the heat input of a building during winter. In a more direct application of solar energy a solar collector may be used to supplement the domestic hot water supply, although a 6 m² collector can supply only about half the needs of an average dwelling and would not be cost-effective unless direct-fired fossil-fuelled heating was not available. The UK climate makes the generation of electricity from wind feasible although rotor diameters of the order of 10 m would be required in light wind areas if the major part of the dwelling heat load was to be met. Both solar and wind energy require either stand-by plant or extensive thermal storage. Geothermal energy represents another possible natural energy source and could supply heat to buildings through a district heating scheme if suitable areas could be found. It is estimated, after allowance for the restrictions placed by the existing building stock, that the use of natural sources of energy could save perhaps one to two per cent on the national primary energy consumption.

Lighting

Potential savings are limited to non-domestic buildings, particularly important areas being offices and schools. Savings from better lighting control, better lighting design to meet task requirements and choice of more efficient luminaires could save over one-half per cent of the national primary energy.

The cost-effectiveness of such measures is under investigation.

Many of the measures listed above interact or are mutually exclusive. After some allowance is made for this fact it may be concluded that in excess of 15 per cent of the national primary energy consumption could ultimately be saved by action in buildings. Not all the measures included in this figure represent a cost-effective use of resources at present costs, but in the domestic sector alone national savings of six per cent, which would be cost-effective when taken over the whole housing stock, can be obtained by improved thermal insulation and the direct use of fossil fuels instead of electricity for heating.

Future supply of energy to buildings

Figure 8 shows the projection of the primary energy consumption of the UK given in evidence to the Select Committee on Science and Technology in 1975, implied by an economic growth rate in the range of 2·7 to 3·3 per cent per annum over the next 15 years. The mix of primary fuels into the UK economy varies considerably within the range of the projections. The underlying assumptions which produce this variability are described more fully in Ref 3. With the notable exception of nuclear power with its characteristically inflexible input because of its long lead time in construction, it is evident that the energy supply can alter considerably in a time-span fairly short compared with the lifetime of a building. This emphasises the need for the designer to avoid decisions which severely limit the building users future options. Keeping options open may not necessarily incur great expense. For example, space can be earmarked for possible storage of fossil fuels even if natural gas is intended to be used initially in a boiler installation. For similar reasons, special attention should be paid to the thermal insulation of those building elements which would later prove extremely expensive to up-grade.

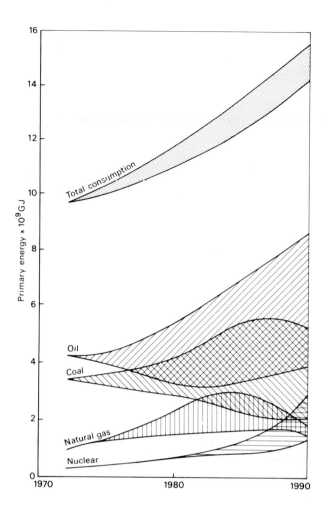

Fig 8 Envelopes of Department of Energy consumption forecasts (1974). The figure shows the range of primary energy inputs under a number of different assumptions for an economic growth rate of between 2·7 per cent and 3·3 per cent per annum

References

1 Energy conservation: a study of energy consumption in buildings and possible means of saving energy in housing. BRE Current Paper CP 56/75
2 Digest of UK Energy Statistics 1974, HMSO
3 Select Committee for Science and Technology (Energy Resources Sub-Committee) Evidence to Session 12 March Annex C

7 External Environment

Wind environment around tall buildings

Tall buildings can deflect wind down towards the ground, producing unpleasantly high speeds in pedestrian areas. Studies at model-scale in wind tunnels and at full-scale around existing buildings have made it possible to estimate the ratios of these speeds to the free, undisturbed wind speed at critical positions for common layouts and shapes of buildings. It is suggested that a design should aim to keep the mean wind speed at pedestrian level below 5 m/s for as much of the time as possible.

Some examples of wind speed ratios are given and the likely occurrence of wind speeds greater than 5 m/s is estimated for particular sites. Remedies used to eliminate high wind speeds from pedestrian areas near to tall buildings are described. Situations are specified under which high priority should be given to the wind environment in the planning of redevelopment schemes. Designs that help to protect pedestrian areas from the downflow from tall buildings are outlined.

The more or less uniformly low buildings in old towns provide shelter from the wind, but in the last 20 years the situation has changed radically with the development of wider, taller buildings that deflect the wind down into previously sheltered areas. The resulting high wind speeds near the ground are unpleasant and sometimes dangerous. Some windy shopping areas, for example, have proved so unpopular with traders and customers that money has been spent to improve conditions. The problem is becoming more widespread with the increase in the number of tall buildings.

Research in the last few years has provided a basis for the recommendations in this digest; wind tunnel tests in appropriate conditions have given qualitative and quantitative information which can be combined with comfort criteria and meteorological wind data to give estimates of the occurrence of unpleasant winds around buildings. Such information has been augmented from the study of practical cases where efforts have been made to improve windy conditions.

The wind and its simulation
The movement of the atmosphere high above the surface of the earth is controlled by large-scale weather patterns, but at lower levels the frictional drag of the surface modifies the flow and produces a boundary layer in which there is a velocity gradient: a decrease in mean wind speed down to zero at the surface. The depth of the boundary layer depends on the roughness of the surface, and varies from about 300 m over open, level country to about 500 m over

212

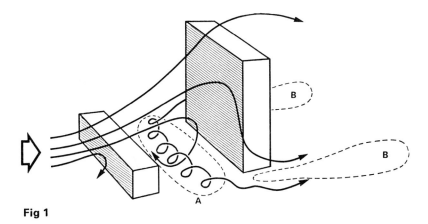

Fig 1

large towns. Conditions within the boundary layer are complex and there is little information available on the detailed turbulent structure of the wind in urban areas.

In the Building Research Station wind tunnel, a typical natural velocity gradient is reproduced, and to improve the simulation of conditions in investigations of particular sites models of the buildings within a radius of 100 to 200 metres around the main building are included. The model scale is usually between 1:100 and 1:250. Under these conditions flow patterns, speeds and pressures around models of particular buildings have been found to agree reasonably well with some full-scale observations.

Flow around buildings

In this digest wind speeds are expressed as a ratio:

$$R = \frac{\text{wind speed at pedestrian height near a building}}{\text{free wind speed at the same height with no buildings present}}$$

When the wind flows over rows of low buildings, as in older parts of a town, pedestrian areas are generally sheltered with wind speed ratios, R, in the region of 0·5 to 0·7.

When the wind meets a building considerably taller than its neighbours a different flow pattern occurs (Fig 1). The wind divides at about two-thirds to three-quarters of the building height. Above this division, the air flows up the face of the building and over the roof. Below, the wind descends to form a vortex in the space in front of the tall building, and sweeps around the windward corners. This produces high wind speeds in regions A and B, where R may be as high as 1·5 and 2·0 respectively. If the building is standing on pilotis or is pierced at ground level by pedestrian ways (Fig 2), some of the descending wind passes beneath the building and produces a region of high wind speed at C, with R values up to about 3·0.

Fig 2

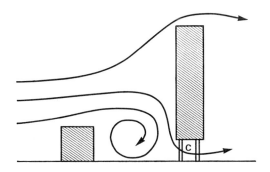

These speeds are dependent on various factors but in particular on the height and breadth of the tall building. The values quoted are for a building four or more times the height of the surrounding low buildings and with a height/breadth ratio of 1 to 2 (slab buildings). When the height is less than twice that of the surroundings, R does not generally exceed 1·0.

The effects of wind on people

The most serious effects are due to the force of the wind, which increases with the square of the wind speed. Thermal effects may also be important, particularly when people are wearing thin clothing, easily penetrated by wind. It is not easy to stipulate what speeds are acceptable from this point of view because of the interaction of sunshine, air temperature, clothing and activity. If people are dressed adequately for the existing air temperature, in windproof clothing, wind force is probably more important than wind chill.

Table 1 describes some effects of wind on people. These descriptions suggest that at a speed of 5 m/s wind begins to be annoying, at 10 m/s it is definitely disagreeable, and at 20 m/s it is likely to be dangerous. The aim, therefore, should be to keep wind speeds below 5 m/s for as much of the time as possible.

Table 1 Effects of wind on people

Beaufort Number	Wind speed m/s	Effect
0, 1	0–1·5	No noticeable wind
2	1·6–3·3	Wind felt on face
3	3·4–5·4	Hair is disturbed, clothing flaps
4	5·5–7·9	Raises dust, dry soil and loose paper
		Hair disarranged
5	8·0–10·7	Force of wind felt on body
		Limit of agreeable wind on land
6	10·8–13·8	Umbrellas used with difficulty
		Difficult to walk steadily
7	13·9–17·1	Inconvenience felt when walking
8	17·2–20·7	Generally impedes progress
9	20·8–24·4	People blown over by gusts

Wind frequency

The frequency of occurrence of wind speeds greater than 5 m/s can be estimated from meteorological data. The Meteorological Office has published wind frequency data for 35 sites in the UK[1] and can supply data for other sites[2].

As an example of the use of such data to indicate orders of magnitude, consider the case of a building pierced at ground level, where the R value is 3·0 for easterly (50° to 130°) and westerly (230° to 310°) winds. The free wind speed which will produce a speed of 5 m/s through the building is $\frac{5}{3\cdot0} = 1\cdot67$ m/s, and meteorological data for a typical inland site show that such a speed could occur from the chosen directions for perhaps 30 per cent of the time. In comparison, in the surrounding town an R value of 0·7 might be appropriate, requiring a wind speed of $\frac{5}{0\cdot7} = 7\cdot1$ m/s, which would occur for only about 5 per cent of the time.

Practical cases

Until the last few years no guidance on wind environment was available to designers and consequently some town centres were developed in a way that gave rise to serious wind problems. The following examples illustrate three cases where conditions were considered bad enough to justify expenditure on remedial measures. Mean speeds of 5 m/s or more were estimated to occur for about 20 per cent of the time in each case.

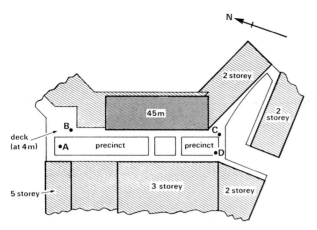

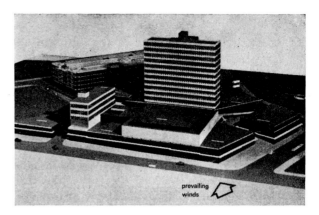

Fig 3 Layout and model of the Leeds centre

Conditions around a 45 m high building adjacent to a shopping precinct in Leeds were reported to be much less pleasant than in the streets in the vicinity. Tests on a model showed that prevailing westerly winds were deflected down into the shopping area, causing high wind speeds at points A, B, C and D (Fig 3), with R values up to 2·2. The addition of a roof over the precinct in the model prevented the wind reaching pedestrian level. A roof was subsequently built and has proved satisfactory in producing a comfortable and pleasant environment.

At Croydon, a 75 m high building spans the end of a shopping mall, with a passage-way 12 m wide and 4 m high beneath the building connecting the mall to the street. Unpleasant conditions were reported in the mall and in the passage-way. Wind tunnel tests showed that a vortex formed on the windward side of the tall building, and that wind then passed under the building at high speed. High-speed easterly and westerly winds were experienced in the passage-way with R values of 2·5 at point A (Fig 4), whilst in the mall, at points B and C, R reached about 1·7. Twenty designs of roof were studied in the wind tunnel, and subsequently a roof was built, carried on existing canopies. Glazed screens were also erected in the passage-way to help particularly in dealing with east winds.

At Edmonton Green, London Borough of Enfield, three tall buildings, 75 m high, adjoin a shopping mall and market square. Each tall building has passage-ways beneath, giving access to the shopping centre. The architects suspected at the design stage that high wind speeds might occur in the shopping areas, and a scale model was tested.

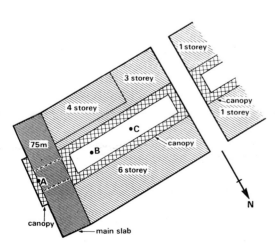
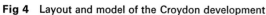

Fig 4 Layout and model of the Croydon development

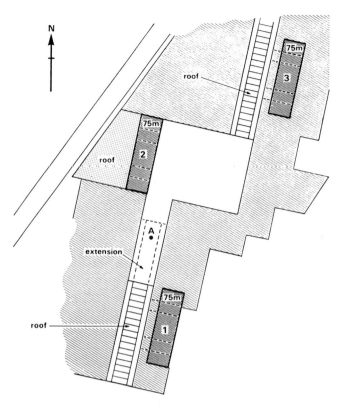

Fig 5 Layout of the Edmonton development

The tests revealed high-speed areas caused by the windward vortex and flow beneath the building, and interaction between two of the tall buildings. With a westerly wind the flow down the face of building 1 (Fig 5) was sucked into the low pressure region in the lee of building 2, resulting in a south to north flow along the mall, with an R value of 2·7 at point A. With easterly winds the flow reversed, blowing north to south at similar speeds. As a result of the test, it was recommended that the mall between buildings 1 and 2 should be roofed. Because of structural difficulties the roof was not initially built to the full length of the mall, and high speeds have been experienced in practice at point A, although conditions beneath the roof are good. Following further consultation the roof has recently been extended as shown in Fig 5, and it is expected that this will prevent high speeds at point A. The square is still uncomfortable with east winds due to vortex formation to windward of building 2, but this is considered acceptable as east winds are somewhat less frequent than westerlies. The cost of roofing the whole square would be prohibitive, but as a result of experience with the existing part of the site, future development of a mall on the north part will be completely roofed.

Approach to design
The procedure outlined—combining meteorological data with wind speed ratios to esimate the likely occurrence of unpleasant winds—may be used for design, and several specific cases have been dealt with in this way at the Building Research Station, but some simpler recommendations can also be made. One approach is to consider the limiting condition where the volume of complaint becomes large enough for money to be spent on remedial measures. This limit was reached in the three cases described, involving slab buildings 75 m high in the suburbs of London and 45 m high in the windier area of Leeds. This suggests a first rough guideline for developers: where suburban redevelopment schemes are to include slab buildings of

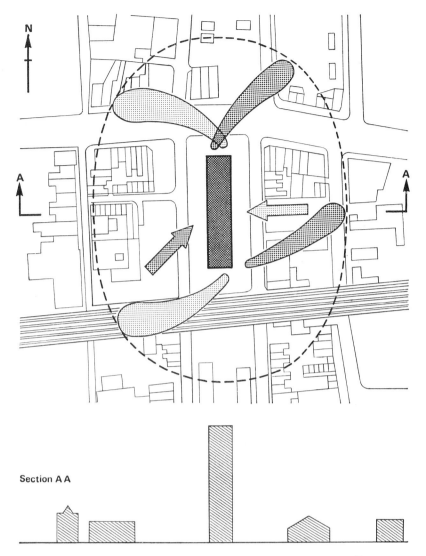

Section A A

Fig 6

these heights or more in similar situations, attention to the wind environment in pedestrian areas should be given high priority in planning and design. This in effect represents a minimum standard based on the only known sites in the UK where extensive remedial measures have been applied after construction. Complaints have arisen with lower buildings but have not been sufficient to persuade developers to take action, and few complaints have been received with buildings less than 25 m high in the suburbs. For buildings of moderate height, therefore, consideration of wind conditions may be given lower priority. The wind environment should not be forgotten, however, if pleasant conditions are to be achieved, for there is likely to be a growing demand for better standards.

Design principles

Figure 6 shows a tall building in part of a town where the average height of surrounding buildings is some 10 to 15 m (3 or 4 storeys). Regions of increased speed extend downwind for a distance roughly equal to the height of the tall building, so that, taking all wind directions into account, the area shown in the figure may be affected.

The following illustrations suggest some designs which help to protect pedestrian areas from the downflow from tall buildings:
a tall building standing on a large podium; the affected areas can be confined to the roof of the podium (Fig 7);

Fig 7

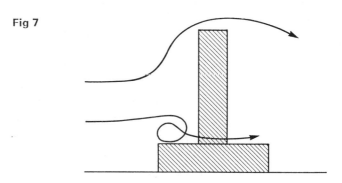

a small podium or a wide canopy, with a vent space above (Fig 8);

Fig 8

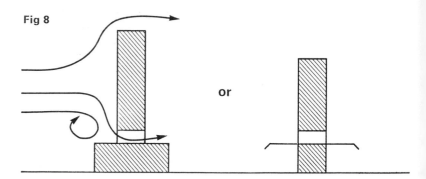

sensitive areas such as shopping precincts can be roofed over (Fig 9);

Fig 9

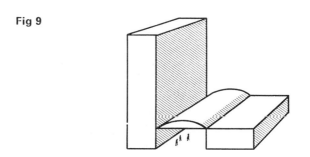

passage-ways beneath buildings can be reduced in area by the erection of screens (Fig 10).

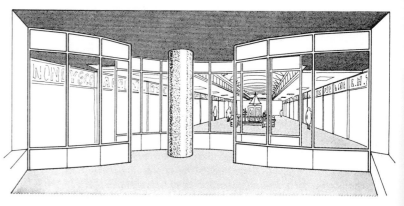

Fig 10

Fig 11

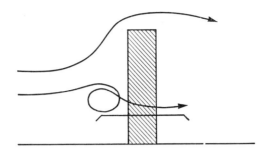

Where buildings prove to be windy after construction, similar remedial treatment to those shown in Figs 9 and 10 can be applied. A large canopy can prevent the windward vortex and corner streams descending to ground level (Fig 11).

The recommendations in this digest apply particularly in cases where the tall building is sufficiently isolated for the flow around it to be unaffected by other tall buildings. If two or more tall buildings are close enough to interact with each other, the situation may be more complicated and require special investigation. Similarly, in the centres of large cities where the general building height may be 25 to 30 m, the influence of other buildings can be considerable. Under these conditions expert advice should be sought to estimate the effect of any proposed tall buildings, and a wind tunnel test may be necessary.

References

1 Tables of surface wind speed and direction over the United Kingdom Meteorological Office, Met. 0.792, HMSO 1968.

2 Advisory offices of the Meteorological Office:

Meteorological Office
Met 0.3 (b)
London Road
Bracknell
RG12 2SZ

Meteorological Office
26 Palmerston Place
Edinburgh
EH12 5AN

Meteorological Office
Tyrone House
Ormeau Avenue
Belfast
BT2 8HH

Further reading

A F E Wise. Wind effects due to groups of buildings. Phil. Trans. R Soc. A.269. 469–485 (1971) (*BRS Current Paper 23/70*)

The assessment of wind loads

The scope of this digest is restricted to the assessment of wind loads on buildings such as dwellings, offices, factories and their associated constructions, including chimneys. It does not deal with the problems of specially flexible structures such as bridges, tall masts and transmission lines.

The recommendations are in line with British Standard Code of Practice CP 3: Chapter V: Part 2: 1972.

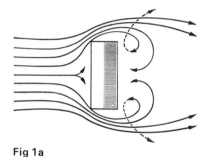

Fig 1a

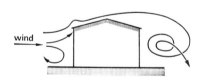

Fig 1b

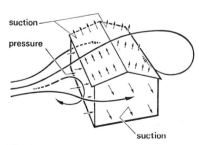

Fig 1c

The generation of pressures and suctions

When the wind blows more or less square-on to a building, it is slowed down against the front face with a consequent build-up of pressure against that face; at the same time it is deflected and accelerated around the end walls and over the roof with a consequent reduction of pressure, ie suction, exerted on these areas. A large eddy is created behind the building which exerts a suction on the rear face. These effects are illustrated in Figs 1a, 1b and 1c.

The greater the speeding up of the wind, the greater will be the suction. Thus the channelling of wind between two buildings can produce severe suction loadings on the sides facing the gap between them. Access openings through and under large slab-like blocks are usually subjected to high wind speeds through them due to the pressure difference between the front and rear faces of the building; the facings of such openings are particularly prone to high suction loadings which may damage the glazing and cladding.

Effect of roof pitch On the windward slope of a roof, the pressure is dependent on the pitch. When the roof angle is less than about 30°, the windward slope can be subjected to severe suction. Roofs steeper than about 35° generally present a sufficient obstruction to the wind for a positive pressure to be developed on their windward slopes, but even with such roofs there is a zone near the ridge where suction is developed and roof coverings may be dislodged if they are not securely fastened. Roofs of all pitches are affected by suction along their windward edges when the wind blows along the direction of the ridge (see Fig 2). Lee slopes are always subject to suction.

The suction over a roof, particularly a low-pitched one, is often the most severe wind load experienced by any part of a building. Under strong winds the uplift on the roof may be far in excess of its dead weight, requiring firm positive anchorage to an adequate foundation to prevent the roof from being lifted and torn from the building.

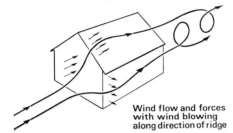

Fig 2

Wind flow and forces with wind blowing along direction of ridge

Variation of pressure over a surface The distribution of pressure, or suction, over a wall or roof surface is generally far from uniform. Pressure tends to be greatest near the centre of a windward wall and falls off towards the edges. The most severe suction is generated at the corners and along the edges of walls and roofs; careful attention must be paid to the fixings at these locations.

Any projecting feature such as a chimney stack or a bell tower will generate eddies in the air flow which increase suction locally in their wake. The roof cladding around projections therefore needs special attention.

Roof overhangs In the assessment of uplift on roofs the effect of roof overhangs is sometimes overlooked. The flow diagrams in Fig 1 show how a windward wall deflects the air flow downwards and upwards. The upward element gives rise to a pressure on the underside of the roof overhang and any other section of the roof, such as the cover to a balcony, to which it has access. This upward pressure on the underside reinforces any suction there may be over the roof and must be taken into account in assessing the roof uplift.

Vortex action on roofs When the wind blows obliquely on to a building it is deflected round and over it. The pressures on the walls are generally less severe than in the previous cases, but strong vortices may be generated as the wind rolls up and over the edges of the roof, as in Fig 3. These give rise to high suctions on the edges of the roof which must be resisted by especially firm fixing of the roof structure

and covering in these areas. Since at most sites the wind can blow from any direction, all edges and corners need special attention (see Fig 4).

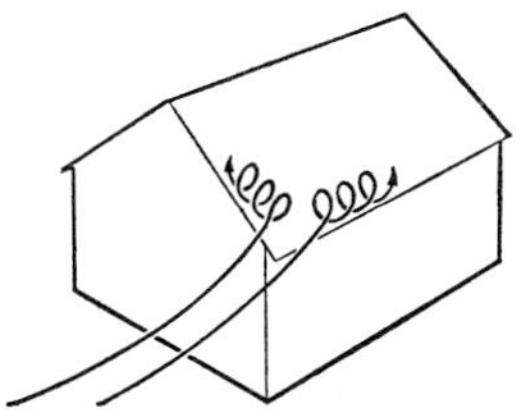

The response of buildings to the wind

Tall chimneys and towers often have natural periods of oscillation of several seconds. For example, the Post Office Tower in London has a natural period of about 6 sec; but most ordinary buildings are relatively stiff with natural periods of less than 1 sec. An ordinary building is thus responsive to short duration loadings and so will respond to short gusts if these are of sufficient spatial dimension to encompass the building. It is therefore important to ensure that deflections under the loading due to a short gust do not exceed the permissible limits. Trouble from excessive deflection has been rare in the past but the situation is being reached where blocks of flats rising to about 30 storeys have been subject to sway and wind vibration which, although not structurally dangerous, has been beyond the limit acceptable for residential occupancy.

Gust loadings are not normally sufficiently frequent or regular to excite a resonant oscillation but it is possible that eddies shed by the structure in a regular pattern may give rise to a resonant oscillation. Precautions which should be taken against this situation are dealt with on page 231.

The nature of the wind

Within the area of a wind storm many local influences modify the general wind flow. There is a thermal bubbling and mixing of the air masses, and a mechanical stirring caused by the friction of the air over the ground. The scale of the turbulence varies over wide limits. Some of the major eddies may be several thousand metres in extent and give rise to squalls lasting several minutes. At the other end of the scale small, though possibly severe, eddies may be due to the passage of the wind past a building or other minor obstruction, and have correspondingly small dimensions and durations measured in fractions of a second. Usually the pattern is complex with small eddies superimposed on larger ones, with the result that wind speeds vary greatly from place to place and from moment to moment. Thus, the result of any measurement of wind speed will depend on the duration over which the sample is taken. A long averaging time allows the inclusion of a large eddy, while a brief averaging time may cover only a small superimposed eddy, but this may have a higher speed.

Vortices produced along edge of roof when wind blows on to a corner

Fig 3

Areas where high suctions must be allowed for on the cladding

Fig 4

222

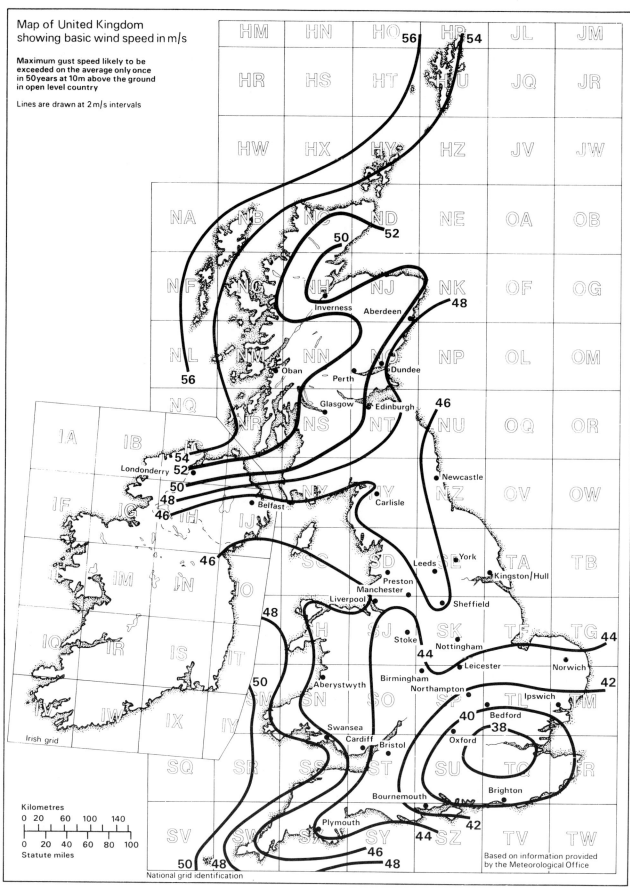

Fig 5

Wind speeds in the United Kingdom

The Meteorological Office records the hourly mean wind speeds and the maximum recorded gust speed each day at a height corresponding to 10 m above open level sites.

Gust speeds are used as the basis of wind load calculations. Maximum gust speeds, at a height of 10 m above ground, likely to be exceeded not more than once in 50 years in open level country are shown by isopleths (lines of equal wind speed) on the map in Fig 5. It should be assumed that the maximum wind may come from any horizontal direction, though near the coast there is a tendency for the strongest winds to blow from the sea.

Design wind speed V_s

The first step in the assessment of wind load is to determine the maximum wind speed appropriate to the structure. This is based on the maximum gust speed V for the locality, as given in the map, but to convert this to the design wind speed V_s three factors must be applied;

S_1 for local topographic influences

S_2 for surface roughness of the environment, gust duration appropriate to the size of the building, height of the structure

S_3 for the design life of the building

Factor for topography S_1 (from Table 1)

The basic wind speeds given on the map take no account of the local effect of hill and valley configuration. They must, therefore, be adjusted in accordance with the factors S_1 shown in Table 1 and described below.

Examination of the wind speed records has shown that the height of the site above sea-level does not by itself affect the basic speed, so unless there are special local effects, the value of S_1 should be taken as 1·0.

Exposed hills rising well above the general level of the surrounding terrain may give rise to accelerated winds; so may some valleys, particularly those shaped so that funnelling occurs, when the wind blows up the valley. Sites so affected are often well known locally for their abnormal winds. For any of these situations, S_1 should be taken as 1·1.

Table 1

Topography	Value of S_1
All cases, except as below	1·0
Very exposed hill slopes and crests where acceleration of the wind is known to occur	1·1
Valleys shaped to produce a funnelling of the wind	
Sites that are known to be abnormally windy due to some local influence	
Steep sided, enclosed valleys, sheltered from all winds	0·9

Table 2

Surface category	1. Open country with no shelter			2. Open country with scattered windbreaks			3. Country with many windbreaks; small towns, outskirts of large cities			4. Surface with large and numerous obstructions eg city centres		
Height above ground level metres	3-sec gust	5-sec gust	15-sec gust	3-sec gust	5-sec gust	15-sec gust	3-sec gust	5-sec gust	15-sec gust	3-sec gust	5-sec gust	15-sec gust
3 or less	0·83	0·78	0·73	0·72	0·67	0·63	0·64	0·60	0·55	0·56	0·52	0·47
5	0·88	0·83	0·78	0·79	0·74	0·70	0·70	0·65	0·60	0·60	0·55	0·50
10	1·00	0·95	0·90	0·93	0·88	0·83	0·78	0·74	0·69	0·67	0·62	0·58
15	1·03	0·99	0·94	1·00	0·95	0·91	0·88	0·83	0·78	0·74	0·69	0·64
20	1·06	1·01	0·96	1·03	0·98	0·94	0·95	0·90	0·85	0·79	0·75	0·70
30	1·09	1·05	1·00	1·07	1·03	0·98	1·01	0·97	0·92	0·90	0·85	0·79
40	1·12	1·08	1·03	1·10	1·06	1·01	1·05	1·01	0·96	0·97	0·93	0·89
50	1·14	1·10	1·06	1·12	1·08	1·04	1·08	1·04	1·00	1·02	0·98	0·94
60	1·15	1·12	1·08	1·14	1·10	1·06	1·10	1·06	1·02	1·05	1·02	0·98
80	1·18	1·15	1·11	1·17	1·13	1·09	1·13	1·10	1·06	1·10	1·07	1·03
100	1·20	1·17	1·13	1·19	1·16	1·12	1·16	1·12	1·09	1·13	1·10	1·07
120	1·22	1·19	1·15	1·21	1·18	1·14	1·18	1·15	1·11	1·15	1·13	1·10
140	1·24	1·20	1·17	1·22	1·19	1·16	1·20	1·17	1·13	1·17	1·15	1·12
160	1·25	1·22	1·19	1·24	1·21	1·18	1·21	1·18	1·15	1·19	1·17	1·14
180	1·26	1·23	1·20	1·25	1·22	1·19	1·23	1·20	1·17	1·20	1·19	1·16
200	1·27	1·24	1·21	1·26	1·24	1·21	1·24	1·21	1·18	1·22	1·21	1·18

224

On the other hand, there are some steep-sided, enclosed valleys where wind speeds will be less than normal. Caution is necessary in applying a reducing factor but for such sites a value of 0·9 may be used for S_1.

More extreme values may be necessary at some especially abnormal sites but until more data are available it is recommended that values outside the range 0·85 to 1·2 should not be used. Local knowledge may help the designer to select the S_1 value, but if he is in doubt the advice of the Meteorological Office* should be sought.

Factor for surface roughness of the environment, gust duration and height of structure, S_2 (from Table 2)
Surface roughness In conditions of strong wind, the wind speed usually increases with height above ground. The rate of increase depends on ground roughness, and also on whether short gusts or mean wind speeds are being considered.

Ground roughness is divided into four categories as follows:

1 Long stretches of open, level or nearly level country with practically no shelter. Examples are flat, coastal fringes, fens, airfields and grassland, moorland or farmland without hedges or walls around the fields.

2 Flat or undulating country with obstructions such as hedges or walls around fields, scattered windbreaks of trees and occasional buildings. Examples are most farmland and country estates with the exception of those parts that are well wooded.

3 Surfaces covered by numerous obstructions. Examples are well-wooded parkland and forest areas, towns and their suburbs, and the outskirts of large cities. The general level of roof-tops and obstructions is assumed to be about 10 m, but the category will include built-up areas generally apart from those that qualify for category 4.

4 Surfaces covered by large and numerous obstructions with a general roof height of about 25 m, or more. This category covers only the centres of large towns and cities where the buildings are not only high, but are also reasonably closely spaced.

Although a classification of ground roughness has been made, it should be recognised that the change from one category to another is necessarily a gradual process. The wind must traverse a certain ground distance before equilibrium is established in a new velocity profile. The change starts first in the layers of wind nearest the ground and the new profile extends to an increasingly deep layer as the distance increases.

It may be assumed that a distance of a kilometre or more is necessary to establish a different roughness category, but within the actual roughness layer, ie below the general level of buildings or obstructions in the windward direction, a lesser distance may apply as follows, depending on the density of buildings and other obstructions on the ground:

Ground coverage:	Required distance:
not less than 10%	500 m
not less than 15%	250 m
not less than 30%	100 m

For a site where the ground roughness varies in different directions, the most severe grading should be used, or, exceptionally, appropriate gradings may be used for different wind directions. For example, the sea-front of a coastal town would generally rank as ground roughness category 1.

Wind speed averaging time The variation of wind speed with height is, however, also dependent on the size of gust that is being considered, ie on the wind speed averaging time. Thus the table gives values of the factors for averaging times of 3 sec, 5 sec and 15 sec.

The recorded gust (a 3-second average) should be used in the design of all units of glazing, cladding and roofing, whatever the size or proportion of the whole building.

For structural design, a 5-second gust should be used for buildings whose largest horizontal or vertical dimension does not exceed 50 m; a 15-second gust should be used for buildings whose largest horizontal or vertical dimension is greater than 50 m.

Height The factor S_2 may be taken as appropriate to the total height of the structure above the level of the surrounding ground; alternatively, the height of the structure may be divided into convenient parts and the wind load calculated on each part, using a factor S_2 which corresponds to the height of the top of that part.

If the structure is on or near to a cliff or escarpment, its effective height H for the purpose of determining the value of S_2 should be measured as follows:

First, find the angle θ, the inclination of the mean slope of the escarpment to the horizontal.

1 If tan θ does not exceed 0·3, the height of the structure should be measured from the natural ground level immediately around the building.

2 If tan θ exceeds 0·3, the height of the structure should be measured from an artificial base Z_c, which is set out by one of the following rules in which

Z_1 is the general level of the ground at the foot of the escarpment

*Advisory offices of the Meteorological Office:
Meteorological Office / Met O.3 (b) / London Road / Bracknall / RG12 2SZ
Meteorological Office / Tyrone House / Ormeau Avenue / Belfast / BT2 8HH
Meteorological Office / 26 Palmerston Place / Edinburgh / EH12 5AN

Z_2 is the general level of the ground at the top of the escarpment

z is the difference of level Z_2-Z_1

(a) If tan θ exceeds 0·3 but is less than 2·0:
Set out the following points:

  A, at the intersection of the level Z_1 with the mean slope of the escarpment

  B, at the intersection of the level Z_2 with the mean slope of the escarpment

  C, such that BC$=z$

  D, such that CD$=3z$

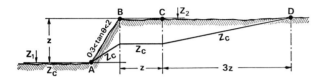

In front of A, $Z_c=Z_1$

from B to C, $Z_c=Z_1+\dfrac{2-\tan\theta}{1\cdot7}z$

behind D, $Z_c=Z_2$
between A and B, and between C and D, Z_c is obtained by linear interpolation.

(b) If tan θ is 2·0 or greater:
Set out the points B, C and D, as in (a)

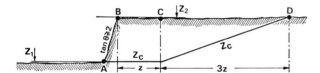

In front of C, $Z_c=Z_1$
behind D, $Z_c=Z_2$
between C and D, Z_c is obtained by linear interpolation

This method of defining the effective height of a building is taken from the French code of practice *Règles NV 65*.

Factor for building life S_3 (from Fig 6)

Whatever the wind speed adopted for design purposes, there is always an element of risk that it may be exceeded in a storm of exceptional violence; the greater the intended life-span of the structure, the greater is the risk. It is in the nature of things that the recorded maxima of natural phenomena tend towards more extreme values with the passing of time, and statistical methods have to be used to estimate the trend of the extremes. For this reason it is not possible to state categorically that a certain maximum value will never be exceeded: probability levels have to be used. The map windspeed is that which is likely to be exceeded on average only once in 50 years. This implies that in any one year there is a 1 in 50 (0·02) probability that the map speed will be exceeded. However, in any period longer than one year there is an increased probability of its being exceeded. It follows statistically that in any one period of 50 years there is a 0·63 probability that it will be exceeded.

Fig 6 shows values of S_3, a factor which takes account of the intended life of the structure and the degree of security that is required. Normally, wind loads should be calculated using $S_3=1\cdot0$, but there

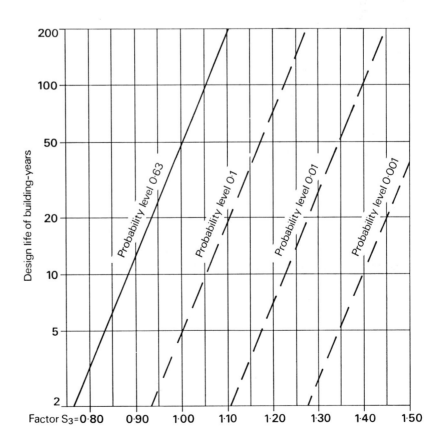

Fig 6 Factor for building life

are exceptions, as follows:

    (a) temporary structures

    (b) structures for which a life of less than 50 years is appropriate.

    (c) structures with an abnormally long intended life

    (d) structures where additional safety is required.

For these special cases both the intended lifetime and the probability level may be varied according to circumstances.

The design wind speed V_s can now be calculated from:

$$V_s = V \times S_1 \times S_2 \times S_3$$

The assessment of load on ordinary buildings

When designing in accordance with codes that use permissible stresses, the loads given by the digest should be related directly to the permissible stresses; in limit-state design, the loads given are the characteristic loads.

For most normal buildings there are two main aspects to be considered in the calculation of wind loads (a) the load on the structural frame taken as a whole; (b) the loads on individual units such as walls and roof, and their elements of cladding. Sometimes (a) may be computed directly and sometimes it must be derived from a summation of the individual loads. In all cases the loads are based on the dynamic pressure of the wind.

The dynamic pressure of the wind q

If the wind is brought to rest against the windward face of an obstacle, all its kinetic energy is transformed to dynamic pressure q, sometimes referred to as the 'stagnation pressure'.

This can be calculated from the formula:

$$q = kV_s^2$$

$k = 0.613$ in SI units (N/m^2 and m/s)

    0.0625 in metric technical units (kgf/m^2 and m/s)

    0.00256 in imperial units (lbf/ft^2 and mile/h)

Because information may be available in various units, a conversion chart covering V in knots, mph and m/s, and q in lbf/ft^2, N/m^2 and kgf/m^2 is given in Fig 7.

Pressure on a surface—the use of pressure coefficients

The pressure on any surface exposed to the wind varies from point to point over the surface, depending on the direction of the wind and the pattern of flow. The pressure p at any point can be expressed in terms of q by the use of a pressure coefficient C_p. Thus:

$$p = C_p . q$$

A negative sign prefixing the coefficient C_p indicates that p is negative, that is, a suction rather than a positive pressure. The pressure or suction at every point acts in a direction normal to the surface.

In the calculation of wind load on any structure or element it is essential to take account of the pressure difference between opposite faces. For clad structures it is therefore necessary to know the internal pressure as well as the external, and it is convenient to use distinguishing pressure coefficients:

$$C_{pe} = \text{external pressure coefficient}$$
$$C_{pi} = \text{internal pressure coefficient}$$

Values of C_p have been determined by experiment on models for a number of building shapes and wind directions. They are expressed in various ways in different compilations of data. The most complete information relative to the pressure on a surface can be given by a contour diagram:

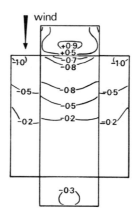 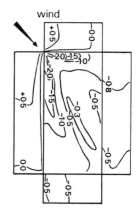

This shows, for a specified wind direction, the distribution of pressure over the surfaces of a building. It will be noticed that, confirming the earlier description of flow, the distribution is far from uniform. This form of presentation gives a detailed picture of the pressure distribution but it is not usually the most convenient for design purposes; a simplified presentation giving mean pressure coefficients over whole surfaces or parts of surfaces is often preferred and has been adopted in Tables 3 and 4.

For each building shape, pressure coefficients are given for each individual surface (or part of a surface) for wind directions that produce critical loading. This enables calculation to be made of the maximum individual load on each surface and, by vectorial summation of individual loads, the total load on the building. It must be understood that, although the wind has been shown blowing against one face and the pressure coefficients given for that wind direction, the wind may, in practice, blow against any face of a building. Therefore each face should, in turn, be considered as the windward one and the pressure calculated from the equation $p = C_{pe}q$. This will indicate the maximum external loading, positive or negative, on each face.

Building height ratio	Building plan ratio	Side elevation	Plan	Wind angle α	C_{pe} for surface				Local C_{pe}
					A	B	C	D	
$\frac{h}{w}<\frac{1}{2}$	$1<\frac{\ell}{w}<\frac{3}{2}$			0°	+0·7	−0·2	−0·5	−0·5	−0·8
				90°	−0·5	−0·5	+0·7	−0·2	
	$\frac{3}{2}<\frac{\ell}{w}<4$			0°	+0·7	−0·25	−0·6	−0·6	−1·0
				90°	−0·5	−0·5	+0·7	−0·1	
$\frac{1}{2}<\frac{h}{w}<\frac{3}{2}$	$1<\frac{\ell}{w}<\frac{3}{2}$			0°	+0·7	−0·25	−0·6	−0·6	−1·1
				90°	−0·6	−0·6	+0·7	−0·25	
	$\frac{3}{2}<\frac{\ell}{w}<4$			0°	+0·7	−0·3	−0·7	−0·7	−1·1
				90°	−0·5	−0·5	+0·7	−0·1	
$\frac{3}{2}<\frac{h}{w}<6$	$1<\frac{\ell}{w}<\frac{3}{2}$			0°	+0·8	−0·25	−0·8	−0·8	−1·2
				90°	−0·8	−0·8	+0·8	−0·25	
	$\frac{3}{2}<\frac{\ell}{w}<4$			0°	+0·7	−0·4	−0·7	−0·7	−1·2
				90°	−0·5	−0·5	+0·8	−0·1	

ℓ - length of major face of building. w - width of building (length of minor face)

Table 3 Pressure coefficients C_{pe} for vertical walls of rectangular clad buildings

Building height ratio		Roof angle degrees	wind angle 0°		wind angle 90°		local coefficients			
			EF	GH	EG	FH				
$\frac{h}{w}<\frac{1}{2}$		0	−0·8	−0·4	−0·8	−0·4	−2·0	−2·0	−2·0	
		5	−0·9	−0·4	−0·8	−0·4	−1·4	−1·2	−1·2	−1·0
		10	−1·2	−0·4	−0·8	−0·6	−1·4	−1·4		−1·2
		20	−0·4	−0·4	−0·7	−0·6	−1·0			−1·2
		30	0	−0·4	−0·7	−0·6	−0·8			−1·1
		45	+0·3	−0·5	−0·7	−0·6				−1·1
$\frac{1}{2}<\frac{h}{w}<\frac{3}{2}$		0	−0·8	−0·6	−1·0	−0·6	−2·0	−2·0	−2·0	
		5	−0·9	−0·6	−0·9	−0·6	−2·0	−2·0	−1·5	−1·0
		10	−1·1	−0·6	−0·8	−0·6	−2·0	−2·0	−1·5	−1·2
		20	−0·7	−0·5	−0·8	−0·6	−1·5	−1·5	−1·5	−1·0
		30	−0·2	−0·5	−0·8	−0·8	−1·0			−1·0
		45	+0·2	−0·5	−0·8	−0·8				
$\frac{3}{2}<\frac{h}{w}<6$		0	−0·7	−0·6	−0·9	−0·7	−2·0	−2·0	−2·0	
		5	−0·7	−0·6	−0·8	−0·8	−2·0	−2·0	−1·5	−1·0
		10	−0·7	−0·6	−0·8	−0·8	−2·0	−2·0	−1·5	−1·2
		20	−0·8	−0·6	−0·8	−0·8	−1·5	−1·5	−1·5	−1·2
		30	−1·0	−0·5	−0·8	−0·7	−1·5			
		40	−0·2	−0·5	−0·8	−0·7	−1·0			
		50	+0·2	−0·5	−0·8	−0·7				

Key plan

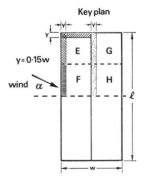

y = 0·15w

wind α

Note: The pressure coefficient on the underside of any roof overhang should be taken as that on the adjoining wall surface
The coefficient for a low–pitch monopitch roof should be taken as −1·0

Table 4 Pressure coefficients C_{pe} on roofs of rectangular clad buildings

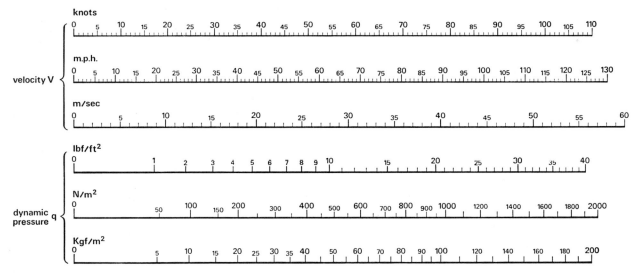

Fig 7 Conversion chart for wind speed and dynamic pressure

Examine for high local pressures The tables also give an indication of the high local pressures that may occur under certain conditions of wind. These should be taken into account in determining the local loads on elements of cladding and their supporting members. They are not to be taken into account in calculating the total structural loads, except that they may influence the pressure inside a building if a major door or window opening occurs in the area affected, as is shown later.

Allow for effect of stacks, etc In addition to the local pressure areas indicated in the diagrams of Tables 3 and 4, any roof or wall surface in the immediate wake of a protuberance such as a chimney stack, may be affected by severe suction caused by vortex action. It is not possible to define exactly the extent of this influence, but it would be prudent to allow for a local C_{pe} of -1.2 for a distance around the obstruction equal to the minor dimension of the obstruction across the wind direction.

Allow for narrow paths A similar local coefficient would also be appropriate for many surface areas that are swept by high-speed air currents such as occur in narrow gaps between buildings and in open passages through buildings.

Effect of internal pressures

The total wind force on a wall or roof depends on the difference of pressure between the outer and inner faces. Open doors, windows or ventilators on the windward side of a building will increase air pressure inside the building and this will increase the loading on those parts of the roof and walls which are subjected to external suction (Fig 8a). Conversely, openings to positions which are experiencing external suction will reduce the pressure inside the building thus increasing total loads on a windward wall, as shown in Fig 8b.

Most buildings have some degree of permeability on each face, through windows, ventilation louvers, leakage gaps around doors and windows, and to some extent through the cladding itself; if there are chimneys, these can provide a low-resistance path for air flow. Permeability, in this context, is measured by the total area of such openings in a face. The problem is to determine the resulting internal balance of all the contributing leakage points and, for design purposes, to assess the worst probable combination of circumstances that may arise.

With all windows nominally closed, the permeability of a house or office block is likely to be in the range of 0·01 per cent to 0·05 per cent of the face area, depending on the degree of draughtproofing.

For a rectangular building of which two opposite faces are equally permeable and the other faces are impermeable, the value of C_{pi} should be taken as $+0.2$ if the wind direction is normal to a permeable face, or -0.3 if the wind direction is normal to an impermeable face, If all four faces are uniformly permeable, C_{pi} should be taken as -0.3. Where the permeability differs from face to face, as when one contains a dominant opening such as an open doorway or window (a window broken by wind force or flying debris during a storm may serve) the value of C_{pi} will change depending on the size and position of the opening in relation to the other

Fig 8a

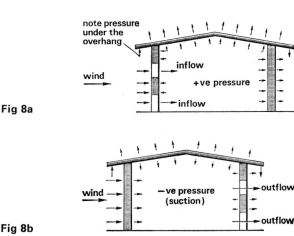

Fig 8b

permeability. The following figures will serve for guidance:

Size and position of dominant opening C_{pi}

(a) on windward face:
if the permeability of the windward face equals the following proportion of the total distributed permeability of all the faces subject to suction

proportion 1	+0·1
1½	+0·3
2	+0·5
3 or more	+0·6

(b) on leeward face: any dominant opening   −0·3

(c) on a face parallel to the wind:

  (i) any dominant opening not in an area of high local C_{pe} −0·4

  (ii) any dominant opening in an area of high local C_{pe}:
if the area of the opening equals the following proportion of the total other distributed permeability of all the faces subject to suction

proportion ¼ or less	−0·4
½	−0·5
¾	−0·6
1	−0·7
1½	−0·8
3 or more	−0·9

Where it is not possible, or is not considered justified, to estimate the range of C_{pi} for a particular building, the coefficient should be based on one of the following paragraphs for any determination of wall or roof loading:

(a) Where there is only a small probability of a dominant opening occurring during a severe storm, C_{pi} should be taken as the more onerous of +0·2 and −0·3.

(b) For situations where a dominant opening is likely to occur, C_{pi} should be taken as 0·75 of the value of C_{pe} outside the opening. The extreme conditions should be determined for the various wind directions that give rise to critical loadings and it should be noted that especially severe internal pressures may be developed if a dominant opening is located in a region of high local external pressure.

The total wind load P on any flat wall or roof panel of area A is: $P = A \cdot q \, (C_{pe} - C_{pi})$.

There is a further complication in a wall or roof element that comprises several layers. For example, a roof may be boarded and felted, and covered with tiles. The pressure difference between outside and inside will then be broken down into steps, across each layer; these steps will depend on the relative permeability of the various layers and the access of air to the spaces between them. Each case needs careful study to ensure that the whole of the wind load is not accidentally transferred to a single membrane such as a thin metal sheet which may not be designed to carry it. Similarly, the pressure difference between windward and leeward faces of a building may be broken down in steps across internal partitions and impose loads on them.

The value of C_{pi} can sometimes be limited or controlled to advantage by the judicious distribution of permeability in the walls and roof or by the deliber-

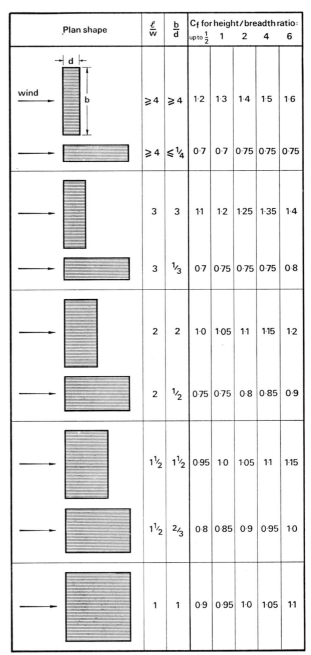

Plan shape	$\frac{\ell}{w}$	$\frac{b}{d}$	C_f for height/breadth ratio: up to $\frac{1}{2}$	1	2	4	6
	≥4	≥4	1·2	1·3	1·4	1·5	1·6
	≥4	≤¼	0·7	0·7	0·75	0·75	0·75
	3	3	1·1	1·2	1·25	1·35	1·4
	3	⅓	0·7	0·75	0·75	0·75	0·8
	2	2	1·0	1·05	1·1	1·15	1·2
	2	½	0·75	0·75	0·8	0·85	0·9
	1½	1½	0·95	1·0	1·05	1·1	1·15
	1½	⅔	0·8	0·85	0·9	0·95	1·0
	1	1	0·9	0·95	1·0	1·05	1·1

b breadth of building across wind direction
d depth of building in direction of wind

Table 5 Force coefficients C_f for rectangular buildings

ate provision of a venting device which can serve as a dominant opening at a position having a suitable external pressure coefficient. An example of such an application is a ridge ventilator on a low pitch roof.

Parapets

Near the top of a building, the pressure coefficient on the windward face is decreasing rapidly from the mean value given in Table 3. It is not therefore appropriate to use this value for the design of parapets. The loading will vary according to the ratio of height of the parapet to the height of the building, but in general the value of the pressure coefficient on the windward side of the parapet will not exceed +0·3. The pressure coefficient on the leeward side of the parapet will, in general not exceed −0·5.

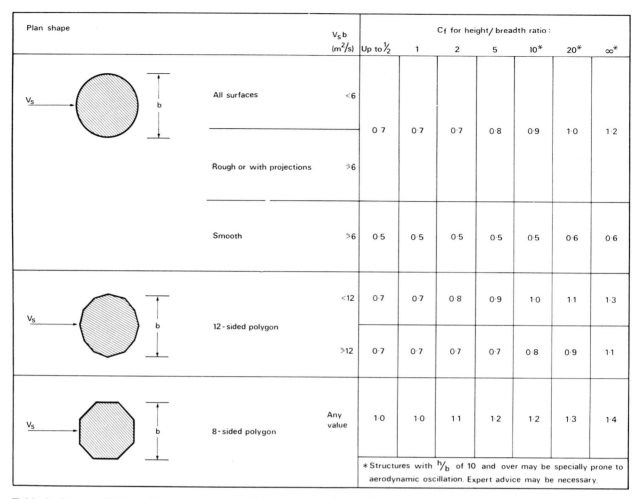

Plan shape		$V_s b$ (m²/s)	Cf for height/breadth ratio:						
			Up to ½	1	2	5	10*	20*	∞*
(circular)	All surfaces	<6							
			0·7	0·7	0·7	0·8	0·9	1·0	1·2
	Rough or with projections	≥6							
	Smooth	≥6	0·5	0·5	0·5	0·5	0·5	0·6	0·6
(12-sided polygon)	12-sided polygon	<12	0·7	0·7	0·8	0·9	1·0	1·1	1·3
		≥12	0·7	0·7	0·7	0·7	0·8	0·9	1·1
(8-sided polygon)	8-sided polygon	Any value	1·0	1·0	1·1	1·2	1·2	1·3	1·4

*Structures with h/b of 10 and over may be specially prone to aerodynamic oscillation. Expert advice may be necessary.

Table 6 Force coefficients C_f for structures of uniform section, such as chimneys

Total wind loads on a structure

The total wind load on a structure may be obtained by the vectorial summation of the loads on the various wall and roof surfaces; in some cases this is the only available method. Care should be taken to summate only those loads which occur simultaneously, ie for a given wind direction, rather than the absolute maxima for the various wall and roof elements.

The use of force coefficients For some structural shapes, total wind forces have been measured directly and force coefficients (see Tables 5 and 6) derived such that:

$$F = A_e q C_f$$

where F is the force in the direction of the wind (usually for a specified direction)

A_e is the effective frontal area of the structure (usually the projected area on a plane normal to the wind direction)

C_f is the force coefficient in the direction of the wind

The rectangular plan shapes of Table 5 can also be used as the overall dimensions of buildings of sub-stantially rectangular plan but having projections or recesses not exceeding ten per cent of the floor area; there will then be high local pressures in the region of all external angles of walling.

In cases where structures are not symmetrical about the wind direction there will usually be a component of load transverse to the wind, but this may in general be neglected as being smaller than other loads that will at some time operate on this axis.

For circular structures such as chimneys (see Table 6), the force coefficient depends on the magnitude of the product $V_s.D$ (where D is the diameter) and on the surface roughness of the structure. Thus there are alternative sets of values for C_f and care must be taken to use the appropriate one.

These force coefficients give overall loads directly and not the loads on individual parts of a building; they offer the most direct means for the determination of stability against overturning. For a quick but approximate calculation, the value of the dynamic pressure q appropriate to the top of the building may be taken. This implies a constant pressure over the whole height of the building and, assuming a constant cross-section at each floor level, a line of action F at a height of $0.5\ H_a$, where H_a is the average height of the top of the area A_e. Because in practice the actual pressure decreases at lower heights, this calculation may give an over-estimate of the overturning moment.

A better estimate can be achieved by dividing the building into height zones and applying to each zone the value of q appropriate to the top of the zone. The overturning moment may be computed by taking the moment of each individual height zone. This method is always necessary in cases where the cross-section of the building varies substantially with height above ground.

Frictional drag

In addition to direct pressure on a surface, there may be some frictional drag. In most cases this will be small, and may be neglected. Where a surface over which the wind is blowing is large in relation to the effective frontal area of the building, a frictional force may be calculated from:

$$F' = 0.025\, A_s . q$$

where

A_s is the area of the surface
F' is the force in the direction of the wind.

Oscillation of slender structures

If the natural frequency of a slender structure is less than the frequency of vortex shedding at the maximum wind velocity and the structural damping is small, then oscillations are likely. The likelihood is increased if two or more similar structures are located in close proximity, for example, at spacings of less than ten of their diameters apart.

The shedding frequency n in cycles per second for slender structures can be determined by:

$$n = \frac{Sv}{b}$$

where

S = 0.20 for structures of circular cross-section
   0.15 for structures of square cross-section and for flat plates
v = wind speed in m/sec
b = breadth of structure across the wind direction in metres.

If oscillations are likely, the amplitude must be restrained by adequate structural damping or by the application of devices to reduce the aerodynamic excitation (see references 4 and 5).

Wind loads during construction

The pressure coefficients which apply to a completed building will generally not be appropriate at all stages of the construction, and care is necessary to guard against adverse conditions. The most common variation is due to extremes of internal pressure when the building is partially clad. This risk should be borne in mind when the construction work is programmed, so that the most vulnerable conditions of partial cladding or dangerous structural shapes are avoided.

A critical loading can occur on framed buildings when the floors are in place but not the walls. The wind load is then the sum of the pressure on the exposed edges of the floors (for which C_f may be taken as 1.2), plus the drag of the exposed columns (for which C_f may be taken as 2.0 for square or rectangular columns or rolled sections), plus the frictional drag on the upper and lower surfaces of the floors. This will depend on surface roughness but may generally be taken as $0.05\, q$ multiplied by the total floor area.

In considering whether columns are exposed to the wind or sheltered by others in front of them, it should be recognised that the worst loading may be due to a wind slightly inclined to the face of a building, causing a maximum exposure.

It is reasonable to assume that the maximum design wind speed V_s will not occur during a short construction period, and a reduced factor S_3 can be used to estimate the probable maximum wind. For normal construction, it is suggested that the S_3 factor for a 2-year building life should be used. The graphs of Fig 6 should not, in any circumstances, be extrapolated to a period of less than two years.

Further reading and references

1 J B Menzies: *Wind damage to buildings in the United Kingdom 1962–1969*. BRE Current Paper CP35/71.

2 C W Newbury, K J Eaton, J R Mayne: *Wind pressures on the Post Office Tower, London* BRE Current Paper CP37/71.

3 *Effect of wind loading on typical buildings*: a package of 40 slides with lecture notes and the above Current Papers: (£4.88, including VAT).

4 The oscillations of large circular stacks in wind. L R Wootton. Proceedings of the Institution of Civil Engineers Paper 7188, Vol. 43, pp 573–598, August 1969.

5 Preventing wind-induced oscillations of structures of circular section. D E Walshe and L R Wootton. Proceedings of the Institution of Civil Engineers Paper 7289. Vol. 47, pp 1–24, September 1970.

An index of exposure to driving rain

R E Lacy B Sc, FR Met Soc

For most of the functional requirements of buildings—loading, heat loss, etc—the basic design data can be expressed in quantitative terms. In considering the exposure of a building to driving rain, however, it has hitherto been necessary to use rather indefinite terms such as 'sheltered', 'moderate' or 'severe'. This digest describes the development of a numerical driving-rain index and suggests how it can be used to give these terms a more quantitative basis. Maps are presented showing how the severity of driving-rain conditions varies over Britain as a whole; variations due to local topography and those occurring over the face of a building are discussed briefly.

'In Spain', we are told, 'the rain falls mainly on the plain.' In Britain, the converse is more nearly true. The map of annual rainfall resembles closely one showing elevation of the ground above the sea, with the contour lines of high rainfall following the contour lines enclosing the hilly areas.

This simple picture is distorted by a number of factors—by distance from the sea, by the shape of the hills and their orientation with respect to prevailing winds, and above all by the fact that most rain-bearing winds blow from directions between roughly south and north-west, so that, other things being equal, the western parts of the country get more rain. In the rainier parts of the country it also rains for longer periods than in drier parts. Around London on average it is raining for 5 per cent of the time: on the western coast of England and Wales this figure rises to 7 or 8 per cent, and in hilly districts to over 10 per cent. In north-west Scotland the proportion is even higher, being more than 15 per cent over a large area.

Technical enquiries arising out of this digest should be addressed to the Building Research Station Advisory Service.

From considerations of rainfall alone, it would be expected that buildings in the west of Britain would experience most trouble from rain penetration. However, in the absence of wind, rain does not wet walls—it is the combination of rain and wind that gives 'driving rain', that is, rain driven along at an angle to the vertical so that it impinges on vertical surfaces. Maps of average wind speed show that highest winds occur near the coasts—and especially near west coasts. Inland regions are relatively sheltered—but with local increases in wind speed on hills. For example, during a three-month spell in one winter it was found that a west-facing driving-rain gauge in a field at Garston, Herts, caught 20 litre/m² of driving rain in the nine worst storms. On the outskirts of a town in South Wales the nine worst storms in the same winter produced 304 litre/m² in a similarly exposed gauge (and about 166 litre/m² on the wall of an adjacent house). But there may be large differences in the degree of exposure over a short distance in hilly country. In the same Welsh town, it was found that the driving rain measured in a park near the town centre was sometimes only 1/15th of that on the outskirts, although the difference in elevation was only about 90 m.

The driving-rain index described below may be taken to apply to a standard building with a constant degree of local exposure. The method of allowing for variations in local exposure is described on page 236.

Measurements made by BRS using rain gauges set in the walls of buildings have shown that the amount of rain driven on to a wall is directly proportional to the product of the rainfall on the ground and the wind speed during the rain. However, the wind speed during periods of rainfall is not usually recorded separately; therefore the product of annual average rainfall and annual average wind speed (both of which are available from published data) is used as an index of driving rain, and hence of likelihood of rain penetration. Such evidence as is available indicates an almost constant relationship between the annual average wind speed and the average wind speed during rain, so that the overall picture is not distorted by this expedient.

Fig 1 shows how an index calculated in this way varies over the country. The map is based on long periods of observation of rainfall and wind and hence indicates the amount of rain which would be driven on to a vertical surface in an average year. The numbered contour lines represent the product of annual rainfall in millimetres and average wind speed in metres per second, divided by 1000; hence the index is expressed in m²/s.

In Fig 1, the indexes are proportional to the total amount of rain that would be driven in one year on to a vertical surface always facing the wind. In many parts of the country, most rain falls with winds blowing from directions between south and west or north-west, but this is far from true in places near the east coasts of Britain. The driving-rain 'roses' in Fig 2 show the amounts of driving-rain index attributable to winds from different directions. The roses are aligned so that, for example, the line representing winds coming from directions 350° to 010° points to true north. The length of each of the twelve 'petals' is proportional to the annual mean driving-rain index from that direction. Because the driving-rain index is proportional to the amount of driving rain, the length of each line is an indication of the amount of wetting likely to be experienced in an average year by walls facing in different directions. It is clear that in western parts of

234

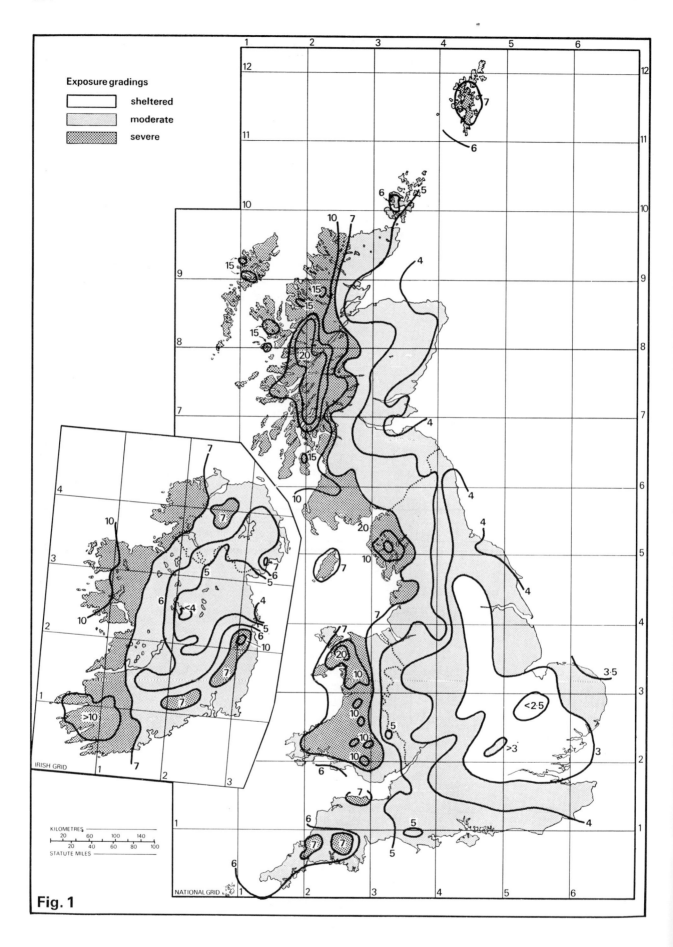

Fig. 1

235

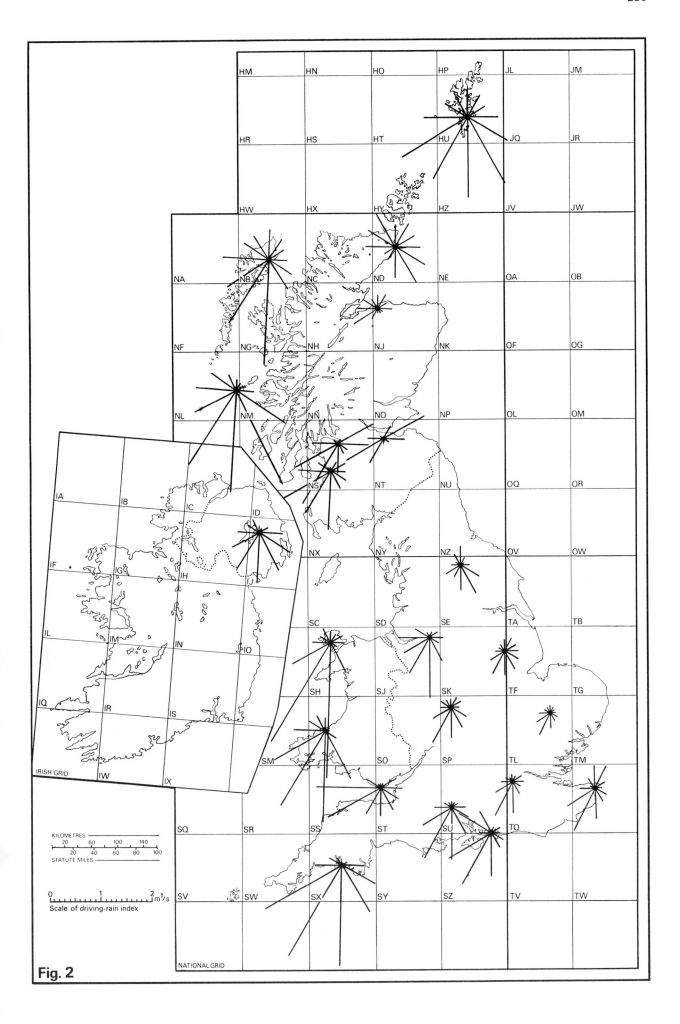

Fig. 2

the country most driving rain reaches walls facing south, south-west or west, and little falls on north-facing walls, so that it is desirable to pay special attention to making the former weathertight. But in some eastern districts, north or north-east walls may be the ones most severely exposed to driving rain.

However, though most driving rain conforms to these patterns, short periods of intense driving rain may be experienced from any direction, as is shown in Table 1.

Most rain penetration of masonry construction probably occurs in a few severe and prolonged storms of driving rain. These tend to be more severe, and to be much more numerous, in wetter areas and where the wind is strongest.

Exposure grading

The contour lines in Fig 1 join places having the same driving rain index; in addition, areas of 'sheltered', 'moderate' and 'severe' conditions are indicated. These hatched areas are based solely on the index figures and may need to be modified according to local conditions, in particular, proximity to the coast, elevation of the site and the height of the building. It is therefore suggested that the exposure grading of the site should be decided by the following rules:

a *Sheltered* conditions obtain in districts where the driving-rain index is 3 or less, excluding areas that lie within 8 km of the sea or large estuaries, where the exposure should be regarded as *moderate* (see also **d** below).

b *Moderate* conditions obtain in districts where the driving rain index is between 3 and 7, except in areas which have an index of 5 or more and which are within 8 km of the sea or large estuaries, in which exposure should be regarded as *severe* (see also **d** below).

c *Severe* conditions obtain in areas with a driving-rain index of 7 or more.

d In areas of sheltered or moderate exposure, high buildings which stand above their surroundings or buildings of any height on hill slopes or hill tops, should be regarded as having an exposure one grade more severe than that indicated by the map.

Even this method of grading cannot take account of all the local circumstances; for example, high ground or a belt of trees on the windward side may make the exposure of the site less than that of its surroundings. Conversely, a site on the windward slope of even a slight hill, or in a valley facing into the strongest rain-bearing winds, may be more severely exposed than the average for the area. The Meteorological Office* may be able to advise in cases of doubt.

*Advisory offices of the Meteorological Office:

Meteorological Office	Meteorological Office	Meteorological Office
Met O.3	Tyrone House	26 Palmerston Place
London Road	Ormeau Avenue	Edinburgh
Bracknell	Belfast	EH12 5AN
RG12 2SZ	BT2 8HH	

Intense driving rain

Most driving rain is caused by more or less prolonged storms, but intense driving rain can occur in heavy showers lasting for a few minutes, or on occasion up to an hour or so.

A recent analysis* of hourly values of driving-rain index at 23 stations has produced the values shown in Table 1. The annual mean driving-rain indexes in the fourth column generally confirm Fig 1, which was produced from maps of annual mean wind speed and rainfall. Where differences occur they can reasonably be attributed to local topographic effects.

The last four columns refer to the highest hourly driving-rain index experienced at each station during 10 years. With one exception the maximum hourly index lies in the range 0·09 to 0·21 m²/s and is independent of the degree of exposure. The exception was the very high value of 0·404 m²/s at Plymouth, when 46 mm of rain fell in 0·9 hour with a wind of about 9 m/s. This fall can be expected about once in 100 years, but the frequency with a wind speed as high as 9 m/s may be appreciably less. It appears that the highest hourly driving-rain index to be expected once in 10 years is about 0·1 to 0·2 m²/s, corresponding to a driving rain amount of about 20 to 40 litre/m² in the hour on the vertical. This rate may be expected on the most exposed parts of the building, near corners; the rates over the façade as a whole might be about half this. A much higher rate,

Table 1   Annual mean driving-rain index at twenty-three places, with greatest hourly driving-rain index (mostly in ten years 1957–66) and corresponding wind direction

Place	Alt. metres	Map ref.	Annual mean DRI m²/s	Amount m²/s	as % of annual mean DRI	Direction of wind deg.	Month of occurrence
				Maximum hourly driving-rain index			
LERWICK	82	HU 453397	10·60	0·117	1·1	180	December
WICK	36	ND 364522	6·37	0·169	2·6	030	August
STORNOWAY	3	NB 459332	9·34	0·104	1·1	240	August
KINLOSS	5	NJ 069625	2·99	0·099	3·3	030	June
EDINBURGH Turnhouse	35	NT 159739	3·68	0·096	2·6	060	September
TIREE	9	NL 999446	10·25	0·181	1·8	150	September
GLASGOW Renfrew	5	NS 480667	5·51	0·205	3·7	060	June
PRESTWICK	16	NS 369261	5·69	0·125	2·2	030	August
DISHFORTH-LEEMING	32	SE 305890	3·39	0·211	6·2	240	June
WADDINGTON	68	SK 988653	3·46	0·107	3·1	120	August
MILDENHALL	5	TL 683779	2·33	0·104	4·5	270	August
BIRMINGHAM Elmdon	97	SP 171837	3·88	0·094	2·4	360	November
LONDON Heathrow	25	TQ 077769	3·37	0·140	4·1	090	July
KEW	5	TQ 171757	2·75	0·078	2·8	030	July
MANSTON	44	TR 335666	4·45	0·086	1·9	240	January
THORNEY ISLAND	3	SU 758026	4·42	0·110	2·5	210	September
BOSCOMBE DOWN	126	SU 172403	4·82	0·147	3·1	300	August
MANCHESTER Ringway	76	SJ 818850	4·87	0·167	3·4	090	September
HOLYHEAD Valley	10	SH 310758	8·18	0·178	2·2	210	September
ABERPORTH	133	SN 242521	7·67	0·146	1·9	240	December
CARDIFF Rhoose	62	ST 060670	5·82	0·199	3·4	360	August
PLYMOUTH Mt Batten	27	SX 492529	9·08	0·404	4·5	090	July
ALDERGROVE	69	IJ 147798	5·74	0·112	2·0	090	December

*This analysis was made by the Meteorological Office and is based on measurements of wind speed, rainfall and rain duration made each hour and at most stations for a period of 10 years. A full report will be published later.

over 100 litres/m²h, has been measured in a period of a few minutes on a penthouse 24 m above ground in Glasgow. The frequency of such high rates is not known, but may be once in 5 or 10 years.

The wind may blow from any direction during the most intense driving rain, and this direction is not related to the average degree of exposure. For example, at Cardiff, Manchester and London the maximum hourly driving-rain index was associated with wind directions (north, east and east respectively) which normally give little driving rain. At these three stations and at most of the other places the maximum occurred during the summer, suggesting that heavy thundery showers were responsible. But even in winter the wind direction during the worst hour could be from north (Birmingham) or east (Aldergrove).

Windows and curtain walls

The annual mean driving-rain index gives a fair indication of the total amount of driving rain to be expected in Britain and is thus helpful in gauging the likely wetness of absorbent walls. It is not likely to be of use in assessing rates of run-off from impervious surfaces or the likelihood of penetration of rainwater through joints in windows or curtain walls. For this the wind pressure acting when water is in contact with the joints is the significant factor. It is planned to issue a further digest on this subject.

Application

A stage has now been reached in the Station's studies of climate in relation to buildings at which the severity of exposure of a site can be classified more precisely than hitherto and, by the use of maps, a fair indication can be obtained of the driving-rain conditions to be expected in any part of Britain. This enables a designer to compare the exposure at one place with that at another with which he is already familiar.

It is doubtful if it will ever be possible to find an exact correlation between the weather conditions and every type of construction. Not only is there an infinite gradation in severity of exposure, but there can be a great variation in the performance of a given type of structure because of slight variations in detailing, in materials, in sizes of components and in workmanship.

However, the exposure grading proposed in this digest does bear some relationship to the performance of current buildings. For example, in the region classed in Fig 1 as *sheltered* it is likely that a one-brick thick solid wall would not suffer from rain penetration (although such a wall would not offer adequate thermal insulation and might be unacceptable for other reasons). The driving-rain index is used as a classification of exposure in the draft proposals for the revision of BS CP 121·101 'Brickwork' and in the Building Standards (Scotland) Regulations 1963. Moreover, in a recent survey of houses and flats in Scotland, tenants' complaints of rain leakage through windows were twice as frequent in areas of severe exposure (driving-rain index greater than 7 m²/s) as in areas of moderate exposure (index 3 to 7 m²/s).

Designers are often worried about the possibility of high rates of run-off of water from the walls of high buildings when there is severe driving rain. In fact, no reports have been seen of any appreciable amounts of rainwater running down façades, although streaks of dirt on weathered buildings show that there is some run-off. It is believed that most water is thrown off by horizontal cills or other elements of the building and falls as a curtain of large drops close to the wall. Observation suggests that gutters are not needed on the face to intercept the rain. The best arrangement is perhaps a canopy at first-floor level to protect pedestrians from the curtain of large drops. The canopy would need to be drained.

Special factors

The information given in this digest relates to the exposure conditions prevailing on a site and takes account of the influence of local topography on the intensity of driving rain. It is recognised, however, that there are at least two other factors, connected with the building itself, which may affect the risk of rain penetration. These do not lend themselves readily to assessment but they may be mentioned briefly.

Variations in exposure on a building The severity of exposure often varies from place to place on the building, even on the same wall. Part of this variation may result from shelter by other buildings or by projections from the wall, but even on a freely exposed face there may be significant differences. As the wind blows past the building the air is deflected from its normal course and is speeded up, especially around corners and over cornices, where its speed may be twice that of the undisturbed wind (*see* Digest No. 119). The rain intensity is increased correspondingly, so that in such places the driving-rain index is abnormally high. While this need not affect the general assessment of the exposure of the building, the possibility that certain parts may be abnormally exposed in this way should be considered in the design.

Evaporation from building materials Rainwater absorbed by porous materials is removed by natural evaporation from the outside surfaces, the significant climatic elements being solar radiation, wind speed and atmospheric humidity (although some of the heat required may be supplied from inside the building). However, the nature of the materials also plays a large part in determining the rate of evaporation of the contained water. With such an infinite variety of factors it is difficult to make precise estimates of evaporation rates from walls, but in the wetter districts they are rather less than those in the drier south-east of England.

The net effect therefore is to intensify the risk of rain penetration in regions of high driving-rain index, for here the walls have less chance of drying out between spells of rain. However, the variation in driving-rain index is so much greater than the variation in evaporation rate that the latter can be ignored and only the driving-rain index need be considered when assessing exposure.

Index

This two-part index consists of a *numerical* index of BRE Digest current titles followed by an *alphabetical* subject index. The numerical section shows the location of each digest by volume and page number, whilst the alphabetical section gives the relevant digest numbers for each subject.

It will be seen that some digests appear in more than one volume, thus ensuring that the subject area of each volume is fully covered and repeating a feature which proved its value to users of the first edition of this series. The index is common to all four volumes, Building Materials, Building Construction, Services and Environmental Engineering, and Building Defects and Maintenance. When a reader is seeking a particular volume which is temporarily unavailable, as, for instance, in an office library, consulting the index in one of the other volumes will show if any of the required digests is to be found elsewhere.

Key M = Building Materials, C = Building Construction, S = Services and Environmental Engineering, D = Building Defects and Maintenance.

Numerical list of current titles

8   Built-up felt roofs,   C118; D105

12   Structural design in architecture,   C226

13   Concrete mix proportioning and control,   M9

15   Pipes and fittings for domestic water supply,   S10; D132

18   Design of timber floors to prevent decay,   M140; C30; D120

27   Rising damp in walls,   C77; D85

33   Sheet and tile flooring made from thermoplastic binders,   M212; C48; D101

35   Shrinkage of natural aggregates in concrete   M30

Estimating daylight in buildings —
41   Part 1   S150
42   Part 2,   S157

Design and appearance —
45   Part 1,   M226; D2
46   Part 2,   M234; D10

47   Granolithic concrete, concrete tiles and terrazzo flooring,   M67

49   Choosing specifications for plastering, M218; C84

53   Project network analysis,   C237

54   Damp-proofing solid floors,   C34; D94

59   Protection against corrosion of reinforcing steel in concrete,   M46; D29

60   Domestic chimneys for independent boilers,   S66

61   Strength of brickwork, blockwork and concrete walls,   M103; C66

Soils and foundations —
63   Part 1,   C2; D40
64   Part 2,   C9; D47
67   Part 3,   C16; D54

69   Durability and application of plastics, M188; D188

70   Painting: Iron and steel,   M176; D167

71   Painting: Non-ferrous metals and coatings,   M183; D174

72   Home-grown softwoods for building,   M116

73   Prevention of decay in external joinery, M144; C99; D124

75   Cracking in buildings,   D21

77   Damp-proof courses,   C73

79   Clay tile flooring,   C45; D98

80   Soil and waste pipe systems for housing,   S2; D139

Alphabetical subject index